METHODS IN COMPUTATIONAL PHYSICS

Advances in Research and Applications

Volume 1

Statistical Physics

METHODS IN COMPUTATIONAL PHYSICS

Advances in Research and Applications

Edited by

BERNI ALDER

Lawrence Radiation Laboratory
Livermore, California

SIDNEY FERNBACH

Lawrence Radiation Laboratory
Livermore, California

MANUEL ROTENBERG

School of Science and Engineering
University of California
La Jolla, California

Volume 1

Statistical Physics

1963

ACADEMIC PRESS

NEW YORK AND LONDON

COPYRIGHT © 1963, BY ACADEMIC PRESS INC.
ALL RIGHTS RESERVED.
NO PART OF THIS BOOK MAY BE REPRODUCED IN ANY FORM,
BY PHOTOSTAT, MICROFILM, OR ANY OTHER MEANS, WITHOUT
WRITTEN PERMISSION FROM THE PUBLISHERS.

ACADEMIC PRESS INC.
111 Fifth Avenue, New York 3, New York

United Kingdom Edition published by
ACADEMIC PRESS INC. (LONDON) LTD.
Berkeley Square House, London W.1

LIBRARY OF CONGRESS CATALOG CARD NUMBER: 63-18406

PRINTED IN THE UNITED STATES OF AMERICA

Contributors

Numbers in parentheses indicate pages on which the authors' contributions begin.

MARTIN J. BERGER, *National Bureau of Standards, Washington, D.C.* (135)

BENGT G. CARLSON, *Los Alamos Scientific Laboratory, Los Alamos, New Mexico* (1)

DONALD H. DAVIS, *Lawrence Radiation Laboratory, University of California, Livermore, California* (67)

JOSEPH A. FLECK, JR., *Lawrence Radiation Laboratory, Livermore, California* (43)

L. D. FOSDICK, *Digital Computer Laboratory, University of Illinois, Urbana, Illinois* (245)

PAUL GANS, *University of Illinois, Urbana, Illinois* (217)

J. M. HAMMERSLEY, *Oxford University, Oxford, England* (281)

FREDERICK T. WALL, *Noyes Chemical Laboratory, University of Illinois, Urbana, Illinois* (217)

STANLEY WINDWER, *Noyes Chemical Laboratory, University of Illinois, Urbana, Illinois* (217)

CLAYTON D. ZERBY, *Neutron Physics Division, Oak Ridge National Laboratory, Oak Ridge, Tennessee* (89)

Preface

The computer has become the laboratory tool of the theoretical scientist. With it he hopes to learn new things which were previously inaccessible, in very much the same way as an experimentalist employs a piece of equipment. This hope depends to a large extent on the development of new methods designed specifically for the computer. The major aim of this series of books is to collect these techniques which were developed in the solution of physical problems.

The theoretical advances which can be made by this new tool appears to be enormous and in many different areas of science. In chemistry and physics, for example, many of the theoretical problems that are faced are mathematical in nature, the physical laws being well established. If, for example, the Schroedinger equation could be solved for more than two particles, molecular chemistry would become quantitative. To solve such problems, the first tendency is to transfer the well established techniques from the desk calculator to the high speed computer. In this way much can be learned by obtaining more accurate and extensive solutions. Alternatively, however, new methods should be thought about which are made practicable only by having access to a machine able to perform arithmetic at high speeds. It is the development of these new methods which we would particularly like to encourage in this series.

It is for this reason that our first volume covers mainly various aspects of the Monte Carlo method. This method is only suited for a high speed computer and permits the accurate numerical solution of many problems which are only approximately soluble by analytical techniques. In fact, the method was used in one of the first applications of computers, namely, the scattering of neutrons by a wall. Chapters *2* through *5* in this volume give the more recent applications and methods to the scattering of various particles. Chapter *1*, by way of contrast gives a competing method for solving the neutron transport problem. The last three chapters are concerned with application of the Monte Carlo method to problems in statistical mechanics. There are many more applications involving variations on the Monte Carlo method and it is hoped that another volume in the future will be devoted to these.

In the effort to make computers a more useful tool to the physical scientist, we feel it desirable to complement the scientific journals by reversing the emphasis between computational detail and physical results. This is because scientific journals, owing to space limitation, discourage detailed exposition of numerical techniques, even though these techniques are time consuming to develop, and frequently crucial to the solution of the problem. The policy has thus been established that a typical article would contain a statement of the physical situation in which the problem arose, previous analytical attempts, if any, for its solution, numerical techniques which were used (even those that were unsuccessful), a flow diagram if the problem is of general interest, the advantage and limitation of the method, the memory requirements of the program, the accuracy and convergence of the method and how they were ascertained, description of the numerical results, and so on.

It is hoped that the detailed description of techniques will lead to the development of even better techniques. By pooling experiences it should also lead to more efficient use of computers. Furthermore, we hope to expose those as yet not familiar with computers to the power of these tools. The results also ought to impress the skeptics who like to employ only analytical methods. Numerical solutions are never as general and as compact as analytical solutions. Nevertheless, the two methods complement each other in that a numerical solution tabulates a function in terms of which other problems can then be analytically expressed and, conversely, much analysis has to be done before a new numerical method is developed.

Much progress and activity can be expected in this field now that large computers have become generally available. In the next volume we hope to collect recent numerical advances in the field of quantum mechanics. The subsequent volume will be devoted to the field of hydrodynamics.

BERNI ALDER
SIDNEY FERNBACH
MANUEL ROTENBERG

April 1963

Contents

CONTRIBUTORS v
PREFACE vii

THE NUMERICAL THEORY OF NEUTRON TRANSPORT
Bengt G. Carlson

I. Introduction 1
II. Coordinate Systems 5
III. Derivation of Difference Equations 17
IV. Transformation of the Difference Equations 27
V. Solution of the Difference Equations 33
    References 42

THE CALCULATION OF NONLINEAR RADIATION TRANSPORT BY A MONTE CARLO METHOD
Joseph A. Fleck, Jr.

I. Introduction 43
II. Description of the Problem 44
III. Finite Difference Methods of Solution 47
IV. The Monte Carlo Method of Solution 51
V. Numerical Results 56
VI. Summary and Conclusions 65
    References 65

CRITICAL-SIZE CALCULATIONS FOR NEUTRON SYSTEMS BY THE MONTE CARLO METHOD
Donald H. Davis

I. Introduction 67
II. Calculation Details 69

III.	Particle Following	70
IV.	Estimates of α and the Equilibrium Distribution	75
V.	Collision Calculation	76
VI.	Examples of Calculations	80
	References	88

A MONTE CARLO CALCULATION OF THE RESPONSE OF GAMMA-RAY SCINTILLATION COUNTERS
Clayton D. Zerby

I.	Introduction	90
II.	Gamma-Ray Scintillation Counters	91
III.	Idealizations and Approximations	96
IV.	Sampling Procedures and Auxiliary Programs	103
V.	Details of the Monte Carlo Procedure	117
VI.	Results of the Calculations	125
	References	133

MONTE CARLO CALCULATION OF THE PENETRATION AND DIFFUSION OF FAST CHARGED PARTICLES
Martin J. Berger

I.	Introduction	135
II.	General Description of the Monte Carlo Method	139
III.	Particular Monte Carlo Schemes	144
IV.	Computational Aspects	157
V.	Solution of Typical Problems	165
VI.	Appendix: Single and Multiple Scattering Theories	202
	References	213

MONTE CARLO METHODS APPLIED TO CONFIGURATIONS OF FLEXIBLE POLYMER MOLECULES
Frederick T. Wall, Stanley Windwer, and Paul J. Gans

I.	Introduction	217
II.	Monte Carlo Methods	220
III.	Results and Conclusions	234
	References	242

MONTE CARLO COMPUTATIONS ON THE ISING LATTICE
L. D. Fosdick

I.	Introduction	245
II.	The Ising Lattice	246
III.	Theory of the Monte Carlo Method for Estimating the Boltzmann Averages	249
IV.	Practical Considerations of the Computation	258
V.	The Square Ising Lattice	261
VI.	The Simple Cubic Lattice	265
VII.	The Body-Centered Cubic Lattice	275
VIII.	Estimation of the Critical Point	278
IX.	Conclusion	279
	References	280

A MONTE CARLO SOLUTION OF PERCOLATION IN THE CUBIC CRYSTAL
J. M. Hammersley

I.	Introduction	281
II.	Schemes for Estimating $P(p)$ by Monte Carlo Methods	283
III.	Choice of Machine	288
IV.	Details of the Mercury Calculation	290
V.	Monte Carlo Refinements and Numerical Results	295
	References	298

AUTHOR INDEX 299

SUBJECT INDEX 302

The Numerical Theory of Neutron Transport

Bengt G. Carlson
LOS ALAMOS SCIENTIFIC LABORATORY

I. Introduction	1
A. General Principles	1
B. Comments on Development of Theory	3
II. Coordinate Systems	5
A. General Description	5
B. Mesh Systems in Time and Space	7
C. Mesh Systems for Direction	9
III. Derivation of Difference Equations	17
A. Difference Equations for Rectangular (x, y) Geometry	17
B. The General Difference Equation	21
C. Balance Equations	25
D. Main Properties of the Transport Equation	26
IV. Transformation of the Difference Equations	27
A. Elimination of the Time Variable	28
B. Treatment of the Source Term	29
C. Treatment of Angular Variation	31
D. Reduction of Dimensionality	32
V. Solution of the Difference Equations	33
A. Directional Evaluation	34
B. Treatment of Source Components	37
C. Scaled Inner Iterations	38
D. Comments on S_n Calculations	40
References	42

I. Introduction

A. GENERAL PRINCIPLES

The basic relations in neutron transport theory derive from simple statements regarding continuity of neutron flow and numerical balance of the number of particles involved. Neutrons moving into [leaving] a region D of space are also, at the instant they cross the boundaries of D, leaving [moving into] adjoining regions. The neutron properties: number, direction of motion, and velocity are not altered in the process. Continuous flow implies, then, the absence of interference at boundaries.

Statements of balance, for D during a time Δt, for example, equate change in neutron population to gains minus losses. Such statements are usually made for a set of neutron beams, each beam having a fixed direction and velocity. Neutrons flowing into D and those released by sources in D constitute the gains. Neutrons flowing out of D and those removed by the material in D make up the losses. By removals we really mean events leading to beam attenuation, that is, collisions between neutrons and the nuclei of the material. There are no collisions in D if D is in a vacuum. The rectilinear motion of neutrons between collisions is referred to as streaming.

It is assumed that very large numbers of neutrons are involved so that only their average behavior is of concern, and not the departures from that average. Also, in general, one can neglect neutron-neutron interactions entirely, which means that the neutrons flowing through a region D need not be counted as part of the material in D. This has the very important consequence that the equations for neutron transport are linear and thus much more amenable to solution than are the non-linear equations found for particle movement in gases or plasmas.

Sources may be of the surface type, neutrons per unit area per unit time, or of the volume type, neutrons per unit volume per unit time. Familiar surface sources are sources incident on the outer boundaries of systems (configurations, arrangements of materials) under study. In numerical treatment, surface sources are placed in the immediate vicinity of boundaries rather than right on them. This is in order to make it clear to what region they belong. Commonly occurring volume sources are the reemission sources, for example, the neutrons which after collision emerge scattered. Collisions are permitted to change the properties of the neutrons, that is, their number, direction, and velocity. A collision without change is not generally counted as a net collision. In combination, collision and volume source terms provide the mechanism for transferring neutrons between the various neutron beams of fixed direction and velocity. Sources external to a system, in time or space, are usually labeled initial values, or boundary values, respectively.

It is further assumed that the materials in the system are specified with respect to macroscopic properties, such as location, density, and isotopic composition, and with respect to microscopic character including interactions with neutrons, such as capture cross section, scattering cross section, and fission cross section. The cross sections depend on the velocity of the incident neutron as well as on the isotope with which the neutron interacts. The cross-section data are often given as functions of energy or lethargy rather than velocity, and include data about the released neutrons, that is, data about number (zero in the event of capture), velocity, and deflection cosine. Together, this information is used to calculate transfer cross sections which after multiplication by a density factor become transfer probabilities (macroscopic cross sections) in terms of events, such as captures, per unit length of neutron travel.

To supply a program of calculation with transfer cross sections is a considerable undertaking in view of the various difficulties with the basic data and the rather elaborate processing required. The cross sections often exhibit a complex resonance structure in energy and also correlation between energy and deflection angle in scattering. The precision of the experimental data is varying and seldom entirely adequate. In some areas data are entirely lacking.

The basic problem in the processing of cross sections is that of reducing continuous and often complicated sets of data to relatively small sets, without losing significantly in realism in the description of materials and their interactions with neutrons. The averaging rules, which one sets up and follows, inevitably make broad assumptions about the applications one has in mind, in particular about the neutron spectrum in energy which one expects within velocity groups. With sets of transfer cross sections are associated, therefore, statements specifying under what general conditions they may be applicable.

The cross-section compilations for the elements in the periodic table are updated from time to time as new or better data accumulate (Hughes and Schwartz, 1958; Argonne National Laboratory, 1958). An increasing number of cross sections are calculated from nuclear models. Also, the techniques for preparing the basic data for use in computation are advancing. Nevertheless, the cross-section problem in transport calculations remains a serious one. It should always be given careful attention.

Finally, it is assumed here that the materials in the configuration are at rest and have properties constant in time, or rather, that changes in these areas, if any, are relatively slow so that the combined problem, neutron transport with material changes, can be replaced by two problems interleaved in time: neutron transport for a short time with fixed material, change of material as indicated by motions, collisions, etc., during that time; then neutron transport for the next time cycle, another change of material, and so forth.

B. Comments on Development of Theory

Transport theory has developed quite swiftly during the last two decades. Much of the stimulus for the development came from problems connected with the design and operation of nuclear reactors. Methods from mathematical analysis were the main tools during the early period. Important basic results were obtained then by S. Chandrasekhar, in connection with the study of radiation flow in stellar atmospheres, and by G. Placzek and co-workers (among them B. Davison) in the Montreal group, mainly concerned with neutron problems. Many of the developments occurred during the time of World War II, in widely separated places at about the same time.

The general objective in transport theory is the determination of neutron flux, i.e., neutron flow per unit time, across unit area, from a difference or differential equation and as a function of time, position in space, direction in space, and neutron energy.

The analytical techniques first used did not really suffice—the problems were usually too complex—but helped along by simplifying assumptions, by ingenuity and innovation, and by judicious choice of calculations, the techniques proved quite effective in a variety of problems and over many years. They created, in fact, a large number of main topics within transport theory: Simple diffusion theory, Spherical harmonics, Age theory, Serber-Wilson technique, Integral theory, and several others. As the problems increased in complexity the analytical approach became increasingly strained. But then—and to some extent as a consequence—the high-speed computer emerged on the scene. With this revolution in computation, the emphasis shifted to numerical methods, first to methods effective within the separate topics of the theory, and then, somewhat later, to methods attacking the problem of transport directly, starting from first principles. Numerical methods are now clearly the most versatile tools in the field, provided that means of rapid computation are fairly accessible.

It is quite clear that many results in transport theory related to the existence of solutions, asymptotic behavior, series expansions, etc., and obtained by analytical means, cannot readily be deduced from difference equations or by the use of difference methods. For this and other obvious reasons, analytical and semianalytical methods will continue to play important, although somewhat changed roles in the theory. They remain important in the numerical context to help answer, for example, these basic questions: How do numerical results depend on the fineness of resolution specified for the variables? How do the various numerical methods compare? Analytical and numerical methods, separately, provide some, usually not very precise answers. Results from physical experiments, on the other hand, such as critical assemblies, are often compared as follows: experimental results to theoretical results to cross-section data used. This comparison is most often used to modify the cross sections until experiment and theoretical model come to reasonable agreement, a use which is quite sensible, generally, but precludes help with the questions just posed. Comparisons between numerical and analytical results for simple problems, where both sets of results can be obtained, remain, therefore, among the better means for gaining information about resolution and for judging numerical methods. Chandrasekhar (1960), Davison (1957) and Case *et al.* (1953), to mention only a few of the analysts, provide numerous opportunities along these lines.

A few basic numerical techniques and most of the methods for simplifying problems and theoretical description were introduced in the early period. The discrete method for handling neutron flux as a function of direction was introduced then. This method was first suggested by G. Wick but was mainly developed by S. Chandrasekhar (see Davison, 1957, pp. 174-182). Variations on it appeared later, an important one is from J. Yvon (see Davison, 1957, pp. 171-173). Early efforts in simplification produced the special topics within transport theory as mentioned previously and also the first methods for eliminating variables, processing cross sections, treating scattering laws, etc.

The main object of this article is to give the most general formulation possible at this time of a numerical theory of neutron transport. From this is derived a general method for solving neutron transport problems numerically, a method around which particular procedures, the topics of numerical transport theory, may group or develop. The basic difference equation of the theory will be derived directly from the principles and assumptions stated at the beginning.

Since the discrete representation of direction remains, the general formulation and method may be regarded as a generalization of the Wick-Chandrasekhar method. It is usually called the discrete S_n method or simply S_n for convenience and in reference to the first formulation which made the generalization possible. An effort has been made to include here the main information needed for the planning and flow diagramming of S_n calculations and codes, and also, either directly or indirectly through the references given, some guidance to what else is required in actual calculations.

II. Coordinate Systems

A. General Description

The later discussion of difference equations will be made in relation to three familiar and basic space symmetries, rectangular, cylindrical, and spherical. The left-side diagrams of Fig. 1 illustrate the coordinate systems corresponding to these symmetries. They name the variables of position (**r**) and direction (**Ω**), specify the components of the former, and relate these, in the cylindrical and spherical cases, to the rectangular coordinates x, y, and z. Shown also, in each left-side diagram, is a sample neutron path starting at position **r** (at point P) and extending some distance (Δs) in the direction **Ω**.

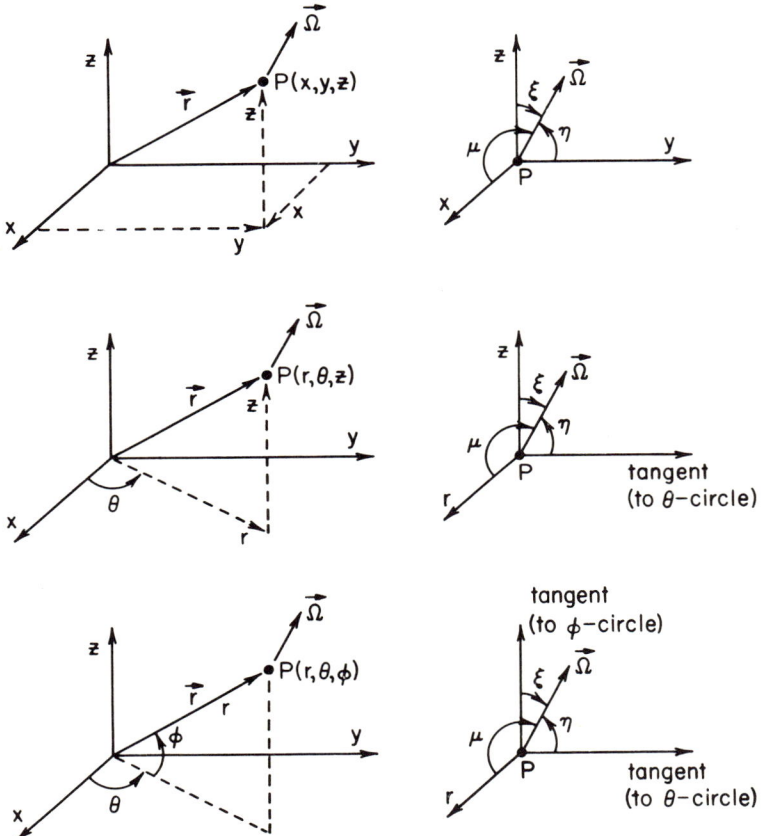

FIG. 1. Coordinate systems; fixed and local frames of reference for rectangular, cylindrical, and spherical symmetries. (Symbols with arrows correspond to boldface symbols in text, and μ, η, and ξ denote cosines.)

In the diagram to the right, again referring to Fig. 1, the components μ, η, and ξ of the unit vector **Ω** are shown. These are measured with respect to a local frame of reference with origin at P, thereby forming a coordinate triplet for specifying direction. The local reference system is lined up with the intersecting tangents touching the space coordinate lines, i.e., the grid lines passing through P. It is clearly a rectangular system since the basic geometries we are dealing with are orthogonal. The components of Ω obey, therefore, the relation $\mu^2 + \eta^2 + \xi^2 = 1$. One usually selects one of the components, say μ, as one of the independent variables, and writes the others, η and ξ, in terms of ω, a second one, as follows: $\eta^2 = (1 - \mu^2) \sin^2 \omega$ and $\xi^2 = (1 - \mu^2) \cos^2 \omega$.

The local frame of reference is here regarded as a moving one, attached to the neutron beam and moving with it along a straight line path. Since this frame is also tied to a space-coordinate system, it may change in orientation as it moves, and generally does this in cylindrical and spherical systems where the space coordinates are curvilinear. Neutrons in the process of streaming must therefore be permitted to change direction coordinates. This is beam attenuation quite apart from that provided by the mechanism of collision discussed earlier. The process will be called angular redistribution. It shifts neutrons in direction cosine, for descriptive purposes really, without affecting their number, basic direction, or velocity. It is the expression of continuity of flow unaffected by changes of variable.

Directions may be depicted as points on the surface of a unit sphere. In what follows it may be useful to refer to Fig. 2, which shows octants of unit spheres, and to Fig. 4, which shows the upper front quadrant of one, in perspective. In rectangular coordinate systems no change of direction cosines is possible during streaming. In cylindrical systems, neutron movement causes higher values of μ always with η first increasing while μ remains negative and then decreasing; ξ is uneffected. In spherical systems of full symmetry, i.e., in systems without θ and φ variation, the shift is also toward higher μ. In the general case, however, this pattern is combined with another one: a shift toward smaller values of ω if φ is positive, toward higher values if φ is negative.

There are some special points and lines (great circles) on the unit sphere. In the cylindrical case there is no angular redistribution along the line $\eta = 0$ and none into or out of directions having $\eta = 0$. In the spherical case there is no flow into or out of the directions $\mu = \pm 1$ and no flow into or out of other directions with $\eta = 0$ except by redistribution along the line $\eta = 0$. The neutron flux can readily be calculated for these special directions, a fact that will be made use of later.

B. Mesh Systems in Time and Space

In numerical treatment, discrete variables take the place of continuous ones and vary over domains that are finite sequences rather than intervals. Thus the time variable t in our case is assumed to take on the values t_s, where $s = 0, 1, 2, ...$, the elements of the sequence $\{t_s\}$, with steps $\Delta t_s = t_s - t_{s-1}$, where $s = 1, 2, 3, \cdots$. Similarly, the space variable x is assumed to range over x_i, where $i = 0, 1, 2, ...$, the elements of $\{x_i\}$, with intervals Δx_i, etc. We now adopt the following conventions. Time is to be represented by t and the subscript s, and position by the coordinate triplets (x, y, z), (r, θ, z), or (r, θ, φ), as the case may be, as well as by subscript triplets (i, j, k), with one-to-one correspondence between elements. Thus, for example, we shall use the terms (r, z) geometry and (i, k) cylindrical geometry interchangeably.

Many factors enter into the choice of geometry and the selection of time and space points. The geometry is most often selected on the basis of the dominating form of the surfaces which define the configuration. Thus, for a right circular cylinder, inside a similar figure and sharing its axis, with no other forms in sight, one obviously chooses cylindrical geometry. Pertinent defining surfaces here are those enclosing the system or separating the materials within it, those carrying surface sources, and those delineating volume sources.

To form a space mesh, that is, to fill all of space with mesh intervals and simultaneously with mesh points, we let the sets $\{x_i\}$, $\{y_i\}$, and $\{z_i\}$, define mutually orthogonal sets of surfaces in space. The mesh interval

TABLE I

AREA AND VOLUME ELEMENTS FOR MESH CELLS IN RECTANGULAR, CYLINDRICAL, AND SPHERICAL COORDINATE SYSTEMS.

Geometry and variables	A	Area elements B	C	Volume element
Rectangular				
x	1	...	...	Δx_i
x, y	Δy_j	Δx_i	...	$\Delta x_i \Delta y_j$
x, y, z	$\Delta y_j \Delta z_k$	$\Delta x_i \Delta z_k$	$\Delta x_i \Delta y_j$	$\Delta x_i \Delta y_j \Delta z_k$
Cylindrical	$C_i = (r_{i+1}^2 - r_i^2)/2$			
r	$2\pi r_i$	...	...	$2\pi C_i$
r, θ	$r_i \Delta \theta_j$	Δr_i	...	$C_i \Delta \theta_j$
r, z	$2\pi r_i \Delta z_k$	...	$2\pi C_i$	$2\pi C_i \Delta z_k$
r, θ, z	$r_i \Delta \theta_j \Delta z_k$	$\Delta r_i \Delta z_k$	$r_i \Delta r_i \Delta \theta_j$	$C_i \Delta \theta_j \Delta z_k$
Spherical	$S_i = (r_{i+1}^3 - r_i^3)/3$,	$S_k = (\sin \varphi_{k+1} - \sin \varphi_k)$		
r	$4\pi r_i^2$	...	...	$4\pi S_i$
r, φ	$2\pi r_i^2 S_k$	...	$2\pi C_i \cos \varphi_k$	$2\pi S_i S_k$
r, θ, φ	$r_i^2 \Delta \theta_j S_k$	$C_i \Delta \varphi_k$	$C_i \Delta \theta_j \cos \varphi_k$	$S_i \Delta \theta_j S_k$

defined by $(x_i, x_{i+1}, y_j, y_{j+1}, z_k, z_{k+1})$ is then a six-sided figure of volume V with areas A_i and A_{i+1} in the x-direction, B_j and B_{j+1} in the y-direction, and C_k and C_{k+1} in the z-direction. Similar statements can be made for the curved geometries. Formulas for the volume V and the areas A, B, and C are given in Table I for the basic coordinate sytems and most of the subsystems.

The defining surfaces of a physical configuration must often be modified in order to accommodate that configuration in one of the available coordinate descriptions. Modifications can occasionally be avoided or reduced by introducing a special geometry. But usually one accepts the fact that physical configurations must be idealized, and, for example, in two-space, that surfaces (here lines) must be pieced together from straight lines and/or circles or pieces thereof. Generally one tries to select abscissas, elements for the sequences $\{x_i\}$, $\{y_i\}$, etc., so that the surfaces of the configuration are on surfaces in the mesh, as far as this is possible and practical. All mesh cells are assumed to be uniform in material and free internally of boundaries of any kind.

In selecting elements for mesh-defining sequences one takes into account a number of other factors as well: the resolution one has in mind for the direction variable, the relative importance of subregions of the configuration, and perhaps others. The main object is to define a calculation which will yield results of about the character and detail one desires without excessive computing or human efforts.

C. Mesh Systems for Direction

In infinite slab (x) and in spherical (r) geometries, where all quantities are independent of the space variables (y, z) and (θ, φ), respectively, the symmetry is such that direction can be described by one cosine, say μ. The points on a fixed μ-level on the unit sphere can be combined, in these cases, into a single point. Since μ varies from -1 to $+1$ and summations over this interval are of fundamental importance, the procedures of mechanical quadrature come to mind. The Gauss abscissas thus became the first discrete representation $\{\mu_l\}$ of μ, and the associated weights w_l the first neighborhoods of μ_l. We shall refer to μ_l as the level cosines and to w_l as the level weights. The order n of the approximation refers to the number of abscissas or μ-levels on the interval $(-1, 1)$. Among many relations satisfied by μ_l and w_l in the Gaussian quadrature set, the most important ones are

$$\sum_l w_l = 1, \quad \sum_l w_l \mu_l = 0, \quad \text{and} \quad \sum_l w_l \mu_l^2 = \tfrac{1}{3}, \tag{1)[1]}$$

---

[1] Σ is used throughout as a symbol of summation, and the subscript specifies the index of summation.

where the summations extend over the n μ-levels. We expect these relations to be satisfied also for other quadrature sets which may be introduced.

The first relation in (1) represents solid-angle normalization. In the general case, the sum extends over one, two, or four quadrants, depending on the number needed, which in turn depends on the geometry, whether the latter has one, two or three space dimensions. The second relation, besides reflecting a degree of symmetry about $\mu = 0$, is of considerable importance, as we shall see later, in connection with angular redistribution (with the coefficients involved) in curved geometries. There are also a number of important aspects to the last relation in (1). One will become apparent in the next few paragraphs, another is encountered in Section III.

The above method for treating direction was introduced nearly twenty years ago. In transport theory it eventually became known as the method of discrete ordinates, the Wick-Chandrasekhar method, or simply "discrete P_{n-1}" (see Davison, 1957, pp. 174-182). To avoid the confusion and difficulties of having directions along mesh lines, quadrature sets with nonzero μ_l are normally chosen. For the standard sets this means even n. The P_1 method, the case with fewest directions, is closely related to what is known as the diffusion approximation or Simple diffusion theory. The largest incremental change of accuracy for a given problem is generally observed when P_1 resolution is replaced by that of P_3. After that the effects of added resolution tend to come slowly.

Considerable gain in precision, holding n fixed, was obtained by placing Gauss abscissas separately on the intervals $(-1, 0)$ and $(0, 1)$. In some applications the gain was quite startling. This variation in procedure is ascribed to J. Yvon (see Davison, 1957, pp. 171-173). It is usually called the Double-P method, in our notation $DP_{n/2-1}$. In the neighborhood of $\mu = 0$ the abscissas of Double-P are denser than those of Single-P. It is conjectured that this and the fact that, in addition to (1), the relation $\Sigma_l w_l \mid \mu_l \mid = \frac{1}{2}$ is satisfied, makes Double-P especially accurate for problems in which boundary effects dominate. This effect has been observed for thin slabs without reflectors, for instance, and also for layer problems, where thin sheets of absorbers and scatterers alternate.

The P_{n-1} and $DP_{n/2-1}$ quadrature sets, together with a third set (S_n) to be derived, are regarded as basic sets in numerical transport theory. All satisfy the relations (1). Several other sets have, however, been introduced. The original S_n set, at the basis of Carlson and Bell (1958), but never tabulated explicitly, had some special features. It accorded weight to $n+1$ directions but the cosines were not symmetric about $\mu = 0$. Although the asymmetry was to advantage in some applications, it generated large flow distortions in others. There are problems which benefit markedly from special sets of directions, however. No one particular quadrature set can necessarily be singled out as the best. A choice of representation is as important here as in the case of the space variables, including choice of resolution as well as placement of points.

As it became possible and practical to carry out calculations in two or more space dimensions, another question arose with regard to distribution of points on the unit sphere. This is concerned with asymptotic behavior in space. If one moves to regions distant from the origin one often has or else approaches the conditions of plane one-dimensional (slab) geometry. In a finite cylinder, for example, the upper strata, slabs by definition, may be quite uniform in material and of sizable diameter. The outer shells of the same cylinder may also be extended and fairly uniform, and far out they have effectively lost their curvature. Since directions with respect to perpendicular axis are the only directions of consequence in extended slab geometry, one would prefer to have representations of direction exhibit projection invariance, that is, yield the same one-dimensional quadrature sets, weights and cosine-levels, independent of the axis on which the general sets were projected.

In the remainder of this section we shall explore the consequences of this invariance with regard to both the directions on the unit sphere and the quadrature weights to be associated with these directions.

Referring to Fig. 2, the lower left point in the diagram for $n = 8$, we find that our symmetry requirement implies

$$\mu_4^2 + \mu_1^2 + \mu_1^2 = 1 , \qquad (2)$$

for that point, or stated otherwise, that $\eta_1 = \xi_1 = \mu_1$. Here, in counting levels with respect to the x-pole, we start near the side opposite x and then move toward the pole, and similarly for levels with respect to y and z. Now, if we move up from the lower left point, the indices of μ, η, and ξ change by -1, 0, and $+1$, respectively, and, if we move sideways instead, by -1, $+1$, and 0. Hence, by an inductive proof:

$$\mu_l^2 + \mu_m^2 + \mu_{n/2-l-m+2}^2 = 1 , \qquad (3)$$

where $l = 1, 2, 3, ..., n/2$; $m = 1, 2, 3, ..., n/2 - l + 1$.

There are $n(n+2)/8$ points per octant in this scheme. An octant exhibits $n/2$ of the μ-levels, a full quadrant all n of them. Examination of the octants in Fig. 2 shows that the points fall into three symmetry types, classes of points with six members, classes with three, and classes of one. For $n = 8$ we have all three types represented. Symmetry requires that points within a given class have identical weights. With this satisfied, it can readily be shown, using Eq. (3), that the average μ^2 equals one-third for each class of points, regardless of symmetry

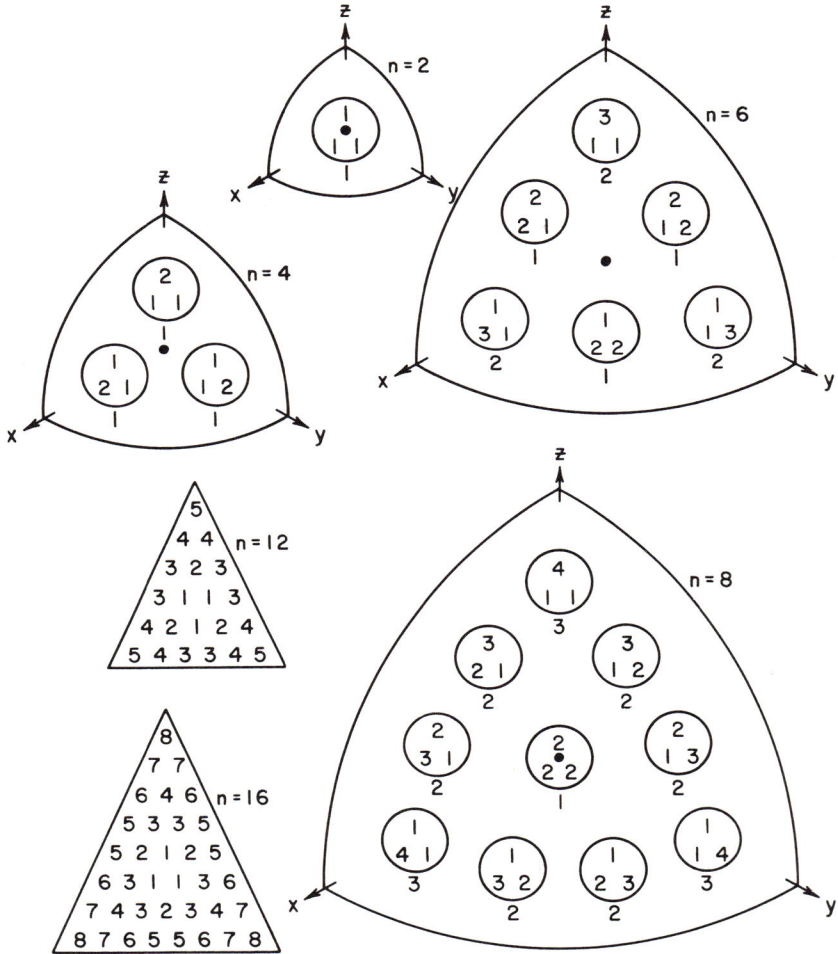

FIG. 2. Unit sphere octants, illustrating the direction mesh for $n = 2, 4, 6, 8, 12$, and 16, giving indices l of the direction cosines (in circles) and index m of the weight (under circles) for each discrete direction. The indices refer to the quantities μ_l and w_m in the tables. For $n = 12$ and 16 only indices of weight are given.

type and pole of reference and hence for all points on the quadrant independent of the separate point weights and the summation of these into level weights. This establishes the last condition of equation (1) for any direction set with projection invariance.

Equation (3) has a very simple solution derived from the fact that the indices sum to $n/2 + 2$, namely,

$$\mu_l^2 = \mu_1^2 + (l-1)\Delta, \qquad l = 1, 2, ..., n/2, \qquad (4)$$

where

$$\Delta = 2(1 - 3\mu_1^2)/(n-2). \qquad (5)$$

Hence, if μ_1 is given, all other μ's can be calculated, the value of μ_1 defining the spread of the points on the octant.

If we now let $W_1 = w_1$, $W_2 = w_1 + w_2$, etc., with

$$W_l = w_1 + w_2 + ... + w_l,$$

we find, making use of Eq. (4) and the third part of Eq. (1), that

$$\sum_l W_l = (n-2)/3, \qquad (6)$$

where the sum extends from $l = 1$ to $n/2 - 1$. We shall have need for this result shortly.

There is probably no unique and best procedure for determining μ_1 and W_l. There are not sufficient degrees of freedom for moment methods to be helpful. Negative numbers among the weights can hardly be tolerated. The following scheme for determining μ_1 and W_l is based on certain analogies with the Gauss case and has the virtue of simplicity.[2] In Gauss quadrature the μ's are approximately equal to the cosines of a set of angles linear in l (Szegö, 1939, p. 188), the weights corresponding to W_l approximately the cosines of the equidistant midpoint angles. In our case we have linearity in μ^2 and the transformation is "square root" rather than "cosine." Hence, by analogy,

$$W_l^2 = W_1^2 + (l-1)\Delta, \qquad l = 1, 2, ..., n/2 - 1. \qquad (7)$$

To progress from here in a simple manner, we anticipate the final result; that is, we let $\Delta = 2/(n-1)$. Next, we substitute this in Eq. (7), then (7) in Eq. (6), and solve for W_1. The numerical value found may

[2] The quadrature sets obtained in this way agree quite well with those calculated by Lee (1961, Chapter 3), using the Area method. In this method, the area on the unit sphere is divided and assigned to the various points, all in a well-defined manner.

be written in this form: $W_1^2 = (4 + \epsilon_n)/3(n-1)$, where ϵ_n turns out to be a small quantity which goes to zero as n approaches infinity. This can be shown using Euler's summation formula.

Continuing the analogy, we write $\mu_1^2 = W_1^2 - \Delta/2$ and this, in combination with (5), gives

$$\mu_1^2 = 1/3(n-1), \qquad \Delta = 2/(n-1), \tag{8}$$

as approximate expressions for μ_1^2 and Δ. Since μ_1^2 (or Δ) is a free parameter with the proviso that $\mu_1^2 = 1/3$ for $n = 2$, we made a good choice above for Δ and the decision is simple. We let μ_1^2 and hence Δ be defined by (8). The following values are found for ϵ_n and $n = 4$, 6, 8, 12, and 16: 0, 0.00416667, 0.00539164, 0.00592059, and 0.00583892.

Given the level weights ($n > 2$) we cannot determine more than $n/2 - 1$ different point or class weights. We have that many for $n = 4$, 6, 8, 10, and 12. After that, the number of point classes grow very rapidly. This number is equivalent to the number of different ways one can construct the sum $n/2 + 2$ using three numbers from the sequence 1, 2, ..., $n/2$. It is given by

$$(1/4)\,[n + (n-4) + (n-12) + (n-16) + (n-24) + ...]$$

if $n/2$ is even and by

$$(1/4)\,[(n+2) + (n-6) + (n-10) + ...]$$

if $n/2$ is odd. In both cases the summation terminates with the last positive term. To get around the indeterminacy for large n, we assume that the point weights, like the point definitions, depend on a more fundamental set, a basis $\{a_l\}$ of weights, $l = 1, 2, ..., n/2$, such that the weights w corresponding to the directions (μ_l, μ_m, μ_p), $p = n/2 - l - m + 2$, may be written $w = a_l + a_m + a_p$, or, if the a's are redefined,

$$w = \bar{a}_l \mu_l^2 + \bar{a}_m \mu_m^2 + \bar{a}_p \mu_p^2, \qquad \bar{a} = a/\mu^2.$$

We find also, upon closer examination, that $a_{n/2}$ may be set to zero without restriction. The above provides then a solution to our problem, a form of analytical continuation for w over the unit sphere.

The above S_n quadrature set is labelled Set A. Many variations are clearly possible here. A second set, called Set B, available for $n = 4$, 6, 8, ..., which is important for several reasons, is developed below. In one sense this set corresponds to "odd" approximations, $n - 1 = 3, 5, 7, ...$; in another, it relates to Set A in the way Double-P relates to Single-P. Since there is no proviso here for $n = 2$, μ_1 is a completely free parameter and may be determined so that the relation

$$\sum_l w_l |\mu_l| = 1/2, \tag{9}$$

supplementary to Eqs. (1), is satisfied. One expects then Set B, for fixed n, to provide better accuracy in applications than Set A.

Set B is constructed by selecting, for a particular n, the point pattern

TABLE II

DIRECTION COSINES AND WEIGHTS, PROJECTION-INVARIANT SETS (S_n).[a]

n and l or m	w_l	μ_l	μ_l^2	w_m	c_m
Set A					
$n = 2$					
$l = 1$	1.00000000	0.57735027	0.33333333	1.00000000	1
$n = 4$					
$l = 1$	0.66666667	0.33333333	0.11111111	0.33333333	3
2	0.33333333	0.88191710	0.77777778	...	
$n = 6$					
$l = 1$	0.51666667	0.25819889	0.06666667	0.15000000	3
2	0.30000000	0.68313005	0.46666667	0.18333333	3
3	0.18333333	0.93094934	0.86666667	...	
$n = 8$					
$l = 1$	0.43672981	0.21821790	0.04761905	0.07075470	1
2	0.25352175	0.57735027	0.33333333	0.09138352	6
3	0.18276705	0.78679579	0.61904761	0.12698139	3
4	0.12698139	0.95118972	0.90476189	...	
Set B					
$n = 4$					
$l = 1$	0.33333333	0.09175171	0.00841837	0.33333333	3
2	0.66666667	0.70412415	0.49579082	...	
$n = 6$					
$l = 1$	0.33333333	0.11417547	0.01303604	0.00000000	1
2	0.33333333	0.57735027	0.33333333	0.16666667	6
3	0.33333333	0.80847425	0.65363062	...	
$n = 8$					
$l = 1$	0.30807338	0.11104445	0.01233087	0.02525996	3
2	0.27921634	0.50307327	0.25308272	0.07937696	3
3	0.18401386	0.70273364	0.49383457	0.11434821	6
4	0.22869642	0.85708017	0.73458642	...	

[a] c_m = number of points of weight w_m.

TABLE III

DIRECTION COSINES AND WEIGHTS, GAUSSIAN SETS.[a]

n and l	w_l	μ_l	μ_l^2
P_{n-1} Set			
$n = 2$			
$l = 1$	1.00000000	0.57735027	0.33333333
$n = 4$			
$l = 1$	0.65214515	0.33998104	0.11558711
2	0.34785485	0.86113631	0.74155574
$n = 6$			
$l = 1$	0.46791394	0.23861919	0.05693912
2	0.36076157	0.66120939	0.43719786
3	0.17132449	0.93246951	0.86949939
$n = 8$			
$l = 1$	0.36268378	0.18343464	0.03364827
2	0.31370665	0.52553241	0.27618431
3	0.22238103	0.79666648	0.63467748
4	0.10122854	0.96028986	0.92215662
$DP_{n/2-1}$ Set			
$n = 4$			
$l = 1$	0.50000000	0.21132487	0.04465820
2	0.50000000	0.78867513	0.62200846
$n = 6$			
$l = 1$	0.27777778	0.11270167	0.01270167
2	0.44444444	0.50000000	0.25000000
3	0.27777778	0.88729833	0.78729833
$n = 8$			
$l = 1$	0.17392742	0.06943184	0.00482078
2	0.32607258	0.33000948	0.10890626
3	0.32607258	0.66999052	0.44888730
4	0.17392742	0.93056816	0.86595710

[a] Data taken from Lowan, Davids, and Levenson (1942). *Bull. Am. Math. Soc.* **48**, 739. 1942.

for $n + 2$ and then deleting the corner points and hence one of the μ-levels. One finds then, corresponding to Eqs. (4), (5), and (6),

$$\mu_l^2 = \mu_1^2 + (l-1)\Delta, \tag{10}$$

$$\Delta = 2(1 - 3\mu_1^2)/n, \tag{11}$$

$$\sum_l W_l = (n-3)/3, \tag{12}$$

and selects, corresponding to (7),

$$W_l^2 = W_1^2 + (l-1)\Delta', \quad \text{with} \quad \Delta' = 2/n. \tag{13}$$

By substitution of (13) in Eq. (12), W_1 can be determined, and with w_l thus made available, μ_1 can be determined by entering (10) in Eq. (9).

The following values are found for W_1 and μ_1 for $n = 4, 6, 8, 12,$ and 16: $W_1 = \frac{1}{3}, \frac{1}{3}, 0.30807338, 0.26479622,$ and 0.23443138; and $\mu_1 = 0.09175171, 0.11417547, 0.1110445, 0.9861006,$ and 0.08811484.

The basic quadrature sets are given in Table II, for the S_n Sets A and B, and in Table III, for Single-P and Double-P.

III. Derivation of Difference Equations

A. Difference Equations for Rectangular (x, y) Geometry

We turn now to the derivation of difference equations and consider first the case of neutron transport in a rectangular (x, y) geometry as a function of time. Here we can derive an important equation and at the same time illustrate the general numerical procedures which are being employed without getting too immersed in details and dimensions. In the next subsection we take up the general case, the case with three space variables in any one of the main geometries, and show that there is one basic difference equation which serves all three geometries. This gives the whole numerical theory a much desired internal consistency. Equations for particular cases are obtained by deleting or specializing terms in the general equation. Equations for different numerical models come from making substitutions in that same equation.

In what follows we shall have frequent occasion to refer to Fig. 3 which illustrates a typical mesh cell in (x, y) geometry and two basic methods for representing the magnitude and variation of neutron flux over that cell. Two principal numerical models in transport theory come from these representations. The first is called the Step model or the step-function method and the second the Diamond model or the line-segment method.

The neutron balance statement for an arbitrary cell in (x, y) geometry embraces the following quantities.

1. Position Dependent Quantities

These quantities are the volume V of the cell in units of length cubed and the surfaces A, B, and C of the cell in units of length squared. For V, A, B, and C, as functions of geometry, see Table I. The surface elements may vary as one moves across the cell from one side to the opposite, e.g., A_{i+1} may be different from A_i, though this is not the case in (x, y) geometry.

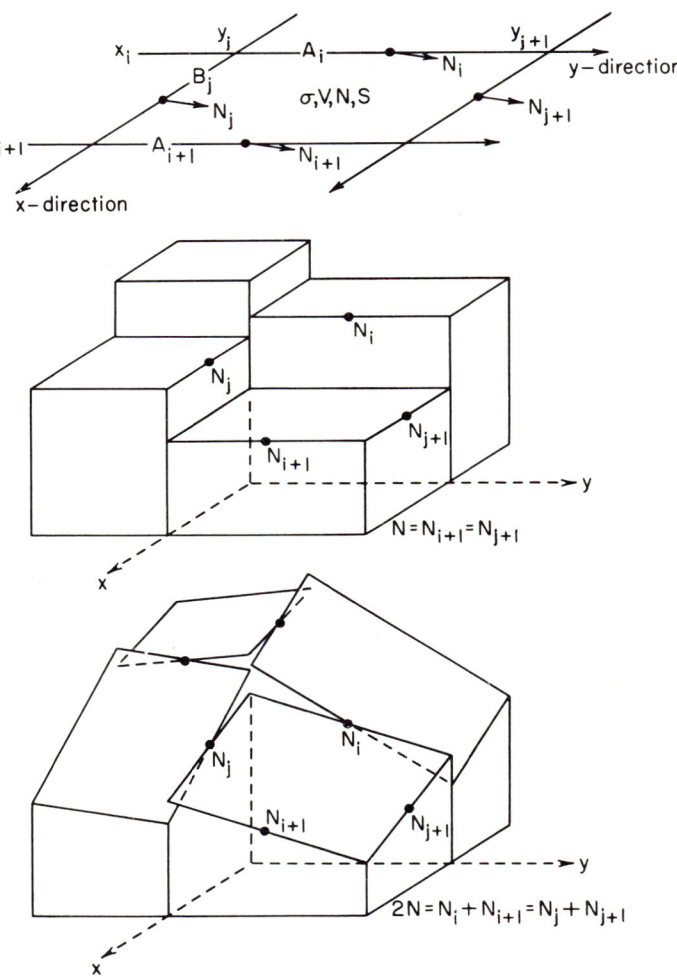

Fig. 3. Illustration of mesh cell for (x, y) geometry and two models for representing the neutron flux N: the Step and Diamond models.

A convenient subscript convention here omits subscripts of no immediate interest and subscripts referring to midpoints of intervals, and retains those (usually none or one) referring to ends of intervals. Thus A_i stands for $A_{i,j+1/2}$, B_j for $B_{i+1/2,j}$, V for $V_{i+1/2,j+1/2}$, etc.

2. *Position and Velocity Dependent Quantities*

These quantities are the collision probabilities σ, per unit length of neutron travel, in the uniform material of the cell. Energy variation is denoted by the group subscript g, when such indication is necessary. The velocity for a particular neutron energy interval of width u is denoted by v and the unit here is length per unit time. Both u and v are independent of position.

3. *Time, Position, Direction, and Velocity Dependent Quantities*

These quantities are the intensities σN and S of respectively neutron collisions and neutrons emitted from the source, in the energy interval u, per unit time, volume, and solid angle. N will in general vary over the cell according to the numerical representation (model) assumed (see Fig. 3). We distinguish between three densities: collision density σN flow density or flux N, and particle density N/v.

Since the intensity of flux N is measured per unit area perpendicular to the direction of flow, the flux crossing a unit area with some other normal direction will differ from N by a projection factor, specifically by the factors μ, η, and ξ in the directions of the axes of the local frame of reference.

We now let N_i, $N_i \equiv N_{i,j+1/2,s+1/2}$, represent the average intensity of flow across the boundary marked A_i in Fig. 3, during the time interval Δt (omitting here subscripts for direction and velocity), with similar interpretations for N_{i+1}, N_j, and N_{j+1} at the other boundaries. Further, we let N_s, $N_s \equiv N_{s,i+1/2,s+1/2}$, represent the average intensity of flow across the cell at time t_s, with the same interpretation of N_{s+1}, except at time t_{s+1}. The neutron balance terms, for a cell during a time Δt (abbreviated Δ below), can now be tabulated for a particular velocity group of width u and a particular direction of weight w:

(1) Number of neutrons in cell at times t_s and t_{s+1}, wVN_s/v and wVN_{s+1}/v.

(2) Number flowing out over areas A_{i+1} and B_{j+1}, $w\Delta \mu A_{i+1} N_{i+1}$ and $w\Delta \eta B_{j+1} N_{j+1}$.[3]

(3) Number flowing in over areas A_i and B_j, $w\Delta\mu A_i N_i$ and $w\Delta\eta B_j N_j$.[3]

---

[3] Whether these terms represent flow in or flow out depends of course also on the signs of μ and η. Negative signs reverse the interpretations given.

(4) Number of collisions in the cell, $w\Delta\sigma VN$.

(5) Number emitted in the cell by the source, $w\Delta VS$.

Combining the above, first dividing by a factor $w\Delta V$, we obtain the following balance equation, the basic difference equation for rectangular (x, y) geometry,

$$\frac{N_{s+1} - N_s}{v\Delta} + \mu \frac{A_{i+1}N_{i+1} - A_i N_i}{V} + \eta \frac{B_{j+1}N_{j+1} - B_j N_j}{V} + \sigma N - S = 0. \tag{14}$$

The above procedure looks after continuity as well as neutron balance. Every term in Eq. (14) has a clear and specific physical interpretation. Any loss term caused by flow, such as those under (2), just given, will become a gain for an adjoining cell unless it represents leakage out of the whole system. Any gain caused by flow is found to have come from an adjoining cell or else from input prescribed by boundary conditions. Similarly, continuity in taking time steps is assured. Here the flow at time t_0 represents initial values. Since continuity is satisfied it is possible to replace certain sums over volume elements, for a given system, by sums extending only over the area elements on the outer surface of that system.

Different models come from different assumptions with regard to the effective or average N over the cell. In the Step model we make the assumption that N equals the boundary N's in the cell at those particular boundaries (half of the total) toward which the flow is directed, in other words that N is constant over the cell. Hence in our example $N = N_{i+1} = N_{j+1} = N_{s+1}$ (see Fig. 3, middle diagram). In the Diamond model we always assume that the sum of N's on opposite cell boundaries are equal and, moreover, equal to twice the average N for the cell. Hence in our example $2N = N_i + N_{i+1} = N_j + N_{j+1} = N_s + N_{s+1}$. It is also feasible to construct mixed Step and Diamond models.

The model relations, that is, the additional relations just discussed, combine with the general equation (14) to define particular transport theory models. They make it possible, moreover, to change (14) into a recursion formula, i.e., to an expression explicitly solved for one of the N's, the unknown N, in terms of the given source and the known N's. The latter come from previous recursive steps, or else from boundary and/or initial conditions. Applying the foregoing model relations to our (x, y) geometry example, we have

$$N_{i+1} = \frac{\frac{\mu}{\Delta x} N_i + \frac{\eta}{\Delta y} N_j + \frac{1}{v\Delta} N_s + S}{\frac{\mu}{\Delta x} + \frac{\eta}{\Delta y} + \frac{1}{v\Delta} + \sigma} \tag{15}$$

for the Step model, and

$$N_{i+1} = \frac{\left(\frac{\mu}{\Delta x} - \frac{\eta}{\Delta y} - \frac{1}{v\Delta} - \frac{\sigma}{2}\right) N_i + \frac{2\eta}{\Delta y} N_j + \frac{2}{v\Delta} N_s + S}{\frac{\mu}{\Delta x} + \frac{\eta}{\Delta y} + \frac{1}{v\Delta} + \frac{\sigma}{2}} \quad (16)$$

for the Diamond model.

A few remarks regarding surface sources are now in order, since the above derivations and equations refer only to volume sources S. A surface source (T) lies, by agreement, inside the configuration and, furthermore, in the immediate vicinity of cell boundaries rather than right on these. The units, variables, and other general assumptions, which have been stated for N, apply also to T. Thus, again referring to Fig. 3, if T_i had been placed just below the line A_i, then N_i would be modified by T_i before it was used in Eq. (15) for the evaluation of N_{i+1}. Similarly, if the source T_{i+1} had been placed just above the line A_{i+1}, then N_{i+1} would be calculated by (15) but modified by the addition of T_{i+1} before there is flow across A_{i+1}.

B. The General Difference Equation

Through the study of (x, y) geometry in the previous subsection, we established, among other things, the form of the difference terms which have to do with movements in time and position. The general difference equation, Eq.(17), contains, therefore, really only two terms which are in need of clarification. These are the somewhat similar terms, involving the coefficients α and γ, terms connected with angular redistribution in curved geometries discussed in Section II, A. The general equation is given by

$$\frac{N_{s+1} - N_s}{v\Delta} + \mu \frac{A_{i+1} N_{i+1} - A_i N_i}{V} + \eta \frac{B_{j+1} N_{j+1} - B_j N_j}{V}$$
$$+ \frac{A_{i+1} - A_i}{2Vw} (\alpha_{l+1} N_{l+1} - \alpha_l N_l) + \xi \frac{C_{k+1} N_{k+1} - C_k N_k}{V} \quad (17)$$
$$+ \frac{C_{k+1} - C_k}{2Vw} (\gamma_{m+1} N_{m+1} - \gamma_m N_m) + \sigma N - S = 0,$$

which we shall also write as

$$D \cdot N + \sigma N \equiv [D + \sigma] \cdot N = S, \quad (18)$$

where D and $[D + \sigma]$ are difference operators defined by Eq. (17).

In Eq. (18), according to the convention set up in the previous subsection, the N's receive all subscripts, and the model relations of Table V extend to cover the new indices, the direction subscripts l and m. We shall see in the following paragraphs that all α and γ are nonnegative. Thus the terms $-\alpha N$ and $-\gamma N$, which have the same signs as S in Eq. (17), represent gains (through redistribution) for the angular interval encompassed by w, whereas the corresponding terms with plus signs represents losses to adjoining intervals. Angular intervals may be visualized as areas on the unit sphere proportional to the w's and situated about the points which define directions and triplets of cosines.

The curvature term in α is clearly associated with the difference (net flow) term involving the A's, the curvature term in γ with the net flow term involving the C's. The quantity $(A_{i+1} - A_i)/2V$ arises as a result of area divergence in the r-direction. It is nonzero for cylinders and spheres, and, in the limit of small r-intervals, equal to $1/2r$ for cylinders and $1/r$ for spheres. The quantity $(C_{k+1} - C_k)/2V$, similarly, results from area divergence in the φ-direction. It is equal to $-\tan \varphi/r$ in the limit. All of these terms are in a sense expressions for average curvature.

Based on a uniformity argument, the curvature terms in Eq. (17) and relations for α and γ are readily established. We regard (17) as a system of equations, for a particular velocity group, containing one equation for each discrete direction. We assume next a configuration specified in such a manner that the neutron flow is uniform. This means that all N are identically equal to some constant and that, for each separate cell and angular interval, the net flow is zero. Equations (17) must, of course, correctly describe this particular situation. From the above, noting that any coordinate system may be used, we find that the curvature terms added to their associate terms must be equal to zero. Hence we deduce for the α-terms,

$$\alpha_{l+1} - \alpha_l = -2w\mu . \qquad (19)$$

Similarly for the γ-terms,

$$\gamma_{m+1} - \gamma_m = -2w\xi . \qquad (20)$$

Note further that the first and last α in any sequence α_l are equal to zero. This is because the outer edges of the first and last angular intervals coincide with the singular loci on the unit sphere, discussed in Section II, A, which do not contribute or remove neutrons in the process called angular redistribution. Similarly, the first and last γ in a sequence of γ_l are zero. If the separate equations in (17) are now multiplied by the appropriate w's and then summed termwise over l and m, we find that the terms in α and γ vanish altogether, i.e., that the net effect of angular redistribution is zero, as it should be.

The coefficients in Eq. (17) are in the limit of small intervals, as it turns out, given by certain projection factors, the derivatives of the variables designated by the subscripts with respect to a distance s measured along the direction of neutron travel. These derivatives are given in Table IV, and the detailed derivations are found in Lee (1961, Chapter 2).

TABLE IV

PROJECTION FACTORS IN THE DIRECTION OF COORDINATE AXES IN RECTANGULAR, CYLINDRICAL, AND SPHERICAL COORDINATE SYSTEMS.[a]

Rectangular	Cylindrical	Spherical
$dt/ds = 1/v$	$dt/ds = 1/v$	$dt/ds = 1/v$
$dx/ds = \mu$	$dr/ds = \mu$	$dr/ds = \mu$
$dy/ds = \eta$	$d\theta/ds = \eta/r$	$d\theta/ds = \eta/r \cos\varphi$
$dz/ds = \xi$	$dz/ds = \xi$	$d\varphi/ds = \xi/r$
$d\mu/ds = 0$	$d\mu/ds = \eta^2/r$	$d\mu/ds = (1 - \mu^2)/r$
$d\eta/ds = 0$	$d\eta/ds = -\mu\eta/r$	$d\eta/ds = (-\mu\eta - \eta\xi \tan\varphi)/r$
$d\xi/ds = 0$	$d\xi/ds = 0$	$d\xi/ds = (-\mu\xi + \eta^2 \tan\varphi)/r$
$d\omega/ds = 0$	$d\omega/ds = -\eta/r$	$d\omega/ds = -\eta \tan\varphi/r$
	$\mu^2 = (1 - \xi^2) \cos^2\omega$	$\eta^2 = (1 - \mu^2) \sin^2\omega$
	$\eta^2 = (1 - \xi^2) \sin^2\omega$	$\xi^2 = (1 - \mu^2) \cos^2\omega$

[a] The quantity ds denotes an infinitesimal distance in the direction of motion.

In the remainder of this subsection we discuss details concerning the curvature terms and the progressions on the unit sphere.

In cylindrical geometry the angular redistribution proceeds along lines of fixed ξ on the unit sphere, α starting with zero and μ with the value $-\sqrt{1 - \xi^2}$ ($w = 0$), then proceeding through the regular points on that ξ-level with $\alpha = \alpha_l$, $\mu = \mu_l$, and $w = w_l$. Here l is a point index rather than a level index. From relation (19), α is a found to be

$$\alpha_l = -2(w_l \mu_l + w_{l-1}\mu_{l-1} + \ldots), \tag{21}$$

where the terms extend back to the starting point ($\alpha_0 = w_0 = 0$).

Here it may be useful to refer to Fig. 4, which shows the upper front quadrant of a unit sphere with directions (points) indicated and level lines drawn for the case $n = 4$. Points labeled Q have negative μ, those labeled P have positive ones. If we choose the lower ξ-level, Q_{01} is the starting point, and Q_{21}, Q_{11}, P_{11}, and P_{21}, the regular points.

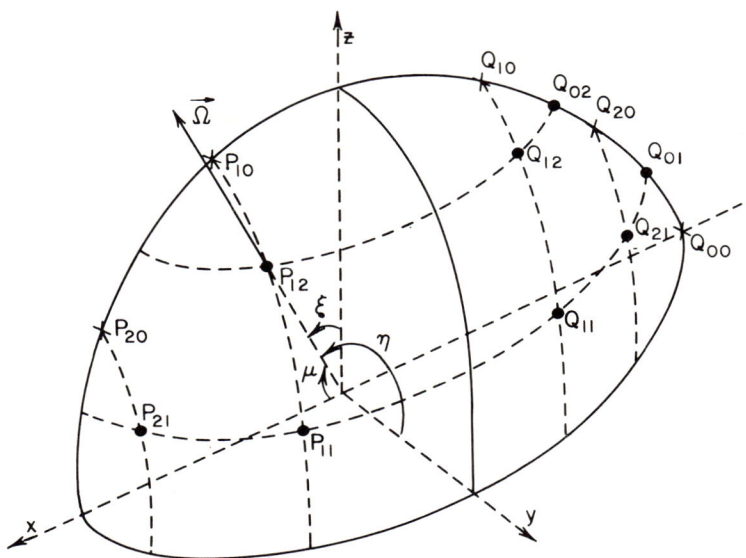

FIG. 4. Direction mesh on the upper front quadrant of unit sphere, for $n = 4$, illustrating sequences of basic and auxiliary (weight zero) points. Lines of constant μ and ξ and typical direction shown. (Symbols with arrows correspond to boldface symbols in text.)

We observe then that α increases with l (moving from right to left through the points in Fig. 4) since we begin summation where μ is negative. This implies that $\Sigma_l w_l \mu_l$, summing over all l, must be zero if the last α in the sequence is to be zero. This should be remembered if one constructs special sets of directions and weights. For the standard sets this condition is observed [see Eqs. (1)].

The maximum α is attained at the halfway mark, between Q_{11} and P_{11} in our reference to Fig. 4. If the points on the fixed ξ-level are symmetrically located about this mark so are, in the numerical sense, w_l and α_l.

For the spherical case, Eq. (21) applies also, except that $\mu = -1$ is the starting point and w_l are level weights. In the limit, as above, α approaches $1 - \mu^2$. The sequence of points Q_{00}, Q_{20}, Q_{10}, P_{10}, and P_{20} illustrates the progression for $n = 4$.

For angular redistribution along lines of constant μ, controlled by the coefficients γ, we find, using relation (20),

$$\gamma_m = -2(w_m \xi_m + w_{m-1}\xi_{m-1} + ...), \qquad (22)$$

where the w's are point weights and the terms extend back to $\xi = -\sqrt{1-\mu^2}$, the starting value for ξ.

Since the set of points on the unit sphere are arranged in triangular fashion, the spherical case with φ-variation is complicated somewhat by the need for "splitting coefficients" for negative μ-levels and "joining coefficients" (the same set of coefficients but used in the reverse order) for positive μ-levels. Splitting coefficients are required when neutrons are passed from one μ-level to another and the number of points per level increase, joining coefficients in the opposite case.

C. Balance Equations

The system of equations (17) for a particular velocity group consists of an equation for each representative neutron beam (discrete direction). The cell balance equation (23), given subsequently, is derived from a combination of the equations under (17). The group balance equation (24) is obtained from (23) by summation over all cells, and the total balance equation from (24) by summation over all groups. The methods for solving the difference equations (17) are closely tied to these balance equations, which are clearly also important in the monitoring of calculations.

To obtain the cell balance equation we multiply (17) by the point weights w_m and by $V\Delta$ and follow this by a sum over m. The result is

$$V(\bar{N}_{s+1} - \bar{N}_s)/v + \Delta(A_{i+1}I_{i+1} - A_i I_i)$$
$$+ \Delta(B_{j+1}J_{j+1} - B_j J_j) + \Delta(C_{k+1}K_{k+1} - C_k K_k) + V\Delta\sigma\bar{N} - V\Delta\bar{S} = 0, \qquad (23)$$

where $\bar{N}$ is the average neutron flux, $\bar{N} = \Sigma_m w_m N_m$; $\bar{S}$ the average source; I the neutron current (net flux) in the i-direction, $I = \Sigma_m w_m \mu_m N_m$, and J and K the currents in the other two directions. The currents represent net flow across cell surfaces. Note that the terms involving the coefficients α and γ in Eq. (17) vanish in the summation.

Positive terms in Eq. (23), as that equation is written, represent losses, the negative terms represent gains. This interpretation may be reversed for terms involving I, J, and K since these quantities may in themselves be negative. Other interpretations include: $V\Delta\sigma\bar{N}$ = the number of neutron-nuclei collisions in the cell during the time interval, $V\Delta\bar{S}$ = the number of neutrons emitted, $\Delta A_i I_i$ = the net number of neutrons flowing across side A_i, $V\bar{N}_s/v$ = the number of neutrons present at the beginning of the time interval, etc.

Now, summing Eq. (23) over all cells, we obtain the group balance equation

$$(\tilde{N}_{s+1} - \tilde{N}_s)/v + (\bar{I}_I - \bar{I}_0) + (\bar{J}_J - \bar{J}_0) + (\bar{K}_K - \bar{K}_0) + \sigma\tilde{N} - \tilde{S} = 0 , \quad (24)$$

where $\sigma\tilde{N}$ is the total number of collisions in the system for a particular velocity group, $\sigma\tilde{N} = \Delta \Sigma_c \sigma_c V_c \bar{N}_c$ with c indexing the cells; $\tilde{S}$ the total source; $\bar{I}$ the leakage in the i-direction ($\bar{I}_I$ on the front side, $\bar{I}_0$ on the back), $\bar{I}_I = \Delta \Sigma_c A_{I,c} \bar{I}_{I,c}$, $\bar{J}$ and $\bar{K}$ the leakages in the other directions (right, left, top, and bottom), and $\tilde{N}_s/v$ the total number of neutrons in the system at the beginning of the time interval.

D. MAIN PROPERTIES OF THE TRANSPORT EQUATION

The difference operator $[D + \sigma]$, operating on N to yield S and defined by (17) and (18), as well as the inverse process illustrated by Eq. (12) and (13) are clearly linear operators. Thus, if S is decomposed into components $S_1, S_2, ...,$ and $N_1, N_2, ...,$ are the corresponding solutions obtained by applying the inverse operator, then $N = a_1 N_1 + a_2 N_2 + ...$ is the solution corresponding to $S = a_1 S_1 + a_2 S_2 + ...$ These linear properties are of major importance in the solution of transport difference equations (see Section V). Note, for instance, that surface sources and boundary conditions may be regarded as components of the source S.

If for the moment we restrict the discussion to problems with one velocity group, then $[D + \sigma]$ can be shown to be a symmetric operator (matrix) provided the set of directions used is reflective. The latter property implies that, if Ω is an arbitrary direction in the set, then the direction opposite to Ω is also in the set.

To establish the symmetry we consider two arbitrary cells C_1 and C_2 with N_1 neutrons flowing into C_1 (on some free side) and N_2 flowing out of C_2. We connect the two cells by an arbitrary "path," by a chain of cells so that some part of N_1 can flow through and join N_2. The contribution to the flow N_2 caused by N_1, i.e., the probability of penetration along this path times N_1, is clearly the product of a string of attenuation factors like the factor multiplying N_i in Eq. (12). The probability of penetration in the opposite direction is the same provided the opposite path is actually provided. The latter follows if the set of directions is reflective. The possibility of collisions, with or without change of number, does not affect the symmetry provided the probabilities involved are independent of the direction of motion along the path.

This is generally assumed to be the case. It is assumed, in addition, that it is possible for neutrons to get from one cell to any other cell in the system, i.e., that the matrix-operator is irreducible.

For problems which are physically realistic and involve more than one velocity group, the operator is generally not self-adjoint, which stems from the physical fact that the matrix of group transfer probabilities is rarely if ever found to be symmetric. One can therefore obtain adjoint (N^*) as well as regular solutions (N). The former are obtained by the same numerical procedures as the latter. The adjoint calculation is preceded by the transposing of the transfer matrix, and followed by an interpretation to the effect that the N formally obtained for a particular direction is actually the N^* for the reflected opposite direction.

Whereas the regular solution N is defined to mean neutron flux, the adjoint solution can be interpreted to mean "neutron importance." For this reason adjoint solutions in combination with regular ones have many uses. They are used in perturbation theory to estimate the effect of relatively small material changes in a particular reactor configuration (Hansen, 1957). To give another example, they are important in the reduction (group collapsing) of transfer cross sections (Bell, 1960, 1962). The process here is one of reducing the order of a matrix of cross sections. The assumptions behind a reduction based on N and N^* are then mainly those implied by the problem which generated these functions. (See also Lee 1961, Chapter 6).

The transport difference equation has been derived here directly from physical principles. This is not the conventional approach. One commonly starts by deriving an analytical model, a linear integro-differential equation in this case, and then proceeds to examine a variety of difference schemes, after which one applies the scheme that appears most promising. This approach may raise some difficult questions with regard to continuity, neutron balance, and interpretation of terms. In numerical model building, one attends to these and related questions from the beginning.

IV. Transformation of the Difference Equations

In some applications it may be possible to eliminate one or more of the variables and make other simplifications toward the end of saving numerical effort without compromising the results which are of primary interest. Simplifying assumptions are almost always made with regard to the form and physics of the source S. Variation in time and in one or more of the space variables can often be separated. For sets of related calculations, so-called parameter studies, a selected few are often performed and the results used, in one way or another, to reduce the remaining computational effort. The number of groups may be reduced (collapsed), the number of intervals cut, S_4 precision replaced by S_2, etc. Since the savings in time and effort may be enormous, such transformations are very important even if they have only the stature of recipes.

A. ELIMINATION OF THE TIME VARIABLE

One may be able to assume that N and S are replaceable by NE and SE where only E varies with time, $E = \exp(\alpha t)$ with α constant, for example. One expects solutions of this type to exist when geometry, cross sections, and extraneous sources are time independent. If substitution is made in Eq. (17) and all terms are divided by $E(t_{s+1/2})$, we find that $N_{s+1} = N \exp(\alpha \Delta/2)$ and $N_s = N \exp(-\alpha \Delta/2)$, and hence,

$$\left[D + \left(\sigma + \frac{2}{v\Delta} \sinh \frac{\alpha \Delta}{2} \right) \right] N = S, \qquad (25)$$

where the operator D no longer involves time. Since time is being eliminated we may as well choose an infinitesimal Δ. Hence,

$$[D + (\sigma + \alpha/v)] N = S. \qquad (26)$$

The quantity α/v is often referred to as "time absorption."

It may happen here that α is negative. The collision probability σ may be small in some regions or for some velocity groups. In any case σ is always zero in a vacuum. Hence it may very well happen that $\sigma + \alpha/v$ is a negative quantity. For reasons that will become clearer as we proceed we shall need another model for N here, called the Alpha model. In this model N is set equal to the average of the cell boundary inflows. Thus, returning to our example in Section III for (x, y) geometry, we let $N = (N_i + N_j)/2$. N_s does not appear here since the time variable has been eliminated. Substituting this expression for N in Eq. (14) we obtain,

$$N_{i+1} = \frac{\frac{\mu}{\Delta x} N_i + \frac{\eta}{\Delta y} N_j - \frac{1}{2} (\sigma + \alpha/v)(N_i + N_j) + S}{\frac{\mu}{\Delta x} + \frac{\eta}{\Delta y}}. \qquad (27)$$

It should be noted here that it is not necessary to use the same difference model throughout a given problem. Equation (27) should only be used for those intervals for which $\sigma + \alpha/v$ is negative. This may be for all intervals for one velocity group, for some or none in others. The point of the Alpha model is to avoid cancellation of significant figures in the denominator, which would occur in the other models if $\sigma + \alpha/v$ were negative. See Eqs. (15) and (16).

B. Treatment of Source Term

If one considers the most general case, the transfer of neutrons between the various beams of different direction and velocity presents a rather complicated picture, both to describe and to reproduce numerically. In principle, a different transfer cross section could be used to describe the scattering of neutrons from one beam into each of many other beams. The following approximate treatment is, however, acceptable in most cases. It is an extension of methods which have been in use for a long time and represents a major simplification. The total cross section is decomposed as follows:

$$\sigma = \sigma^c + \sigma^f + \sigma^i + \sigma^e, \tag{28}$$

where

σ^c = cross section for capture,

σ^f = cross section for fission.
 Here 1, 2, 3, or more neutrons may emerge per event. The average number is denoted by ν. The spectrum of the fission neutrons is given by χ_g where $\Sigma_g \chi_g = 1$. The quantity ν may depend on the incident neutron velocity and on the target nucleus. The spectrum χ_g is normally assumed to be independent of these factors.

σ^i = cross section for scattering with change of velocity.
 This quantity is usually divided, among several groups, generally of lower velocity than the incident neutrons. Each group component of σ^i may be further divided into components as defined below under σ^e.

σ^e = cross section for scattering without change of velocity, i.e., no change of velocity group.

$\sigma^e = \sigma^s + \sigma^+ + \sigma^-, \qquad \sigma^e(\mu) = \sigma^s + 2\sigma^+\delta(\mu - 1) + 2\sigma^-\delta(\mu - 1).$

σ^s = the isotropic component, deflection with equal probability in all directions.

σ^+ = the forward component, no deflection.

σ^- = the backward component, 180° deflection.

The components σ^s, σ^+, and σ^- may be obtained by the procedure described below. This is the simplest one. One can, of course, devise other rules.

Let $\sigma^e(\mu)$ be the probability of deflection through an angle whose cosine is given by μ, with normalization as follows:

$$M_0 = \sigma^e = \frac{1}{2} \int_{-1}^{1} \sigma^e(\mu) \, d\mu ,\tag{29}$$

and let M_1 and M_2 be defined by

$$M_1 = \frac{1}{2} \int_{-1}^{1} \mu \sigma^e(\mu) \, d\mu , \qquad M_2 = \frac{1}{2} \int_{-1}^{1} \mu^2 \sigma^e(\mu) \, d\mu .\tag{30}$$

By substitution of the δ-function form of $\sigma^e(\mu)$ in Eqs. (30), these formulas follow:

$$\sigma^s = 3(M_0 - M_2)/2 , \quad \sigma^+ - \sigma^- = M_1 , \quad \sigma^+ + \sigma^- = (3M_2 - M_0)/2 .\tag{31}$$

If the relation $3M_2 - M_0 = 2M_1$ or $\sigma^- = 0$ is assumed, then $\sigma^s = M_0 - M_1$ and $\sigma^+ = M_1$.

The function $\sigma^e(\mu)$ for angular scattering in some particular velocity group is often considered to be a step function defined as nonzero over the interval (μ_0, μ_1) and as zero elsewhere, normalized so that M_0 is equal to σ^e or to components of σ^i, as the case may be. M_1 is then given by $\sigma^e(\mu_0 + \mu_1)/2$ and M_2 by $\sigma^e(\mu_0^2 + \mu_0\mu_1 + \mu_1^2)/3$.

The two simplest treatments of the source S in Eq. (17) will now be described. In the isotropic treatment one assumes that scattering without change in velocity (self-scattering) is anisotropic with two components, σ^s and σ^+, and that all other scattering (σ^i scattering) is purely isotropic. In this case the term of S with σ^+ ($\sigma^+ N$) is transposed to the left-hand side of Eq. (17) which leads to replacement there of σN by $(\sigma - \sigma^+)N$, all in accord with agreement that collision without change is no net collision. Hence S is, in this case, given by Eq. (35) below omitting the second summation. The quantity $\sigma - \sigma^+$ is usually called the transport cross section.

In the simple anisotropic treatment one assumes that self-scattering is anisotropic with three components, σ^s, σ^+, and σ^-; and σ^i scattering with two, σ^s and σ^+. Here the term of S with σ^+, all or part of it, may be transposed as above, likewise, if $N(-\Omega) + N(\Omega) = 2\bar{N}$ is assumed, the term with σ^-. If both components are entirely transposed, then σN is replaced by $(\sigma - \sigma^+ + \sigma^-)N$.

The self-scattering part S_e of the source term S in Eq. (17) can now be written as follows:

$$S_e = \sigma^s \bar{N} + \sigma^+ N(\Omega) ,\tag{32}$$

where $\bar{N}$ is the flux $N(\Omega)$ averaged over Ω and σ^s and σ^+ are revised, that is, adjusted for the transpositions actually made. The final reduction is based on approximating N by its constant and linear terms in expansion with respect to the direction cosines:

$$N = \bar{N} + 3\mu I + 3\eta J + 3\xi K, \qquad (33)$$

which then changes (32) to

$$S_e = \sigma^s \bar{N} + 3\sigma^+(\mu I + \eta J + \xi K), \qquad (34)$$

where σ^s is again revised, now denoting the previous σ^s plus σ^+. The final expression for S for the case of simple anisotropy, omitting any fixed sources which may be present, can now be written as follows:

$$S = \chi_g F + \sum_h \sigma^s_{hg} \bar{N}_h + 3 \sum_h \sigma^+_{hg}(\mu I_h + \eta J_h + \xi K_h), \qquad (35)$$

where the summation over h extends over all groups and where F is the fission neutron source given by

$$F = \sum_h \nu_h \sigma_h{}^f \bar{N}_h. \qquad (36)$$

The above treatment of scattering may appear to be a very crude approximation. Actually it is quite good. One must remember that it is used only in the source term where the scattering function is averaged with the neutron flux. The latter is seldom extremely anisotropic. If it is isotropic, or nearly so, then it clearly does not matter how one approximates $\sigma^e(\mu)$ as long as $\sigma^e = M_0$.

In a general treatment of scattering it might be convenient to start with given functions $\sigma_g{}^e(\mu)$ and $E_g(\mu)$, where the latter specifies the neutron energy or velocity after deflection by μ. These functions might be given in the form of tables permitting interpolation, and would be used in a calculation to generate cross sections as required and in the right form and quantity the treatment would include a few adjustable coefficients so that the main properties of σ and E (certain averages) could be preserved.

C. Treatment of Angular Variation

Since one is seldom interested in N as a function of angle except for the purpose of calculating $\bar{N}$ and the currents, one endeavors naturally to hold the angular description to a minimum of separate directions, to the S_2 and S_4 resolutions most of the time. One goes to higher order n only if the effect of angular resolution needs to be investigated further or if one is interested in the angular variation of

N per se. In case of extreme flux anisotropy one should certainly investigate the possibilities of special sets of directions rather than to go to higher n in one of the standard sets, beyond $n = 8$ say.

A number of important simplifications come from combinations of the S_2 equations. There are eight equations, one for each octant of the unit sphere, for each velocity group in the most general case. We derive the four S_2 diffusion equations given below as follows: Multiply Eqs. (17) by w_m and sum over m. That gives the first relation, the cell balance equation, except for a factor. Next, multiply by $w_m \mu_m$ and sum, then by $w_m \eta_m$ and $w_m \xi_m$, thereby obtaining the other three relations, the gradient equations,

$$\frac{\bar{N}_{s+1} - \bar{N}_s}{v\Delta} + \frac{A_{i+1}I_{i+1} - A_i I_i}{V} + \frac{B_{j+1}J_{j+1} - B_j J_j}{V} + \frac{C_{k+1}K_{k+1} - C_k K_k}{V}$$
$$+ (\sigma - \sigma^s)\bar{N} - \bar{S} = 0, \qquad (37)$$

and

$$\left.\begin{array}{l} \dfrac{I_{s+1} - I_s}{v\Delta} + \dfrac{A_{i+1} + A_i}{6V}(\bar{N}_{i+1} - \bar{N}_i) + \sigma I = \overline{\mu S} \\[6pt] \dfrac{J_{s+1} - J_s}{v\Delta} + \dfrac{B_{j+1} + B_j}{6V}(\bar{N}_{j+1} - \bar{N}_j) + \sigma J = \overline{\eta S} \\[6pt] \dfrac{K_{s+1} - K_s}{v\Delta} + \dfrac{C_{k+1} + C_k}{6V}(\bar{N}_{k+1} - \bar{N}_k) + \sigma K = \overline{\xi S}, \end{array}\right\} \qquad (38)$$

where in Eqs. (38) σ actually represents $\sigma - \sigma^+$, and

$$\bar{S} = \chi_g F + \sum_h \sigma^s_{hg} \bar{N}_h, \quad \overline{\mu S} = \sum_h \sigma^+_{hg} I_h, \quad \text{etc.}, \quad h \neq g. \qquad (39)$$

The three equations (38) may, after some alteration, be substituted in Eq. (37) which is then converted into a difference equation of second order. This is of practical significance for steady-state problems with isotropic scattering ($\sigma^+ = 0$), in which case the second order equation (the Simple diffusion equation) can be solved by one of the many elegant relaxation methods developed for this purpose. See, for instance, Davison (1957), Lee (1961), and Wachspress *et al.* (1958). The majority of reactor calculations can, as experience has shown, be adequately handled within the assumptions of the simplified theory.

D. REDUCTION OF DIMENSIONALITY

One may use the equations (38) of the previous subsection as a basis for reducing the dimensionality of space, especially in the case of steady-state problems with isotropic scattering. After reduction, which in the simplest case is a quick analytic process, the problem may be solved by using S_2, S_n, or simple diffusion methods, as one chooses.

As an illustration we return to (x, y) geometry and assume, with the intent of eliminating y, that I, $\bar{N}$, and $\bar{S}$ may be replaced by IE, $\bar{N}E$, and $\bar{S}E$, where only E depends on y. On assuming a steady state, an isotropic source, and infinitesimal y-intervals, the second equation in (38) and Eq. (37) may be written as follows:

$$J = -\frac{1}{3\sigma}\frac{\partial \bar{N}E}{\partial y} \tag{40}$$

$$\frac{I_{i+1} - I_i}{\Delta x} E + \frac{\partial J}{\partial y} + (\sigma - \sigma^s) \bar{N}E = \bar{S}E . \tag{41}$$

If it is now assumed that $E = \cos(\pi y/2b)$, where b is equal to the half-thickness of the system in the y-direction, then

$$\frac{\partial J}{\partial y} \equiv \sigma\beta\bar{N}E = \frac{\sigma}{3}\left(\frac{\pi}{2\sigma b}\right)^2 \bar{N} \cos(\pi y/2b) \tag{42}$$

and hence (41) becomes

$$\frac{I_{i+1} - I_i}{\Delta x} + [\sigma(1 + \beta) - \sigma^s] \bar{N} = \bar{S} , \tag{43}$$

where, explicitly, $\beta = (\pi/2\sigma b)^2/3$.

It is not necessary here, or in the elimination of time, to choose E so that the separation is "pointwise" in the variable. If, after substitution in (41) from (40), terms in y in (41) do not cancel, one may simply average over y, i.e., integrate over y from $y = 0$ to $y = b$, term by term. Thus, if E is chosen to be $1 - (y/b)^2$ instead of the above cosine function, the absorption correction β is equal to $(1/\sigma b)^2$ instead of $(\pi^2/12)/(\sigma b)^2$.

V. Solution of the Difference Equations

The solution of the transport difference equations (17) or substitute equations, i.e., subcases or transforms, is based on a few general rules and three principal calculational procedures. There are, of course, a large number of special methods for special problems. These will not be discussed in this article. In computer codes one endeavors in general to keep specialized methods, incompatibilities, and multiplicity of treatment to a minimum.

All problems, whether steady-state or time dependent, are treated as if inhomogeneous in the neutron flux N. This is one of the general rules. Some kind of fixed source, internal or external to the system, is in other words assumed to be present or else one is introduced by normalizing part or all of the homogeneous source. Most homogeneous problems, now referring to steady-state problems, involve a source F of fission neutrons. A fission source F is supplied, directly or indirectly at the beginning of the calculation, normalized to unity or some other "power level" whereupon the inhomogeneous problem is solved.

The calculation does not stop here, however; F is now recomputed, integrated over all space, and normalized to the fixed power level. The constant of normalization, the divisor λ, is the growth factor for fission. This quantity is used to divide χ_g, χ_g/λ replacing χ_g in Eq. (35) whereupon a revised inhomogeneous problem is solved. These so-called "outer iterations" or power iterations continue until λ converges to unity by some prescribed test. The final divisor of χ_g, called the reactivity k, is a principal eigenvalue and is usually taken as the reference eigenvalue. The time absorption α may also be used as such, and usually is used if no multiplicative process like fission is present.

One is often not interested in the reference eigenvalue per se but rather in some parameter of the problem which is taken to vary continuously with the eigenvalue. Examples of such parameters are: "critical dimensions" in the system and "critical concentrations" of particular isotopes. In parameter problems one usually supplies a guess for the parameter, proceeds to calculate the reactivity as outlined above, i.e., the k corresponding to the parameter, repeating this for other guesses or interpolations until the parameter corresponding to $k = 1$, if one exists, has been determined.

The principal procedures for solving the difference equations numerically are (1) directional evaluation of recursion formulas, (2) component treatment of the source S, and (3) scaled inner iterations.

A. Directional Evaluation

The difference equations are solved in the general direction of neutron travel, in energy and in space. The slowing down of neutrons by scattering is the typical process in neutron diffusion. Calculations begin, therefore, at the high velocity end and progress in monotone fashion toward the group of lowest velocity. As one moves from one group to the next, one makes use of the most recent data about $\bar{N}$ and the associated currents in the computation of S, i.e., of continuous updating of S. Thermal activity and fission are the main processes which upgrade neutrons in energy. The fission neutrons, however, are given the special treatment already described. Upscattering of thermal neutrons may possibly upset the technique based on continuous updating and therefore require some special attention. The progression towards groups of lower velocity is, however, seldom tampered with. It is reversed in one case, for obvious reasons; this is when adjoint rather than regular solutions are sought.

We turn now to the equations for a particular group. These are solved in the general direction of neutron travel as specified by the (μ, η, ξ) triplet. The difference equations are, however, first converted to recursion formulas. The conversion process makes use of the model relations to eliminate all unknown N's except one. There may be unknown N's concealed in S but this problem is handled separately (see the next subsection). The following test for proper substitution and solution should be satisfied: All terms in the denominator of the recursion formula should have positive signs [see, for instance, Eq. (15), (16), and (27)].

TABLE V

BASIC SUBSTITUTIONS AND RULES OF SOLUTION FOR THE TRANSPORT DIFFERENCE EQUATIONS.

Geometry and indices	If μ η ξ	solve for	in Step Model[a] after setting	in Diamond Model[a] after setting
1-dim.	$-$	N_i	$N = N_i$	$2N = N_i + N_{i+1}$
(i)	$+$	N_{i+1}	$N = N_{i+1}$	$2N = N_i + N_{i+1}$
2-dim.	$-\ -$	N_i	$N = N_j = N_i$	$2N = N_i + N_{i+1}$
(i, j)	$+\ -$	N_{i+1}	$N = N_j = N_{i+1}$	$= N_j + N_{j+1}$
	$-\ +$	N_i	$N = N_{j+1} = N_i$	in all cases
	$+\ +$	N_{i+1}	$N = N_{j+1} = N_{i+1}$	
2-dim.	$-\ \ \ -$	N_i	$N = N_k = N_i$	$2N = N_i + N_{i+1}$
(i, k)	$+\ \ \ -$	N_{i+1}	$N = N_k = N_{i+1}$	$= N_k + N_{k+1}$
	$-\ \ \ +$	N_i	$N = N_{k+1} = N_i$	in all cases
	$+\ \ \ +$	N_{i+1}	$N = N_{k+1} = N_{i+1}$	
3-dim.	$-\ -\ -$	N_i	$N = N_k = N_j = N_i$	$2N = N_i + N_{i+1}$
(i, j, k)	$+\ -\ -$	N_{i+1}	$N = N_k = N_j = N_{i+1}$	$= N_j + N_{j+1}$
	$-\ +\ -$	N_i	$N = N_k = N_{j+1} = N_i$	$= N_k + N_{k+1}$
	$+\ +\ -$	N_{i+1}	$N = N_k = N_{j+1} = N_{i+1}$	in all cases
	$-\ -\ +$	N_i	$N = N_{k+1} = N_j = N_i$	
	$+\ -\ +$	N_{i+1}	$N = N_{k+1} = N_j = N_{i+1}$	
	$-\ +\ +$	N_i	$N = N_{k+1} = N_{j+1} = N_i$	
	$+\ +\ +$	N_{i+1}	$N = N_{k+1} = N_{j+1} = N_{i+1}$	

[1] Equate also N_{s+1} to N in Step method, $N_s + N_{s+1}$ to $2N$ in Diamond method, for time dependent cases. Extend similarly for subscripts l and m in curved geometries.

There may be more than one way of solving the equations directionally. The choice is usually made according to some convention. Ease of application of certain boundary conditions may on occasion suggest a choice different than the conventional one. Table V lists one set of substitutions and rules of solution for the Step and Diamond models.

Following the procedures described, we obtain recursion formulas linear in the N's and in S and its components, with attenuation coefficients (numbers less than or equal to unity in absolute value) for all N's and a positive coefficient for S. This means that we can progress with evaluations with little or no chance of losing significant figures. In practice one is limited to relatively few figures; 8 to 15 significant figures are carried on present-day computing machines. Any process other than directional evaluation will in most cases lead to quick numerical disaster.

The step model is particularly attractive, since the terms in the numerator are also positive. The Diamond model with negative terms in the numerator may lead to N's which show bounded, usually damped oscillations, especially if intervals and cross sections are large. If the Diamond model is used, it is generally a good idea to be ready to switch momentarily to the Step model, depending on some test. One might repeat a step by this model if N goes negative or changes too rapidly.

The calculations for a particular velocity group consist of a number of major sweeps through the mesh. The number is, in fact, 2 raised to a power equal to the number of space dimensions. In (x, y, z) geometry, for example, one goes (1) in and out in the x-direction for fixed y and z, then (2) to the left in y to the next stack of cells leaving z unchanged and repeating step (1), etc., until it is time (3) to turn in the opposite direction, to the right, still with the same z and performing step (1). Upon completion of one z-level we have made four passes through each cell on that level. We then progress down in z repeating steps (1), (2), and (3), and finally up in z doing the same. The last pass, up in z, changes the number of visits per cell from four to eight.

At the beginning of every major sweep one has to apply boundary conditions. These will be discussed in some detail in the next subsection. Every cell involves, however, boundary inputs and outputs, inputs from cells that have already been visited and outputs to be calculated and saved, as inputs for cells further on in the sequence.

Every sweep involves one or more separate neutron directions, $n/2$ for one-dimensional slabs and spheres, otherwise $n(n+2)/8$. In curved geometries one has to progress directionally among these, as has been discussed in Sections II and III. Additional directions, these without weights and representing starting points for angular redistribution, are needed for the Diamond model and curved geometries. This is because in model relation like $N_{i+1} + N_i = N_{l+1} + N_l$ a first N_l is needed. Thus for a symmetric sphere and S_2 calculations, the first sweep involves $\mu = -1$ ($w = 0$) and $\mu = -\sqrt{3}/3$ ($w = 1/2$), and the second sweep $\mu = \sqrt{3}/3$ ($w = 1/2$). For an infinite cylinder and S_4 the first sweep involves $\mu = -0.943, -0.882, -0.333, -0.471$, and -0.333, with $w = 0, \frac{1}{6}, \frac{1}{6}, 0$, and $\frac{1}{6}$, and $\xi = 0.333, 0.333, 0.333, 0.882$, and 0.882. The second sweep involves $\mu = 0.333, 0.882$, and 0.333, with $w = \frac{1}{6}$, $\frac{1}{6}$, and $\frac{1}{6}$, and $\xi = 0.333, 0.333$, and 0.882. The calculation is usually done in four sweeps, in and out for each ξ-level. The figures used directly above are taken from Table II, Set A excepting those with $w = 0$. The direction scheme is illustrated in Fig. 4.

The special equations needed for directions of zero weight are obtained from Eq. (17) by processes involving limits. For cylindrical geometry, as an example, one sets $w = \epsilon$, and $N_{l+1} = N_l = (N_{i+1} + N_i)/2$ and lets μ approach its starting value. The attention here is on the first curvature term in (17). Since $\alpha_l = 0$ and α_{l+1}/w approaches -2μ, the curvature term reduces to $-\mu((A_{i+1} - A_i)/2V)(N_{i+1} + N_i)$. This combines with the second term of (17) replacing that term by $\mu((A_{i+1} + A_i)/2V)(N_{i+1} - N_i)$.

The method of directional evaluation is of course closely related to the basic procedure in Monte Carlo calculations. In both instances one traces particles through certain paths. Here we follow a deterministic scheme where the errors tend to be systematic. In the Monte Carlo method, which is based on a random walk scheme, the errors one makes are statistical.

B. Treatment of Source Components

Boundary conditions and other components of S, for a particular group, often involve flux terms N for that group which are about to be calculated but not yet available. The boundary conditions of main concern here are the reflective and periodic ones and similar conditions. The reflective conditions are a problem only at boundaries where evaluation sweeps begin, not where they turn around. The volume sources which have to be attended to are the self-scattering components of S, in particular $\sigma^s \bar{N}$. It is the implicit situations just referred to which necessitate so-called boundary and scatter iterations. Together these are called group or inner iterations. All other components of S, fixed or handled as fixed, can now be treated as one and be denoted by S_f.

The boundary iterations deal with homogeneous neutron input at outside boundaries. For reflective conditions $N(-\Omega)$ at a boundary is set equal to $N(\Omega)$ for all Ω directed out of the system. At boundaries limiting the x-variation, $-\Omega$ is Ω with the sign of μ changed. At boundaries limiting y and z, reversing direction changes the sign of η and ξ, respectively. For periodic conditions, $N(\Omega)$ at a boundary, with Ω directed inward, is set equal to $N(\Omega)$ at the opposite boundary. Zero inflow conditions create no problems. In such cases $N(\Omega)$ is equated to zero for all inward Ω.

We solve now a number of separate problems, one based on $S = S_f$ and a zero boundary condition, one on $S = \sigma^s \bar{N}$ also with this condition, and none, one, or more based on S equal to the separate implicit boundary sources, all problems to be combined later under the rules of scaled iterations (see next subsection). We label the solutions N_f, N_s, N_r (reflective condition), N_p (periodic), etc. Here, to obtain N_f, we use $S = S_f$ which is given, and to obtain N_s we use $S = \sigma^s \bar{N}'$, with $\bar{N}' = \bar{N}$ as last calculated in inner or outer iterations. If no $\bar{N}'$ is available we set $\bar{N}' = \bar{N}_f$. To obtain N_r we use $N(-\Omega) = N'(\Omega)$, where $N'(\Omega)$ is the last information available about $N(\Omega)$. If none exists we set $N(-\Omega) = N_f(\Omega)$.

C. SCALED INNER ITERATIONS

Throughout the development of numerical transport methods, those techniques have been carefully avoided which in themselves would create or destroy neutrons. Without this attention on the continuity principle, the physical model would be altered in unpredictable ways. Below is described a method of performing inner iterations which looks after neutron balance in the presence of implicit sources.

To the separate equations determining N_f, N_s, and N_r [with sources S_f, $\sigma^s_{gg} \bar{N}'$, and the boundary source $N(\Omega)$] are attached the scaling coefficients $R_f = 1$, R_s, and R_r. The final solution, because of the linearity of the difference equations, is then given by

$$N = N_f + R_s N_s + R_r N_r, \qquad (44)$$

where the undetermined coefficients R_s and R_r are to be determined by conservation conditions. Since, for a given velocity group, self-scattering must not alter the neutron population, the following condition is imposed upon the coefficients:

$$R_s \tilde{N}'(r) = \tilde{N}_f(r) + R_s \tilde{N}_s(r) + R_r \tilde{N}_r(r), \qquad (45)$$

where the tilde and r indicate summation over all space cells, with weight $\sigma^s_c V_c$, of the average flux. The above equation states that the initial source used to compute N_s is the same, in an averaged sense, as the source of the same type at the end of an iteration.

In the case of reflective type boundary sources continuity of neutron currents requires that

$$R_r \tilde{N}'(\Omega) = \tilde{N}_f(\Omega) + R_s \tilde{N}_s(\Omega) + R_r \tilde{N}_r(\Omega), \qquad (46)$$

where the tilde and Ω denote summation over boundaries and over Ω with weights $\mu_m w_m$. Equation (46) states that the net current assumed for the boundary source, $R_r \tilde{N}'(\Omega)$, at the beginning of the iteration is equal to the net current at the end of an iteration.

Both conditions (45) and (46) are integral conditions over the system, and in the early stages of iteration local errors in the scattering or current balances will be present. The scaling factors R_s and R_r are used to adjust successive approximations to the scattering and boundary sources. Experience has shown that as iteration continues the local errors diminish. It has also shown that scaled iteration as described is considerably faster than direct iteration on the source S.

If either the self-scattering source or the boundary source is zero, then the solution for R_s or R_r is immediate from Eqs. (45) and (46). Otherwise, R_s is found from the following equation obtained by eliminating R_r:

$$[(\tilde{N}'(r) - \tilde{N}_s(r)) \cdot (\tilde{N}'(\Omega) - \tilde{N}_r(\Omega)) - \tilde{N}_r(r)\,\tilde{N}_s(\Omega)]\,R_s \qquad (47)$$
$$= \tilde{N}_f(r)\,(\tilde{N}'(\Omega) - \tilde{N}_r(\Omega)) + \tilde{N}_r(r)\,\tilde{N}_f(\Omega).$$

If two or more boundary sources are present it may be possible to add them and treat them as one. More important, the fixed source S_f and the source $\sigma^s \bar{N}$ may be combined by using a method first suggested by G. Bell. In this method the calculation of N needs to be made only for the combined source. Let N now be the solution corresponding to $S = S_f + \sigma^s \bar{N}'$ and multiply all terms in the difference equation (17) by R_f. The new source $R_f S = R_f S_f + R_f \sigma^s \bar{N}'$ is then written, after adding and subtracting equal terms, as

$$R_f S = S_f + R_f \sigma^s \bar{N} + (R_f - 1) S_f + R_f \sigma^s (\bar{N}' - \bar{N}). \qquad (48)$$

The first two terms on the right-hand side are clearly the terms appropriate for S in Eq. (17), since this substitution leaves the fixed part of S unchanged and makes the self-scattering part of S, $R_f \sigma^s \bar{N}$, correspond to the part of $R_f \sigma N$ (σN in Eq. (17) after scaling) which is scattered without change of velocity group. The remaining terms in (48) must therefore vanish, at least in the mean, i.e., when summed over all cells weighted with the volume elements. Consequently,

$$R_f = \left(\sum_c V S_f\right) \div \left(\sum_c V S_f + \tilde{N}'(r) - \tilde{N}(r)\right). \qquad (49)$$

If an N_r component is present because of a reflective boundary condition, we find first that R_r is given by $R_r/R_f = \tilde{N}(\Omega)/(\tilde{N}'(\Omega) - \tilde{N}_r(\Omega))$, and second that the denominator of Eq. (49) must be modified by $-\tilde{N}_r(\Omega) R_r/R_f$.

Above is a brief description of scaled iterations. The various steps involved are repeated until a test for convergence has been met. The simplest test is based on the source error E defined by $E = (R_s - 1)\sigma^s \bar{N}'$, which in the scaling process is effectively added to both sides of the transport equation. This test requires that the sum of $V_c E_c$ over the system, or, if desired, that each partial sum defined and computed, be reduced below a prescribed level for each velocity group.

The method of source components also offers, in certain cases, the possibility of improving the accuracy in calculation without refining the mesh used. In simple spherical geometry, for example, one may split the source for inward integrations (negative μ) into two components and in this way provide for adjustment in numerical model so that the flux is made isotropic at the origin, exactly—as it should be—rather than approximately. For the outward integrations a discontinuity component of N at $\mu = 0$ can be carried along in the calculations and so determined, for each r-interval, that the extrapolated flux at $\mu = 1$, using the model relations, agrees with the flux at $\mu = 1$ calculated directly (as for $\mu = -1$). With this refinement one obtains a more symmetric treatment and hence a good basis for a Double-S_n procedure in curved geometries.

There are only minor differences in procedure between steady-state and time dependent problems. The calculations for each particular time interval represents essentially a full steady-state problem with outer and inner scaled iterations. The initial conditions or the previous time step supply the sources entering at the beginning of the new interval and estimates for those needed at the midpoint time. If the time interval is short, one outer iteration may be sufficient and relatively few inner ones. If the interval is increased, more repetitive cycles will be required.

D. Comments on S_n Calculations

S_n calculations are quite time consuming because of the many inner and outer iterations usually required, which is mainly a reflection of the "follow the neutron" character of the whole numerical process. Neutron velocities in reactor systems vary from the visinity of 10^9 cm/sec (1/2 Mev) typical of fission neutrons to 2.2×10^5 cm/sec (1/40 ev) for the average thermal neutron. Since most materials promote the slowing down of neutrons, large numbers of intermediate and thermal neutrons are usually generated. At the same time one finds that the neutron mean free path, the average distance $1/\sigma$ between collisions, usually decreases as the neutron energy decreases. For these and related reasons, S_n calculations are more rapid for a fast reactor system than for a thermal one. On holding other factors fixed, the calculation time for a typical reactor problem generally increases if the effective size of the system increases. Here size is measured in $1/\sigma$ units, where σ is averaged over space and energy.

Since S_n calculations tend to be slow, it is quite important to reach the vicinity of the correct solution as quickly as possible. Favorable starting data may be obtained either from a previous and similar calculation or from a quick preliminary run based on S_2 or Simple diffusion theory. In either case the computational effort for a subsequent S_n run is much reduced.

No comprehensive numerical study for the completed S_n theory has been made as yet. Some comparisons for simple systems are presented in Carlson and Bell (1958), Carlson (1961), and Lee (1961). These comparisons are important but far from sufficient. A satisfactory investigation will have to cover a very large number of cases, in which difference model, problem setting, and mesh resolution are varied.

Long experience with S_n codes have established this, however, that S_4 calculations in most cases will yield satisfactory precision. In the calculation of critical dimension, for instance, the critical size is usually determined to better accuracy than 1%. If predominantly thermal systems are studied, then S_2 accuracy is sufficient.

In conclusion, the S_n techniques provide in relation to particle transport and similar problems, a very much enlarged area of activity for numerical evaluation and investigation. Generality has been achieved both in respect to formulation and methods of solution. With respect to convergence some methods are not yet rigorously established. Here a number of questions belonging to mathematical analysis need to be answered. The most important relate to convergence of the numerical solutions in the limit of very fine mesh resolution.

Acknowledgments

The author wishes to express his thanks to the following members of the staff of the Los Alamos Scientific Laboratory[4]: George I. Bell, Gordon E. Hansen, Robert M. Kiehn, Carroll B. Mills, and Clarence E. Lee who, over a number of years, through fresh points of view, suggestions and ideas, and careful examination of calculations and experiments, have contributed greatly to an early completion of a satisfactory numerical theory of neutron transport.

Clarence E. Lee and William J. Worlton have during the last few years furnished a great deal of direct assistance to the author, which is gratefully acknowledged, and Lee has, in his recent report: *The Discrete S_n Approximation to Transport Theory*, Los Alamos Report LA-2595, treated many aspects of the theory, only touched upon in this article, in considerably more detail.

The author also wishes to express his sincere thanks to Sue Vandervoort, Josephine Powers, and Alice Luders for their help in preparing the manuscript.

References

BELL, G. I. (1960, 1962). Internal memoranda. Los Alamos Scientific Laboratory.

CARLSON, B. G. (1961). Numerical solution of neutron transport problems, *Proc. Symposia Appl. Math.* **2**, 219-232.

CARLSON, B. G., and BELL, G. I. (1958). Solution of the Transport equation by the S_n method. *Proc. 2nd Intern. Conf. Peaceful Uses Atomic Energy, Geneva.* P/2386.

CASE, K. M., DE HOFFMANN, F., and PLACZEK, G. (1953). "Introduction to the Theory of Neutron Diffusion." Vol. I. U. S. Government Printing Office, Washington D. C.

CHANDRASEKHAR, S. (1960). "Radiative Transfer." Dover, New York.

DAVISON, B. (1957). "Neutron Transport Theory." Oxford Univ. Press, London and New York.

HANSEN, G. E. (1957). Internal memorandum, #N-2-256. Los Alamos Scientific Laboratory.

HUGHES, D. J., and SCHWARTZ, R. B. (1958). Neutron Cross Sections. Brookhaven National Laboratory, BNL-325 and BNL-325 Supplement (1960).

LEE, C. E. (1961). The Discrete S_n Approximation to Transport Theory. Los Alamos Scientific Laboratory Report LA-2595.

WACHSPRESS, E. L., STONE, P. M., and LEE, C. E. (1958). Mathematical techniques in two-space-dimension multigroup calculations. *Proc. 2nd Intern. Conf. Peaceful Uses Atomic Energy, Geneva.* P/633.

SZEGÖ, G. (1939). "Orthogonal Polynomials." American Mathematical Society. Colloquium Publications. Vol. XXIII.

Reactor Physics Constants (1958). Argonne National Laboratory Report ANL-5800.

---

[4] This article was written under the auspices of the Los Alamos Scientific Laboratory. It is dedicated in memoriam to B. Davison and G. Placzek, formerly of this laboratory and among the founders and main contributors to transport theory.

The Calculation of Nonlinear Radiation Transport by a Monte Carlo Method

JOSEPH A. FLECK, JR.

LAWRENCE RADIATION LABORATORY
LIVERMORE, CALIFORNIA

I. Introduction 43
II. Description of the Problem 44
III. Finite Difference Methods of Solution 47
IV. The Monte Carlo Method of Solution 51
V. Numerical Results 56
VI. Summary and Conclusions 65
References 65

I. Introduction

THE AREAS OF APPLICATION of the Monte Carlo method in physics are numerous. In nearly all cases the technique is applied under conditions which may be regarded as linear. One area of frequent application is neutron transport. The transport of particles in this case may be assumed to depend on the geometry of the system in question, the scattering and absorbing properties of the media which it contains, but to be independent of the particle densities. The transport equation which describes this phenomenon is of course linear. The transport of dense gases on the other hand involves collisions between the gas molecules themselves and gives rise to a nonlinear problem. Further examples of nonlinear transport phenomena are found in the field of radiative transfer. Since the transport of radiation through a medium affects its absorption properties, the transport of individual photons is indirectly affected by the density of the other photons present. The resulting problem is therefore nonlinear.

In the usual applications of the Monte Carlo technique to linear transport problems, one or more quantities pertaining to a given system are computed for each of a reasonably large number of independent particle histories. The desired numbers are obtained by averaging the results over all of the histories. In the case of nonlinear transport phenomena, the motion of individual particles is influenced by the motion of all of the other particles. One cannot, therefore, calculate the behavior of the system simply by taking averages over independent particle histories.

One possible approach to the solution of nonlinear transport problems is to set up a model containing a reasonably large number of mutually interacting particles and to determine their trajectories by simultaneously solving their equations of motion. Alder and Wainwright (1957) have applied this method to problems relating to the theory of liquids. Following a large system of correlated particles can, however, be vastly more complicated and time-consuming than following an equivalent number of uncorrelated particles.

In the transport of radiation there exists a single parameter, temperature, which completely characterizes the nonlinearity of the problem. This simplifies the problem and suggests the following scheme for its solution. From the initial temperature distribution a source distribution is constructed and a population of independent photons generated. These photons are followed and periodically examined. The results of these examinations determine new temperature and source distributions from which new photons are generated. The continual absorption of photons limits the number of particles that must be followed at any time. The calculation is time dependent, but steady-state solutions which are of interest in astrophysical applications may be obtained by considering the solutions at large time.

II. Description of the Problem

The equations that govern the transport of radiation through matter in one dimension are the transport equation,

$$\frac{1}{c}\frac{\partial I_\nu}{\partial t} + \mu \frac{\partial I_\nu}{\partial x} = \kappa'_\nu \rho (2\pi B_\nu(T) - I_\nu)$$

$$- \sigma_s I_\nu + \int_0^{2\pi} d\phi \int_{-1}^1 p(\cos \Theta) I_\nu(x, t, \mu') \, d\mu' , \quad (1)$$

and the equation of energy conservation,

$$\rho c_v \frac{\partial T}{\partial t} = - 2\pi \int_0^\infty \kappa_\nu' \rho B_\nu \, d\nu + \int_{-1}^1 d\mu \int_0^\infty \kappa_\nu' \rho I_\nu \, d\nu . \quad (2)$$

In these equations $I_\nu(x, t, \mu)$ is the radiation intensity at position x, time t, and in direction μ, per unit value of μ. The quantity μ is the cosine of the angle between the direction in question and the x axis. See Fig. 1. The intensity I_ν, it will be noted, is 2π times the intensity per unit solid angle. The following relations also hold for the intensity I_ν:

$$J_\nu(x, t) = \frac{1}{2} \int_{-1}^{1} I_\nu(x, t, \mu) \, d\mu \qquad (3a)$$

$$u(x, t) = \frac{1}{c} \int_0^\infty d\nu \int_{-1}^{1} I_\nu(x, t, \mu) \, d\mu , \qquad (3b)$$

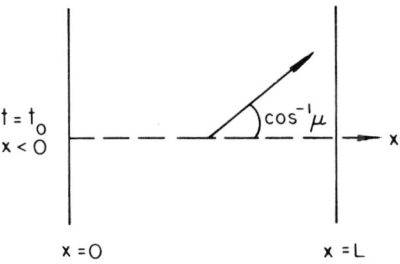

Fig. 1.

where $J_\nu(x, t)$ is defined as the mean intensity, and $u(x, t)$ is the radiation energy density. The function $B_\nu(T)$ is the Planck function at temperature T,

$$B_\nu(T) = \frac{2h\nu^3}{c^2} \frac{1}{e^{h\nu/kT} - 1}, \qquad (4)$$

σ_s is the radiation scattering cross section, and c_v is the specific heat of the medium. The mass absorption coefficient κ_ν' is obtained from the "true" mass absorption coefficient κ_ν by correcting for induced emission:

$$\kappa_\nu' = (1 - e^{-h\nu/kT}) \, \kappa_\nu . \qquad (5)$$

The function $p(\cos \Theta)$ is the phase function which gives the distribution of $\cos \Theta$, where Θ is the angle of scattering of the light quanta. For Thomson scattering of light quanta by electrons, the phase function is

$$p(\Theta) = \frac{3}{16\pi} (1 + \cos^2 \Theta) . \qquad (6)$$

The quantity $\cos \Theta$ can be expressed as

$$\cos \Theta = \mu\mu' + (1 - \mu^2)^{1/2} (1 - \mu'^2)^{1/2} \cos (\phi - \phi'), \qquad (7)$$

where μ and μ' are the values of μ and μ' before and after the collision, and ϕ and ϕ' are the values of the azimuthal angle made by the photon directions before and after the collision. If expression (6) is substituted into Eq. (1), the result is

$$\frac{1}{c}\frac{\partial I_\nu}{\partial t} + \mu \frac{\partial I_\nu}{\partial x} = \kappa_\nu' \rho (2\pi B_\nu(T) - I_\nu) - \sigma_s I_\nu + \frac{3}{16}\sigma_s$$
$$\left[(3-\mu^2) \int_{-1}^{1} I_\nu(\mu')\, d\mu' + (3\mu^2 - 1) \int_{-1}^{1} I_\nu(\mu')\, \mu'^2\, d\mu' \right]. \quad (8)$$

Equation (8) applies to the frequency-dependent or nongrey case. We wish for the present to restrict our attention to the grey case, considering only the intensity I which has been integrated over frequency:

$$I = \int_0^\infty I_\nu\, d\nu. \quad (9)$$

This requires that we find some suitable average for the absorption coefficient κ_ν'. Without attempting further justification we assume that

$$\kappa = \frac{\int_0^\infty \kappa_\nu' B_\nu(T)\, d\nu}{\int_0^\infty B_\nu\, d\nu}. \quad (10)$$

The resulting integrated forms of Eqs. (1) and (2) are

$$\frac{1}{c}\frac{\partial I}{\partial t} + \mu \frac{\partial I}{\partial x} = \kappa \rho \left(\frac{caT^4}{2} - I \right) - \sigma_s I + \frac{3}{16}\sigma_s$$
$$\left[(3-\mu^2) \int_{-1}^{1} I(\mu')\, d\mu' + (3\mu^2 - 1) \int_{-1}^{1} I(\mu')\, \mu'^2\, d\mu' \right], \quad (11)$$

$$\rho c_v \frac{\partial T}{\partial t} = -\kappa \rho (caT^4 - \int_{-1}^{1} I\, d\mu). \quad (12)$$

In developing a Monte Carlo method of solution for Eqs. (11) and (12), rather than for the Eqs. (1) and (2) which apply to the frequency-dependent case, there is actually no important loss in generality. In generalizing the method to the nongrey case one need only consider photons which are sampled from a source distribution depending on frequency as well as on position.

The specific problem which we shall consider is the irradiation of a slab of finite thickness, extending from $x = 0$ to $x = L$, by a hohlraum held at a fixed temperature $T = T_0$ and situated at $x = 0$. See Fig. 1.

III. Finite Difference Methods of Solution

The more conventional method of solving Eqs. (11) and (12) involves reducing them to finite difference form or to a combination of finite difference and semianalytic form. A certain amount of discussion will be devoted to this approach since it suggests how the Monte Carlo technique can be applied to the problem and since it also furnishes numerical solutions which can be compared with the Monte Carlo solutions.

Let us define M discrete values of μ by

$$\mu_m = -1 + \frac{2m-1}{M}, \quad m = 1, 2, \ldots M \tag{13}$$

and the corresponding fluxes by

$$I_m(x, t) = I(\mu_m, x, t). \tag{14}$$

Equations (11) and (12) may now be approximated by

$$\frac{1}{c}\frac{\partial I_m}{\partial t} + \mu_m \frac{\partial I_m}{\partial x} = \kappa\rho\left(\frac{caT^4}{2} - I_m\right) - \sigma_s I_m + \frac{3}{16}\sigma_s \\ \left[(3-\mu_m^2)\frac{2}{M}\sum_{m=1}^{M} I_m + (3\mu_m^2 - 1)\frac{2}{M}\sum_{m=1}^{M} I_m \mu_m^2\right], \tag{15}$$

$$\rho c_v \frac{\partial T}{\partial t} = -\kappa\rho\left(caT^4 - \frac{2}{M}\sum_{m=1}^{M} I_m\right). \tag{16}$$

Equations (14)-(16) represent a form of discrete flux approximation to Eqs. (11) and (12). A steady-state solution to another discrete flux approximation of these equations was first obtained by Wick (1943) and Chandrasekhar (1950). In their method the directions μ_m were chosen so that the integral over intensity could be approximated by the Gauss quadrature formula. The discrete directions in Eq. (14) have been selected only for reasons of convenience. This choice of directions has also been employed by Carlson (1959) in a method of neutron transport calculation known as the DS_n method.

Equations (14) and (15) can be further approximated in finite difference form and solved. This method has been used by Barfield *et al.* (1954) One could also solve Eq. (16) by a finite difference approximation and solve the system of hyperbolic equations (15) in a semianalytic form by integrating along characteristics. The latter method was first suggested by Keller and Wendroff (1956), although no numerical results of the method have been published. The "characteristics" S_n method has been selected for numerical comparison with the Monte Carlo method. It will be referred to in the remainder of this article as simply the "S_n" method. A brief description of the method is now presented.

Consider the slab divided into regions bounded by the planes $x = x_j = j\Delta x$, $j = 0, 1, \ldots J$, $J\Delta x = L$. Consider also the time divided into intervals of duration Δt. Thus a two-dimensional mesh in space and time may be constructed (see Fig. 2). The differential operator on the

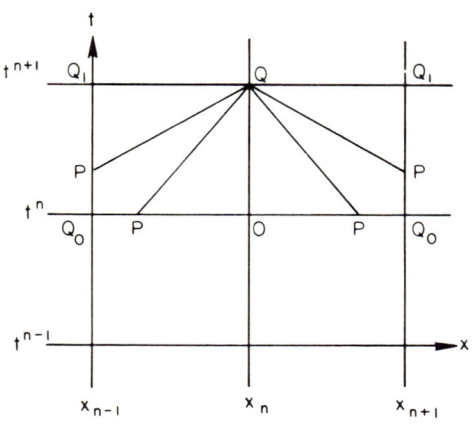

FIG. 2.

left-hand side of Eqs. (15) may be replaced by a directional derivative,

$$\frac{1}{c}\frac{\partial}{\partial t} + \mu_m \frac{\partial}{\partial x} = D_m \frac{d}{ds}, \qquad (17)$$

where d/ds signifies differentiation along a line whose direction cosines with the x and t axes are, respectively,

$$1/cD_m, \mu_m/D_m, \qquad (18)$$

where

$$D_m = [(1/c^2) + \mu_m^2]^{1/2}. \qquad (19)$$

Equations (15) may now be written

$$D_m \frac{dI_m}{ds} + (\kappa\rho + \sigma_s) I_m = S_m, \qquad m = 1, 2 \ldots M \qquad (20)$$

where S_m is the sum of the temperature-dependent source and the scattered particle source. The solution to Eqs. (19) at a mesh point Q in terms of values of I_m and S_m at an earlier time but not necessarily at a mesh point is

$$I_{mQ} = I_{mP} \exp\left[-\frac{1}{D_m} \int_0^s (\kappa\rho + \sigma)\,ds\right]$$
$$+ \frac{1}{D_m} \int_0^s S_m(s') \exp\left[-\frac{1}{D_m} \int_{s'}^s (\kappa\rho + \sigma_s)\,ds''\right] ds'. \quad (21)$$

In Eq. (21) and in what follows the point P and Q refer to Fig. 2. If we approximate the source function by

$$S_m \approx \tfrac{1}{2}(S_P + S_Q) \quad (22)$$

and regard the temperature in a given mesh rectangle as a constant equal to its value at the center of the rectangle, i.e.,

$$T = T[(n + \tfrac{1}{2})\Delta t,\, x_j \pm \Delta x/2] = T^{n+1/2}_{j\pm 1/2}, \quad (23)$$

then Eq. (20) can be approximated as

$$I_{mQ} = I_{mP} \exp\left[-\sigma^{n+1/2}_{j\pm 1/2}\,\Delta s/D_m\right] + \frac{S_P + S_Q}{2\sigma^{n+1/2}_{j\pm 1/2}}\left(1 - \exp\left[-\sigma^{n+1/2}_{j\pm 1/2}\,\Delta s/D_m\right]\right). \quad (24)$$

In Eqs. (23) and (24) the plus and minus signs correspond to the cases where the characteristic has negative and positive slopes, respectively. also

$$\sigma^{n+1/2}_{j\pm 1/2} = \kappa^{n+1/2}_{j\pm 1/2}\rho + \sigma_s, \quad (25)$$

and Δs is the length of the line segment PQ. The quantities I_{mP} and S_P are determined by linear interpolation from values of I_m and S expressed at mesh points. In this computation two cases must be distinguished:

Case I: $c\Delta t\,|\mu_m| < \Delta x$

$$I_{mQ} = I_{mP} \exp\left[-\sigma^{n+1/2}_{j\pm 1/2}\,c\Delta t\right] + \frac{S_P + S_Q}{2\sigma^{n+1/2}_{j\pm 1/2}}\left(1 - \exp\left[-\sigma^{n+1/2}_{j\pm 1/2}\,c\Delta t\right]\right) \quad (26)$$

$$I_{mP} = \frac{1 - c\Delta t\mu_m}{x_j - x_{j\pm 1}} I_{mO} + \frac{c\Delta t\mu_m}{x_j - x_{j\pm 1}} I_{mQ_0} \quad (27)$$

$$S_{mP} = \frac{1 - c\Delta t\mu_m}{x_j - x_{j\pm 1}} S_{mO} + \frac{c\Delta t\mu_m}{x_j - x_{j\pm 1}} S_{mQ_0} \quad (28)$$

Case II: $c\Delta t |\mu_m| > \Delta x$

$$I_{mQ} = I_{mP} \exp[-\sigma_{j\pm 1/2}^{n+1/2} \Delta x/\mu_m] + \frac{S_P + S_Q}{2\sigma_{j\pm 1/2}^{n+1/2}} (1 - \exp[-\sigma_{j\pm 1/2}^{n+1/2} \Delta x/\mu_m]), \quad (29)$$

$$I_{mP} = \left[1 - \frac{x_j - x_{j\pm 1}}{c\Delta t \mu_m}\right] I_{mQ_l} + \frac{x_j - x_{j\pm 1}}{c\Delta t \mu_m} I_{mQ_0}, \quad (30)$$

$$S_{mP} = \left[1 - \frac{x_j - x_{j\pm 1}}{c\Delta t \mu_m}\right] S_{mQ_l} + \frac{x_j - x_{j\pm 1}}{c\Delta t \mu_m} S_{mQ_0}. \quad (31)$$

Actually more accurate approximations of the source function $S_m(s')$ may be made by keeping several terms in a power series expansion in s'. Such approximations might be needed in situations where drastic variations in the absorption coefficient are encountered. The simpler approximation will suffice for the purposes of this paper.

The order in which the solution proceeds is in the direction of increasing x for μ_m positive and in the direction of decreasing x for negative μ_m. The boundary conditions are

$$\begin{aligned} I_m(x=0) &= aT_0^4/2 & \mu_m > 0 \\ I_m(x=L) &= 0 & \mu_m < 0. \end{aligned} \quad (32)$$

In computing the scattering contribution to S_Q in Eqs. (24) to (29) it is necessary to know the values of I_m which it is desired to compute. An iterative procedure may be used whereby initially the scattering source is computed using values of I_m at time t^n. The values of I_m at time t^{n+1} are then computed and the scattering source at Q is improved.

In the calculation as described thus far it has been assumed that the radiation intensities I_m are calculated at times $t^n = n\Delta t$ and positions $x_j = j\Delta x$ and that the temperatures are calculated at times $t^{n\pm 1/2} = (n \pm \frac{1}{2})\Delta t$ and positions $x_{j\pm 1/2} = (j \pm \frac{1}{2})\Delta x$. One could equally well use Eqs. (26)-(32) to advance the radiation intensities I_m from time $t^{n-1/2}$ to time $t^{n+1/2}$ using values of temperature at time t^n to calculate absorption coefficients and source terms and then use the following difference equation derived from Eq. (16) to advance the temperature from time t^n to time t^{n+1}.

$$T_{j-1/2}^{n+1} = T_{j-1/2}^n - \frac{c\Delta t}{\rho c_v} \kappa(T_{j-1/2}^{n+1/2}) \rho \left[a(T_{j-1/2}^{n+1/2})^4 - \frac{2}{M}\sum_{m=1}^M I_{mj-1/2}^{n+1/2}\right], \quad (33)$$

where

$$\begin{aligned} T_{j-1/2}^{n+1/2} &= (T_{j-1/2}^{n+1} + T_{j-1/2}^n)/2 \\ I_{j-1/2}^{n+1/2} &= (I_{mj}^{n+1/2} + I_{mj-1}^{n+1/2})/2. \end{aligned} \quad (34)$$

Note that Eq. (33) is centered at $t = t^{n+1/2}$. The procedure can be started by assuming that the values of I_m at time $t = 0$ hold at $t^{-1/2}$.

IV. The Monte Carlo Method of Solution

We may summarize the calculation procedure just described as follows.

1. The transport equation is solved assuming a temperature distribution which is constant during a given time cycle and throughout a given spatial zone. In so doing the radiation intensities are advanced from time $t^{n-1/2}$ to time $t^{n+1/2}$.

2. The radiation intensities obtained from step 1 are used to calculate radiation absorption rates at time $t^{n+1/2}$ which in turn are substituted into finite difference equations. The latter are used to advance the temperatures from time t^n to t^{n+1}. The entire cycle is then repeated using the new temperatures.

The essential feature of step 1 is that the calculation is performed under the assumption of temperature and medium properties constant in time. We could also have solved this part of the problem by a conventional Monte Carlo procedure. It would be required to generate photons or, more appropriately, energy bundles from a known temperature distribution. These "particles" would be followed in time and space while appropriate account is taken of absorption and scattering processes. At census time, i.e., at the end of a time cycle, the radiation energy is tallied in each zone. This in turn is used as the input to step 2, described above.

In the problem under discussion there will be two types of sources. The first type is the surface source,

$$S_0 = \sigma T_0^4, \qquad (35)$$

where $\sigma = ac/4 =$ Stephan's constant. The second type is the "volume" source,

$$S_{j-1/2}^n = 4\sigma(T_{j-1/2}^n)^4 \Delta x, \qquad j = 1, \ldots J. \qquad (36)$$

Equations (35) and (36) describe the rates at which the surface $x = 0$ and the various zones of heated material radiate energy. Let N_1 be an input number which cannot be exceeded by the number of source particles generated in a given time cycle. Then the number of surface source particles N_0^n and the number $N_{j-1/2}^n$ of volume source particles generated in a given time cycle may be determined from the following equations:

$$N_0^n = \text{Grint}\left[\frac{T_0^4 N_1}{T_0^4 + 4\Delta x \sum_{j=1}^{J} \rho \kappa_{j-1/2}^n (T_{j-1/2}^n)^4}\right]$$

$$N_{j-1/2}^n = \text{Grint}\left[\frac{4\Delta x \rho \kappa_{j-1/2}^n (T_{j-1/2}^n)^4 N_1}{T_0^4 + 4\Delta x \sum_{j=1}^{J1} \rho \kappa_{j-1/2}^n (T_{j-1/2}^n)^4}\right]$$

(37)

In Eqs. (37) "Grint" signifies the greatest integer less than the argument. The energy of the various source particles is given by

$$E_0^n = \Delta t \sigma T_0^4 / N_0^n$$

$$E_{j-1/2}^n = \Delta t \Delta x 4 \rho \kappa_{j-1/2}^n T_{j-1/2}^4 / N_{j-1/2}^n .$$

(38)

The use of Eqs. (37) and (38) insures that energy is conserved and that the most particles are generated where the most energy is radiated. Since a nearly constant number of new particles is generated during each time cycle, the total number of particles being followed in a given time cycle will increase at first. A steady state, however, is eventually reached, and if the number N is judiciously chosen, the total number of particles reaching census in any specified time cycle can be kept within a specified bound determined by the storage capacity of the machine.

To be on the safe side, one may use the following procedure with a varying N_1. Let N be an upper bound to the number of particles which can be allowed to reach census, and let N_2 be the number of particles which reached census at the end of the previous time cycle. Then N_1 for the current time cycle may be chosen so that

$$N_1 + N_2 \leqslant N .$$

(39)

Once the energy, number, and location of the source particles has been decided, they are assigned starting position and time coordinates x_0, t_0 as well as directions μ_0. The two types of source particles are handled according to the following two schemes:

Surface particles:

$$x_0 = 0 + \delta \text{ (δ a small positive number)} \quad (40a)$$

$$t_0 = (n - 1/2 + r) \Delta t \quad (40b)$$

$$\mu_0 = \max(r_1, r_2) \quad (40c)$$

Volume particles, zone j; $j = 1, 2, \ldots J$

$$x_0 = (j - 1 + r) \Delta x \tag{41a}$$
$$t_0 = (n - 1/2 + r) \Delta t \tag{41b}$$
$$\mu_0 = 2r - 1 \tag{41c}$$

In the above equations as well as in what follows r stands for a random number between 0 and 1. Since the particles in the hohlraum adjacent to the surface $x = 0$ are considered isotropic, the actual number crossing this surface from the hohlraum in a given direction will be weighted by the cosine of the angle which this direction makes with the x-axis, i.e.,

$$f(\mu_0) = 2\mu_0, \tag{42}$$

where $f(\mu)$ is the distribution function of the direction μ_0. This is equivalent to the first of the boundary conditions expressed in Eq. (32). The volume particles are emitted isotropically, i.e., uniformly distributed in μ.

Once the particle data are determined, the geometry routine is entered. The latter begins with the determination of the distance to boundary d_B, the distance to collision d_{col}, and the distance to census d_{cen}, where

$$d_B = \begin{cases} [(j+1)\Delta x - x]/\mu, & \text{if } \mu \geq 0 \\ (x - j\Delta x)/\mu, & \text{if } \mu < 0 \end{cases} \tag{43}$$

$$d_{\text{col}} = \lambda_{j-1/2}^n |\ln r| \tag{44}$$

$$\lambda_{j-1/2}^n = (\sigma_s + \kappa_{j-1/2}^n \rho)^{-1} \tag{45}$$

$$d_{\text{cen}} = c[(n+1)\Delta t - t]. \tag{46}$$

Here x and t represent the position and time of the particle. The next step is determined by which of these distances is smallest. If the smallest distance d turns out to be d_{cen}, the particle is advanced in position and time,

$$\begin{aligned} x' &= x + \mu d \\ t' &= t + d/c \\ \mu' &= \mu, \end{aligned} \tag{47}$$

the particle data is stored, and the energy is tallied for the appropriate zone. If the smallest distance turns out to be d_B, the particle is advanced to

$$\begin{aligned} x' &= x + \mu d' \\ t' &= t + d'/c \\ d' &= d_B + \delta \ (\delta \text{ a small number}), \end{aligned} \tag{48}$$

and the new zone number j of the particle is determined by

$$j = \text{Grint}(x'/\Delta x). \tag{49}$$

If $j > J$ the particle is lost from the system and the history is ended. Otherwise the calculation and selection using Eqs. (40)-(43) is repeated. However, after a boundary crossing, Eq. (44) may be replaced by

$$d'_{\text{col}} = \frac{\lambda_{j'-1/2}}{\lambda_{j-1/2}}(d_{\text{col}} - d_B), \tag{50}$$

where d_{col} and d_B are retained from the previous use of Eqs. (43)-(46).

In case $d = d_{\text{col}}$ is selected, the specific collision type must be determined. The process is absorption if

$$r \leqslant \kappa^n_{j-1/2}\rho/(\kappa^n_{j-1/2}\rho + \sigma_s) \tag{51}$$

and scattering otherwise. If the process is absorption, the history is terminated. If the process is scattering, the particle is first advanced,

$$\begin{aligned} x' &= x + \mu d \\ t' &= t + d/c, \end{aligned} \tag{52}$$

and a new direction is selected using the phase function for Thomson scattering [Eq. (6)]. The phase function Eq. (6), gives the distribution in the cosine of the scattered angle Θ per unit solid angle. If expression (7) for $\cos \Theta$ is substituted into the phase function, the following expression is obtained for the distribution in final direction per unit solid angle:

$$\begin{aligned} p(\mu, \mu', \phi, \phi') = \frac{3}{16\pi} [&1 + \mu^2\mu'^2 + 2\mu\mu'(1 - \mu^2)^{1/2}(1 - \mu'^2)^{1/2} \cos(\phi - \phi') \\ &+ (1 - \mu^2)(1 - \mu'^2) \cos^2(\phi - \phi')]. \end{aligned} \tag{53}$$

If Eq. (43) is integrated over azimuth, i.e., over ϕ for fixed ϕ', the result is

$$\begin{aligned} p(\mu, \mu') &= p(\mu', \mu) \\ &= \frac{3}{16}[(3 - \mu^2) + (3\mu^2 - 1)\mu'^2], \end{aligned} \tag{54}$$

where $p(\mu, \mu')$ is the probability of scattering from μ to μ' per unit value of μ in $d\mu$. Note that

$$\int_{-1}^{1} p(\mu, \mu') \, d\mu' = \int_{-1}^{1} p(\mu', \mu) \, d\mu = 1. \tag{55}$$

Figure 3 shows the function $p(\mu, \mu')$ for three different values of μ. For a given μ the value of μ' can be determined by the following rejection technique:

Select r_1 and r_2
Case 1: $\mu^2 \leqslant 1/3$
Let $x = 2r_1 - 1$.
If $(3 - \mu^2) + (3\mu^2 - 1) x^2 \leqslant (3 - \mu^2) r_2$, $\mu' = x$.

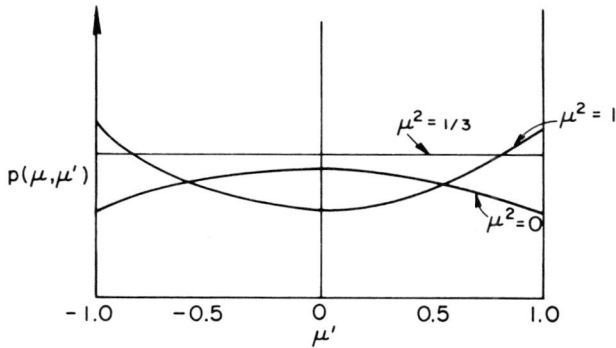

Fig. 3. The distribution $p(\mu, \mu')$ of the scattered photon direction.

Otherwise two new random numbers are selected and the procedure is repeated.

Case 2: $\mu^2 > 1/3$
Let $x = 2r_1 - 1$.
If $(3 - \mu^2) + (3\mu^2 - 1) x^2 \leqslant 2r_2(1 - \mu^2)$, $\mu' = x$.

Otherwise two new random numbers are selected and the procedure is repeated.

The residue particles, i.e., particles generated in previous time cycles which have reached census at time $t^{n-1/2}$, are treated in all respects like the source particles, except that their starting position and time coordinates as well as direction are those associated with the particle at the last census. Finally, the energy $E_{j-1/2}^{n+1/2}$ of all particles reaching census in a given zone is summed. From these sums the intensity integrated over angle in that zone is obtained.

$$\int_{-1}^{1} I_{j-1/2}^{n+1/2} d\mu = c \sum_{i} E_{ij-1/2}^{n+1/2}/\Delta x \qquad (56)$$

The index i designates the particle history. The temperatures in the various zones are then obtained from the following difference version of Eq. (12):

$$T^{n+1}_{j-1/2} = T^n_{j-1/2} + \frac{c\Delta t}{\rho c_v}\kappa(T^{n+1/2}_{j-1/2})\rho\left[a(T^{n+1/2}_{j-1/2})^4 - \sum_i E^{n+1/2}_{ij-1/2}/\Delta x\right], \quad (57)$$

where $T^{n+1/2}_{j-1/2}$ is defined as in Eq. (13).

Equation (57) may be solved by a suitable iteration method, e.g., the Newton-Raphson method. The entire Monte Carlo method of solving the transport and energy equations in briefly summarized by the flow diagram, Fig. 4.

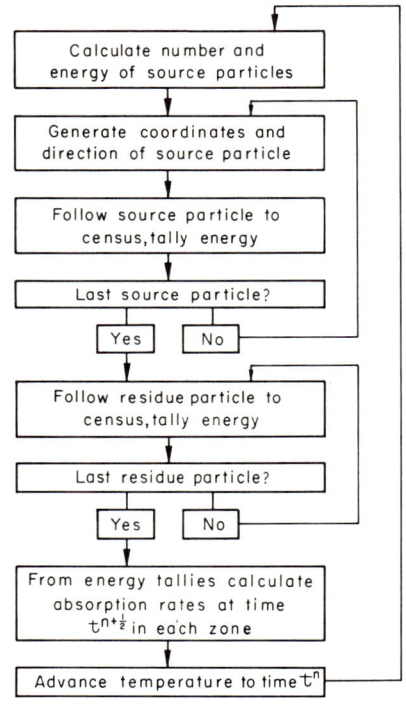

Fig. 4.

V. Numerical Results

There are four questions connected with the Monte Carlo calculation which merit special consideration.

1. The first is the question of convergence. Because of the rather unique combination of a Monte Carlo with a finite difference calculation there is no assurance that the numerical solutions will converge to the actual solutions of the differential equations.

2. The second is the question of statistical accuracy. An important part of any Monte Carlo calculation is estimating the variance of the quantities which are being averaged over the population of histories. As a result of the intricate way in which the Monte Carlo histories calculated during a given time cycle depend on those calculated during previous time cycles, a meaningful variance estimate valid beyond the first time cycle seems out of the question.

3. Thirdly there is the question of errors other than statistical errors, i.e., truncation errors. These errors will in all likelihood be influenced by statistical errors.

4. Finally the question of stability needs to be considered. The situation is complicated in this respect by the addition of statistical errors to round-off errors. In addition to the usual possibility of the growth of roundoff errors, the possibility also exists that statistical inaccuracies may contribute to errors which grow with time.

An attempt will not be made to answer these questions on a rigorous basis. Instead a purely heuristic approach will be taken. We have at our disposal an alternate numerical method, the convergence and stability properties of which are reasonably well understood [see Keller and Wendroff (1956)] and which can be expected to furnish fairly accurate solutions. These solutions in turn can be compared with the Monte Carlo solutions to the same problems. This procedure will furnish answers to the preceding questions in an *ad hoc* way. We can also apply a consistency check to both types of solution. This will further establish the accuracy of solutions obtained by both methods. The consistency check is derived as follows.

If Eq. (11) is integrated over μ, the result is

$$\frac{\partial u}{\partial t} + \frac{\partial F}{\partial x} = \kappa \rho (acT^4 - \int_{-1}^{1} I \, d\mu), \tag{58}$$

where $u(x, t)$ is the radiation energy density and $F(x, t)$ is the radiation flux.

$$u = \frac{1}{c} \int_{-1}^{1} I \, d\mu$$

$$F = \int_{-1}^{1} I\mu \, d\mu. \tag{59}$$

If Eq. (58) is combined with Eq. (12), the result is

$$\frac{\partial}{\partial t}(u + \rho c_v T) = -\frac{\partial F}{\partial x}.\tag{60}$$

The integration of Eq. (60) over the volume of the slab and over time yeilds

$$\int_0^L (u + \rho c_v T)\, dx = \int_0^t [F(0, t) - F(L, t)]\, dt,$$
$$= \epsilon_P(t),\tag{61}$$

which equates the energy content of the slab with the net energy transferred through the faces. The quantity

$$\epsilon = \frac{\epsilon_P^{n+1/2} - \sum_{j=1}^{J}(u_{j-1/2}^{n+1/2} + \rho c_v T_{j-1/2}^{n+1/2})}{\epsilon_P^{n+1/2}}\tag{62}$$

called the "energy check" represents the fractional deviation of the calculated energy in the slab from the calculated net energy pumped in, and provides a useful figure of merit for the accuracy of a calculation. This applies both to the Monte Carlo and to the S_n method.

The "characteristic" S_n method has been coded for an IBM 7090 computer and the Monte Carlo method for the LARC computer at Livermore. In the latter code the LARC's buffering facilities are used to read in and out of fast memory from drum storage the particle data. Blocks of 2500 words may be read into and out of internal storage while calculation is in progress. Thus even though the particle data are stored in drum storage, the access time is that of fast memory.

Results are exhibited in Figs. 5 through 15. In these calculations

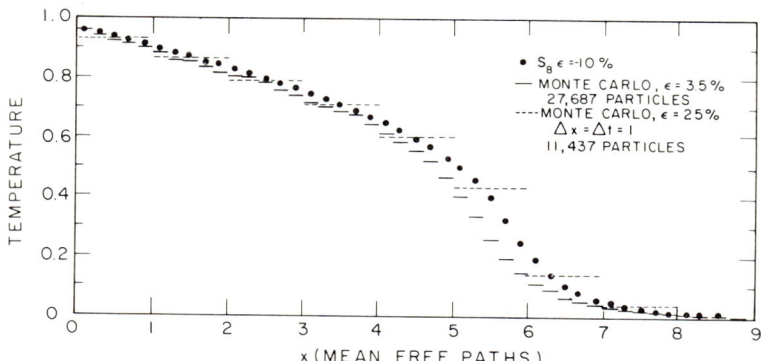

Fig. 5. Comparison of S_8 and Monte Carlo temperature distributions for constant mean free path.

the units of length are cm, the units of time are such that $c = 1$, the units of energy and temperature are such that $a = 1$, $\rho c_v = 0.5917$, and $T_0 = 1$. Figures 5 and 6 show a comparison between an S_n calcula-

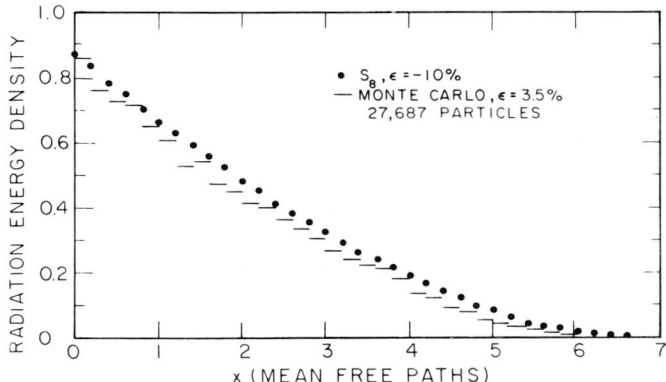

FIG. 6. Comparison of radiation energy density for S_8 and Monte Carlo at constant mean free path.

tion with eight directions (S_8) and a Monte Carlo calculation. Exhibited are the temperature and radiation energy density u as functions of position at time $t = 40$. In this problem $\rho \kappa = 1$, $\sigma_s = 0$, $\Delta x = \Delta t = 0.2$, $L = 20$. This problem in the following discussion will be referred to as problem A.

Both figures show that the S_8 calculation allows energy to penetrate farther into the slab than does the Monte Carlo calculation for the same Δx and Δt. This fact is consistent with the negative value of ϵ (too

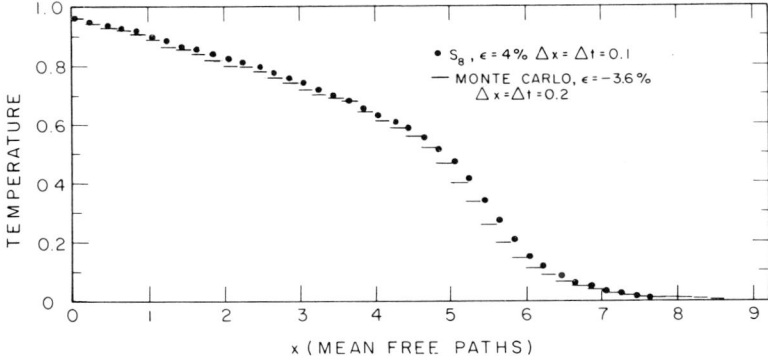

FIG. 7. Temperature distributions for S_8 and Monte Carlo.

much energy) for S_8 and the positive ϵ (too little energy) for the Monte Carlo calculations. One would conclude that the Monte Carlo results are more accurate on the basis of the smaller magnitude of the energy check associated with it. This conclusion is further substantiated when the Δt and Δx are refined to 0.1 and 0.1 for the S_8 calculation. In this case the ϵ is reduced to -4% at $t = 40$, and the two curves are brought closer together (see Fig. 7). Also shown in Fig. 5 is the temperature

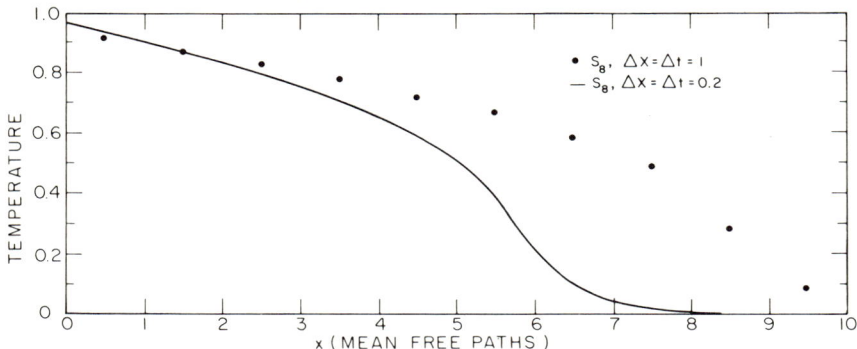

FIG. 8. Comparison of two S_8 runs with different zoning.

distribution for $\Delta x = \Delta t = 1$. The larger zoning and time cycle lead to an ϵ of 25%. Figure 8 shows a comparison of two S_8 calculations, one for $\Delta t = \Delta x = 0.2$ the other for $\Delta t = \Delta x = 1$. The coarser-zoned

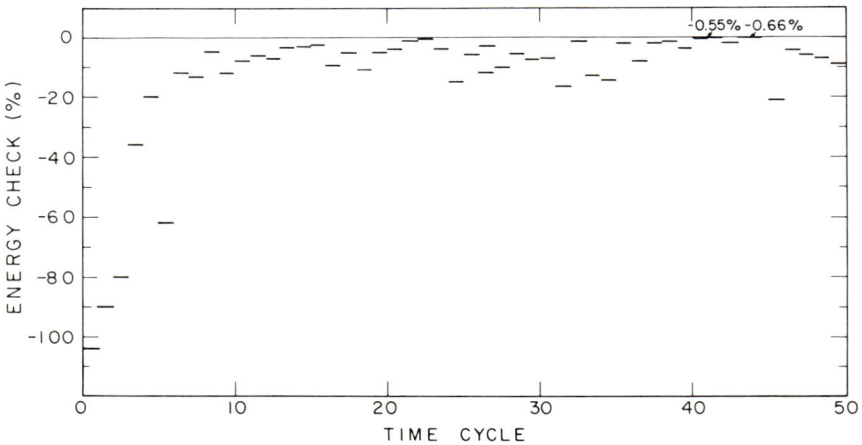

FIG. 9. Energy check as a function of time cycle for 1000 source particles.

case results in $\epsilon = 86\%$. The loss of accuracy is evident in the plotted points.

The superior accuracy of the Monte Carlo method over the S_n method is due to the fact that the Monte Carlo method conserves energy regardless of zone size, whereas in the "characteristics" S_n method energy is not very well conserved except for small zones and short time cycles.

It has already been mentioned that variance estimation appears to be out of the question. However, some qualitative insight into the statistical accuracy of the Monte Carlo calculation can be gained by comparing solutions to the same problem obtained with different sets of random numbers and different numbers of histories. Figures 9 and 10 show the

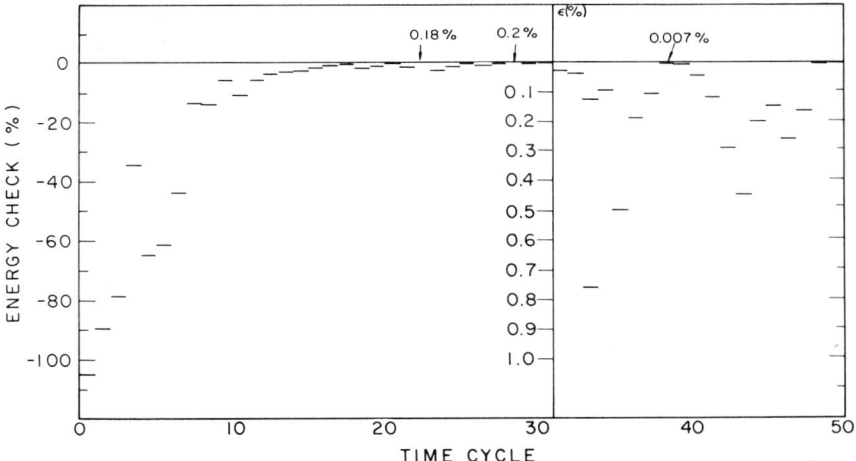

FIG. 10. Energy check as a function of time cycle for 9000 source particles.

energy checks for problem A using 1000 and 5000 source particles per time cycle respectively for the first 50 cycles. It will be noticed that not only is the spread in ϵ for the case of 5000 source particles smaller, but ϵ is generally smaller in magnitude as well. This superiority however was only temporary. The energy checks for the 5000-source-particle case after 170 cycles changed sign and increased in magnitude to an average value of about 4% by cycle 200. The energy checks in the 1000-source-particle case (not shown) past cycle 200 were also of the same order of magnitude but exhibited somewhat greater spread.

A comparison of the temperature distributions for two Monte Carlo runs on problem A using 5000 source particles per cycle is shown in

Fig. 11. The time corresponds to 200 cycles. The agreement is good and it appears that if smooth curves were drawn through the two sets of data they would be almost indistinguishable. The energy penetration E_p defined by

$$E_p = \int_0^t [F(0, t) - F(L, t)] \, dt \tag{63}$$

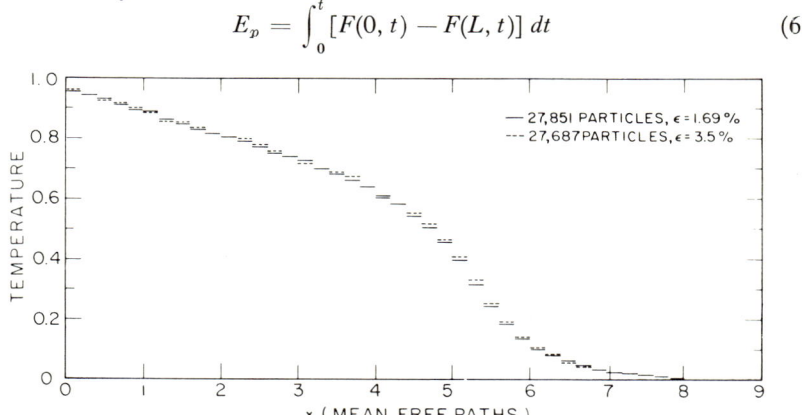

FIG. 11. Comparison of two Monte Carlo runs with different sets of random numbers.

has values of 4.6398 and 4.6159 (in arbitrary units) for the two cases—which is agreement within $\frac{1}{2}\%$. Energy penetrations as a function of time are shown in Fig. 12 for four different cases involving different

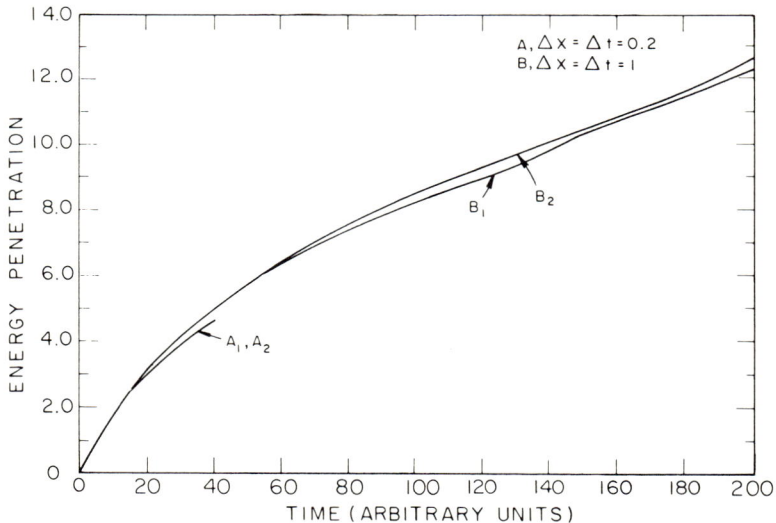

FIG. 12. Comparison of energy penetration for four Monte Carlo calculations involving different random numbers and zoning.

sets of random numbers: the two runs of problem A with 5000 source particles per cycle and the two runs for problem B which differs from problem A in setting $\Delta x = \Delta t = 1$. At time 40 the two curves for problem B agree to within $\frac{3}{4}\%$. However, $t = 40$ corresponds to only 40 cycles for problem B. After this time the gap between the two B curves widens substantially. This gap also shows fluctuations in size. An interesting feature of the B curves is that they never cross. Once the trend is established it does not seem to reverse itself. The same is true of the A curves although this effect cannot be distinguished on the figure.

The effect of introducing scattering is shown in Fig. 13. In this

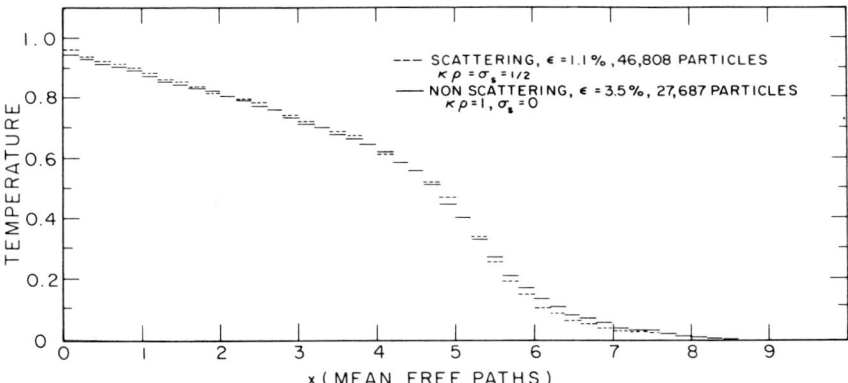

FIG. 13. Comparison of Monte Carlo runs for constant mean free path with and without scattering.

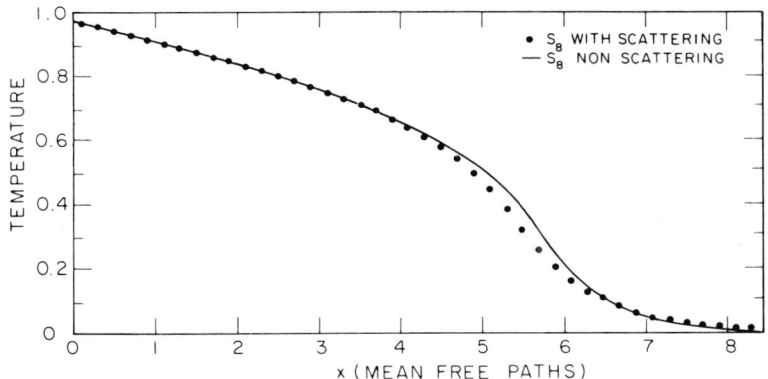

FIG. 14. Comparison of S_n runs for constant mean free path with and without scattering.

problem $\kappa\rho = \frac{1}{2}$, $\sigma_s = \frac{1}{2}$, $\Delta x = 0.2$. The total mean free path is thus the same as in problem A. Differences from the results of problem A are difficult to distinguish and appear to lie within the bounds of statistical error. It does appear, however, that at least below the knee of the curve the scattering results are lower at a given value of x. In Fig. 14, which shows a comparison of the same solutions obtained by the S_n method, the effect is clearly noticeable.

In Fig. 15 the temperature distribution has been allowed to come to

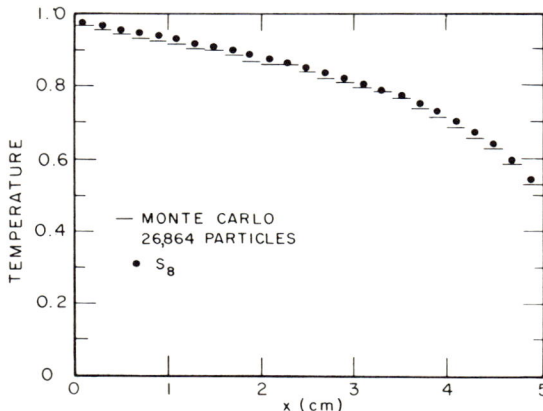

FIG. 15. Monte Carlo and S_n runs for mean free path depending on temperature

equilibrium for the following problem: $\Delta x = 0.2$, $L = 5$, $\sigma_s = 0.5$, $\kappa\rho = 1/(T + 0.1)$. This problem might be thought of as applying to a stellar atmosphere with both scattering and absorbing properties. Once the equilibrium temperature distribution has been determined in this grey case, the source function $B_\nu(T)$ which appears in Eq. (11) can be calculated. The steady-state transport equation with a known source $B_\nu(T)$ can be solved by known techniques (see ref. 3) for the frequency-dependent radiation intensity I_ν. The importance of the Monte Carlo method in the present application is that it can, at least in principle, be readily generalized to take into account the effect of a frequency-dependent absorption coefficient $\kappa_\nu\rho$ on the temperature distribution and the resulting intensity I_ν. It should also be of use when detailed processes, e.g., the polarization of light, become important.

VI. Summary and Conclusions

A method has been described for calculating both time-dependent and steady-state nonlinear radiative transfer by means of a Monte Carlo method. The method is presumably applicable to other nonlinear transport problems provided that there exist suitable variables whose changes can be calculated from finite difference equations coupled in a suitable way to the Monte Carlo calculations. In testing the method it has been found to yield good agreement with purely finite difference methods. Although no method has been devised to estimate variances for the method, it has been possible to achieve a reasonable degree of reproducibility using different sets of random numbers. Furthermore, it has been demonstrated that problems of reasonable length can be run without an appreciable buildup of errors, statistical or otherwise.

REFERENCES

ALDER, B. J., and WAINWRIGHT, T. (1957). Molecular Dynamics by Electronic Computers. *In Proc. Intern. Symposium on Transport Processes in Statistical Mechanics*, Wiley (Interscience), New York, pp. 97-131.

BARFIELD, W. D., VON HOLDT, R., and ZACHARIASEN, F. (1954). A Comparison of Diffusion Theory and Transport Theory Results for the Penetration of Radiation into Plane Semi-Infinite Slabs. U. S. AEC Report No. AECD-3653. Available from Office of Technical Services, U. S. Dept. of Commerce, Washington, D. C.

CARLSON, B. (1959). Numerical Solution of Transient and Steady-State Neutron Transport Problems. U. S. AEC Report No. LA-2260. Available from Office of Technical Services, U. S. Dept. of Commerce, Washington, D. C.

CHANDRASEKHAR, S. (1950). "Radiative Transfer." Oxford Univ. Press, London and New York.

KELLER, H. B., and WENDROFF, B. (1956). On the Formulation and Analysis of Numerical Methods for Time-Dependent Transport Equations. U. S. AEC Report No. NYO-7694. Available from Office of Technical Services, U. S. Dept. of Commerce, Washington, D. C.

WICK, G. C. (1943). *Z. Physik.* **121**, 702.

Critical-Size Calculations for Neutron Systems by the Monte Carlo Method

Donald H. Davis

UNIVERSITY OF CALIFORNIA, LAWRENCE RADIATION LABORATORY
LIVERMORE, CALIFORNIA

I. Introduction	67
II. Calculation Details	69
III. Particle Following	73
IV. Estimates of α and the Equilibrium Distribution	75
V. Collision Calculation	76
A. Elastic Scattering	78
B. Inelastic Scattering (n, n^1) from a Single Level	79
C. Inelastic Scattering Using the Statistical Model	79
D. Fission	80
VI. Examples of Calculations	80
A. One-Velocity Problem	80
B. Godiva Critical Assembly	85
Acknowledgments	87
References	88

I. Introduction

THIS ARTICLE CONCERNS a problem in neutron transport theory. The parts of the problem are the calculations of (a) the critical size (or critical composition) of a system in which neutron multiplication occurs, and (b) the distribution of the neutrons both in energy and space in the system. When more than one neutron, on the average, results from a collision, there is a possibility for the neutron population to increase without bound. When this occurs the neutron population approaches an equilibrium, that is, in the sense that the neutron distribution is given by

$$n(\bar{x}, \bar{v}, t + t_0) = n(\bar{x}, \bar{v}, t_0) e^{\alpha t} ; \qquad (1)$$

where $n(\bar{x}, \bar{v}, t)$ is the number of neutrons per unit volume per unit velocity at the space-velocity point $(\bar{x}, \bar{v})$ and at the time t after a time t_0. The quantity α, the multiplication rate, depends only on the composition and configuration of the system. When α is zero, the system is said to be critical. The fundamental parts of the problem of this article, therefore, are (1) to find the configuration (or composition) of a system for which α equals zero and (2) to find the distribution of the neutrons in the system.

The quantity α depends in a complex way on the composition and configuration. In order to simplify the problem, only those systems will be considered in which a single parameter w can vary. The problem will then be to find the value of the parameter which makes α equal to zero, i.e., to find the solution to the equation

$$\alpha(w) = 0.$$

The distribution $n(\bar{x}, \bar{v}, t)$ can be represented by

$$n(\bar{x}, \bar{v}, t) = cf(\bar{x}, \bar{v}),$$

where

$$\int_s f(\bar{x}, \bar{v}) \, d\bar{x} \, d\bar{v} = 1.$$

The integration is over the all-material parts of the system. The constant c depends on the previous history of the neutron population growth.

The purpose of this article is to present in a computationally complete form a simple solution of the critical problem. Many theoretical questions are examined only in a heuristic way, if at all. For example, for theoretical completeness the question of the existence of the solution expressed in Eq. (1) should be discussed, but is considered to be out of the scope of this article. Also, the computational question of comparative efficiency of the many ways of solving this problem are of necessity inadequately discussed. The existence question, however, is answered in part by comparing calculations to experiment. The second question is answered by examining the machine and coding time.

Many analytical techniques are available for solving the critical problem. There are also several computational techniques which require fast computers. Examples of these are the S_n method (Carlson 1955, 1959; Carlson and Bell 1958) the diffusion method (Davison, 1957) and the Monte Carlo method, (Davis, 1959; Goad and Johnston, 1959; Kaplan, 1958; Morton, 1956). The primary concern of this article is the Monte Carlo method.

The Monte Carlo calculations have several advantages. The most important is that the Monte Carlo method permits the handling of the elementary neutronic processes within existing knowledge of nuclear data without using large amounts of computation time. The second most important advantage is that the method can be used for a wide variety of problems. Disadvantages are that accuracy is limited and quite often machine programs are difficult to construct.

II. Calculation Details

The calculation as described here is a rudimentary one-dimensional one-zone calculation that provides a basis for extensions to more complex problems as well as for extensions to calculations making use of variance reducing techniques.

To provide flexibility in Monte Carlo code development, a set of subroutines has been written. These subroutines are connected with a Fortran program. By a suitably chosen sequence the development that uses these subroutines can proceed from a simple code to a complex code with minimum effort.

In the critical-size problem one of the difficulties in a Monte Carlo calculation is the generation of samples from the equilibrium distribution. If a model sampling technique is used, the Monte Carlo particle population will either increase or decrease exponentially, depending on whether the system is supercritical or subcritical. In either case the number of particles followed could be too large for reasonable calculation times before an equilibrium distribution is attained. To overcome this difficulty a small number of particles are followed for a short time (cycle time, t_0). If at the end of this time the population is larger than at the start, a sample having a size equal to the initial population is selected for the next cycle. If the population is smaller, some of the particles are used more than once as source particles. This process is then repeated for a large number of cycles.

The cycle time t_0 has to be small enough so that population changes during cycles are small, but large enough so that the particles can undergo a reasonable change of position and energy in a cycle. The number of cycles q_{max} has to be large enough so that ($q_{max} \times t_0$) is long compared to the time for equilibrium to be established. Both of these numbers as well as the initial population size I are chosen before the calculation is made. The calculation is, therefore, in a sense subjective. If these numbers are chosen too large, the calculation time can be too large for the accuracy required. If these numbers are too small, the sampling will not be from the equilibrium distribution. To help answer the question of whether equilibrium has been established, provision is made for repeating the calculation using the particles from the last cycle of the previous calculation for the new initial source.

III. Particle Following

The flow diagram (Figs. 1a, 1b, and 1c) illustrates the steps in the calculation. Following is a description of the calculation for one cycle.

A new cycle is started with the coordinates of M_f source particles from the previous cycle. These coordinates include the radial position r of the particle, the cosine β of the angle between the radial vector and the direction vector of the particle, the speed v of the particle, and the time t until the particle reaches census. At the beginning of a cycle the time coordinate is equal to t_0, the cycle time.

The expected number of times ν a particle is to be used as a source particle is given by the equation,

$$\nu = I/M_f,$$

where I is the number of particles to be followed and M_f is the number of particles from the previous cycle. However the actual number of times a particle is used as a source has to be integral. Therefore the number of times a particle is followed is equal to either the integer just smaller than ν or the integer just larger than ν. The number of times the mth particle is used as a source is given by the equation,

$$N_m = \text{Grint } \rho_m \qquad m = 1, 2, \ldots M_f$$

where

$$\rho_1 = u + \nu$$

and

$$\rho_m = \rho_{m-1} - N_{m-1} + \nu.$$

The function "Grint x" means the greatest integer less than x. u is a random number.[1] The above process uses each of the particles from the previous census at least Grint ν times. If ν is not an integer, the process selects the remaining $(I - M_f \text{ Grint } \nu)$ particles from the M_f particles with equal chance.

After the selection of the coordinates of the particle, the next step is the calculation of the distance to a collision. The macroscopic total cross section (the reciprocal of the mean free path) is needed for this calculation. The cross section is approximated by a continuous and piecewise linear function of the neutron velocity. The cross section is

$$\sigma = \sigma_l + (\sigma_{l+1} - \sigma_l)\delta$$

---

[1] The symbols $u, u_1, u_2, u_3 \ldots$ are used throughout the paper to denote values of random variables, independently and uniformly distributed in the interval (0 to 1).

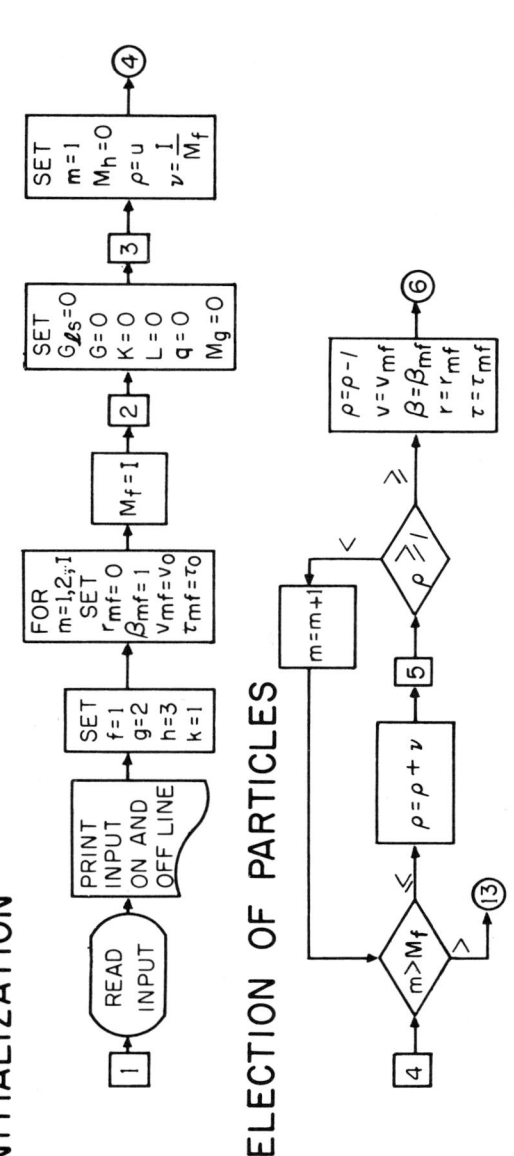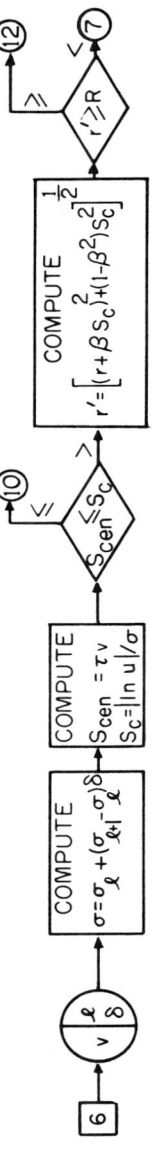

Fig. 1a. Flow diagram.

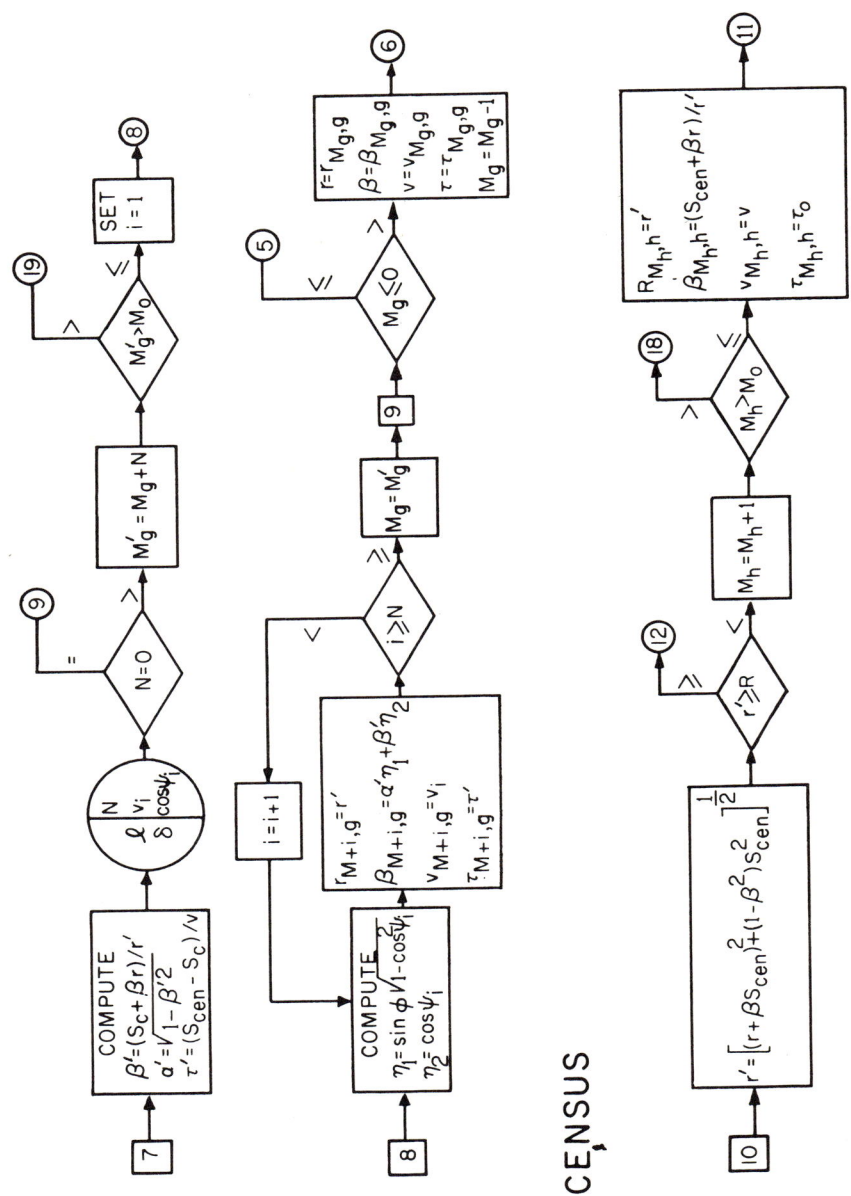

Fig. 1b. Flow diagram (continued).

CRITICAL-SIZE CALCULATIONS

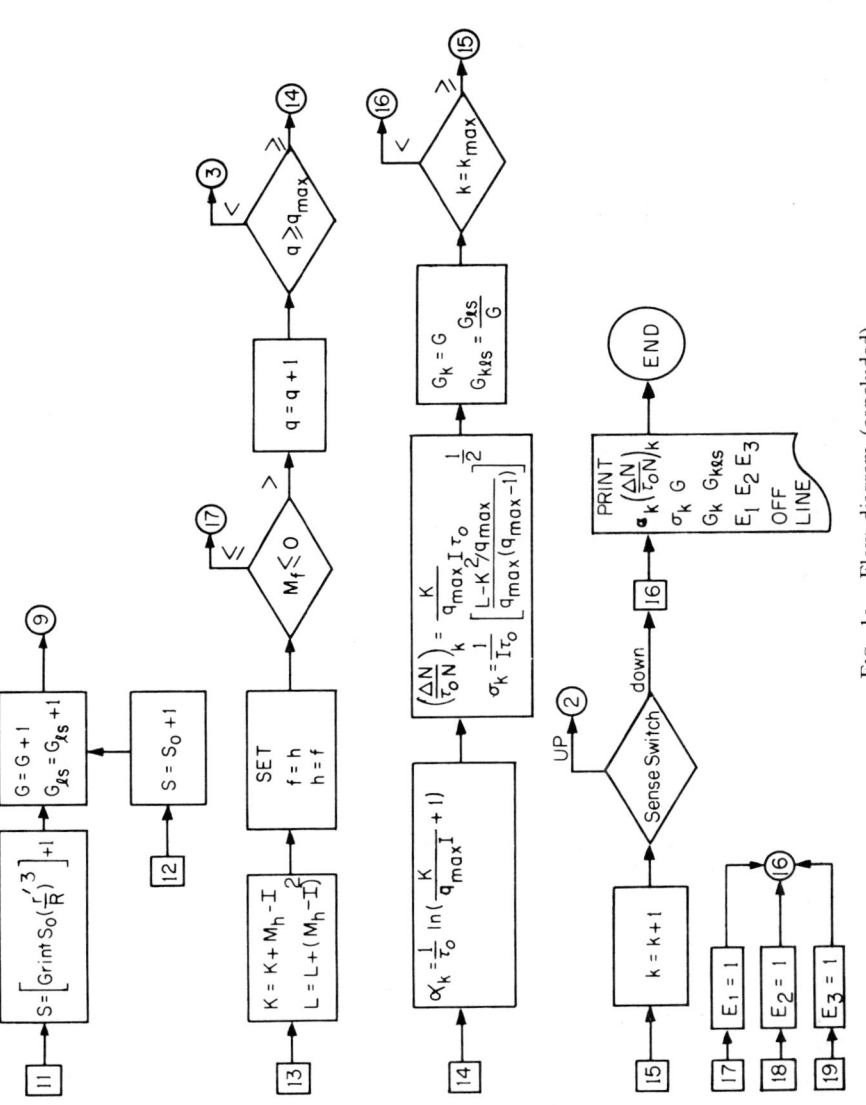

Fig. 1c. Flow diagram (concluded).

where the index l and the interpolation number δ are determined by the velocity v. The values σ_l and σ_{l+1} are the cross sections at velocity v_l and v_{l+1}. These values are stored in a table. The quantities v, v_l, v_{l+1}, and δ are related by

$$v = v_l + (v_{l+1} - v_l)\delta \qquad 0 \leqslant \delta \leqslant 1.$$

The distance to a collision is

$$s_c = |\ln u|/\sigma,$$

where u is a random number. This is equivalent to saying that the probability density for s_c is

$$f(s_c) = \sigma e^{-\sigma s_c}.$$

The distance to a census is given by

$$s_{\text{cen}} = tv,$$

where t is the time to census. If the distance to census is smaller than the distance to collision, the coordinates at census time are computed and stored for use in the next cycle. If the distance to collision is smaller, the coordinates at the collision point are computed for use in the collision calculation. In either case the radial position r is checked to see if the particle is in the system. If the particle is not in the system, neither the census nor the collision calculation is made and another particle is selected for following. In both cases the new radial position r' and the new cosine β' are

$$r' = [(r + \beta s)^2 + (1 - \beta^2) s^2]^{1/2}$$

and

$$\beta' = (s + \beta r)/r',$$

where r and β are the old coordinates and s is the distance to collision or census. For a collision the new time until census is

$$t' = (s_{\text{cen}} - s_c)/v.$$

For a census the time until census is set equal to the cycle time t_0 for use in the next cycle.

The collision calculation consists of two parts. The first part is the neutronics calculation in which the number N of particles resulting from the collision, their velocities v_i ($i = 1, 2, ..., N$), and the cosines $\cos \psi_i$ ($i = 1, 2, ..., N$) of the angles between the direction of the incoming particle and the direction of the outgoing particles, are calculated. The second part is the calculation of the new values of β. The coordinates of the newly generated particles are stored in an intermediate storage table. The neutronics calculation is described in another section of this paper. The new value of β is

$$\beta = (1 - \beta'^2)^{1/2} n_1 + \beta' n_2,$$

where

$$n_1 = (1 - \cos^2 \psi)^{1/2} \sin \phi$$

and

$$n_2 = \cos \psi.$$

The angle ϕ is the azimuthal angle. It is taken to be uniformly distributed in the interval $(0 - 2\pi)$. The $\sin \phi$ is generated by the following process.

$$\sin \phi = [(2u_1 - 1)^2 - u_2^2]/[(2u_1 - 1) + u_2^2].$$

If $(2u_1 - 1)^2 + u_2^2 \geq 1$, two new random numbers are selected.

The last particle whose coordinates are stored in the intermediate storage table is then followed to collision or census by the same series of calculations as just described. When the particles from intermediate storages are exhausted, a new source particle is selected. When all particles reach census or leave the system, a new cycle is started in which the census particles just calculated are used as source particles.

IV. Estimates of α and the Equilibrium Distribution

The multiplication rate α of the system is estimated by

$$\alpha = \frac{1}{t_0} \ln \left(\frac{K}{q_{max} I} + 1 \right),$$

where t_0 is the cycle time, q_{max} is the number of cycles, I is the number of source particles followed per cycle, and

$$K = \sum_{q=1}^{q_{max}} (M_{hq} - I).$$

The quantity M_{hq} is the number of particles which reach census in the qth cycle. Therefore $K/q_{max} I$ is the average increase (if K is positive) or decrease (if K is negative) in the number of neutrons in the system in time t_0 per neutron in the system.

The average rate of increase per neutron is

$$\frac{\Delta N}{t_0 N} = \frac{K}{q_{max} I t_0}.$$

When $K/q_{max} I \ll 1$, α is approximately equal to the rate of increase.

Some estimate of the error in the calculation is needed. An unbiased estimate of the standard deviation of the rate of increase is

$$\sigma = \frac{1}{I t_0} \left[\frac{L - K^2/q_{max}}{q_{max}(q_{max} - 1)} \right]^{1/2},$$

where

$$L = \sum_{q=1}^{q_{max}} (M_{hq} - I)_q^2.$$

This estimate is based on the assumption that the source particles used at the beginning of each cycle are distributed according to the equilibrium distribution and are independent from the source particles of other cycles. Neither of these two assumptions are correct. However, experience has shown that if $It_0 q_{max}$ is large enough σ is reasonably consistent with the variation observed in the estimated α values for identical problems with different random number sequences.

The equilibrium distribution of neutrons in the system is estimated by G_{ls}/G, where G_{ls} is the number of particles at the end of all the cycles in a velocity interval (v_l to v_{l+1}) and a radial space interval (r_s to r_{s+1}). The space intervals are spherical zones with equal volume. G is the total number of particles which reach census and which escape the system. Again if the assumption is made that these particles are independent and are distributed according to the equilibrium distribution, then an estimate of the standard deviation of G_{ls}/G is

$$\sigma = \left[\frac{(G_{ls}/G)(1 - G_{ls}/G)}{G} \right]^{1/2}.$$

V. Collision Calculation

The collision calculation is made in a subroutine. Its purpose is to calculate the number N of neutrons resulting from a collision, the velocity v_i ($i = 1, 2, ..., N$), and the cosine $\cos \psi_i$ ($i = 1, 2, ..., N$) of the scattering angle in the laboratory system of each of the resulting neutrons. The subroutine is entered with the value of the velocity v

of the neutron making the collision, the velocity index l and interpolation number δ.

$$v = v_l + (v_{l+1} - v_l)\delta.$$

Neutron reactions are classified according to the method for calculating the velocity and cosine of the scattering angle of a neutron resulting from a collision. To each type of reaction there corresponds an index j. The remaining information which is needed to make the collision calculation is a table of values v_l and corresponding values of j. The quantity v_{lj} is the expected number of neutrons which come from a collision of a neutron with velocity v_l and which come from a reaction j. These values are calculated at the beginning of a problem from cross section data and the isotopic composition of the material of the system.

$$v_{lj} = \sum_{I(j)} A f_I (v\sigma)_{Ilj}/\sigma_l,$$

where A is the atom density of the material with units of atoms per centimeter-barn, f_I is the fraction of atoms in the material of isotope I, σ_{Ilj} is the microscopic cross section (with units of barns) of isotope I for a reaction of type j for a neutron of velocity v_l, v_{Ilj} is the expected number of neutrons resulting from such a reaction, and σ_l is the total macroscopic cross section of the material at velocity v_l (with units of reciprocal centimeters). The sum is over all isotopes in the material which can have a reaction of type j at velocity v_l. The total macroscopic cross section is given by

$$\sigma_l = \sum_I A f_i \sigma_{Il},$$

where σ_{Il} is the total microscopic cross section in barns for isotope I at a neutron velocity of v_l. The sum is over all isotopes in the material. The values of σ_l are stored in a table for use in the distance-to-collision calculation.

The first step in the calculation is the determination of the number N_j of neutrons from a reaction of type j and the index j. These values are determined by the following process.

$$N_{j_k} = \text{Grint } \rho_{j_k} \qquad k = 1, 2, \ldots n,$$

where

$$\rho_{j_1} = u + v_{j_1}$$
$$\rho_{j_k} = \rho_{j_{k-1}} - N_{j_{k-1}} + v_{j_k} \qquad k = 2, \ldots n$$
$$v_{j_k} = v_{lj_k} + (v_{l+1,j_k} - v_{lj_k})\delta.$$

This process for choosing values of N_j depends on the random number u and is similar to the process used for selecting source particles described in the preceding section. This process is not a model of a real neutron process in that it is possible to generate, for example, one neutron as if it came from an elastic collision and another in the same collision as if it came from a fission. However, the process has the property that if many collisions are made the average number of neutrons from elastic, fission, etc., collisions will be the same as for the real neutron process.

The number N of neutrons resulting from the collision is then

$$N = \sum_{k=1}^{n} N_{j_k}.$$

The method for generating values of v_i and $\cos \psi_i$ ($i = 1, 2, ..., N$) depends on the index j. For each different method there corresponds a different value of j. A method is different not only if the formulas are different but also if the constants used in the formulas are different. Some of the formulas for the various ways of calculating v_i and $\cos \psi_i$ are given in the following sections.

A. Elastic Scattering

Three basic ways of making an elastic-scattering calculation are used. These are

(1) $$v_i = \frac{v}{A+1}(1 + A^2 + 2A \cos \theta)^{1/2}$$

$$\cos \psi_i = (1 + A \cos \theta)/(1 + A^2 + 2A \cos \theta)^{1/2}$$

(2) $$v_i = \frac{v}{A+1}\left[A - \frac{1}{3A(A+1)} + \cos \theta\right]$$

$$\cos \psi_i = (1 + A \cos \theta)/(A + \cos \theta)$$

(3) $$v_i = v$$

$$\cos \psi_i = \cos \theta$$

where v is the velocity of the colliding neutron, A is the mass of the struck nucleus relative to the neutron mass, and $\cos \theta$ is the cosine of the scattering angle in the center-of-mass system.

The last two ways are approximations to decrease calculation time. The first way is appropriate for light isotopes, the second for intermediate-mass isotopes, and the third for heavy isotopes.

The value of $\cos\theta$ can be calculated in two ways.

(1) $\qquad\cos\theta = 2u - 1$

(2) $\qquad\cos\theta = M_l + (M_{l+1} - M_l)\,\delta$,

where

$$M_l = \mu_{lm} + (\mu_{l,m+1} - \mu_{lm})\,\Delta$$

$$m = \text{Grint } 16u$$

$$\Delta = 16u - m,$$

where μ_{lm} are tabulated values. Both ways depend on a random number u. The first way corresponds to an isotropic scattering in the center-of-mass system. The second way corresponds to anisotropic scattering in the center-of-mass system where the differential cross section is represented by a histogram with 16 equal-area intervals.

B. INELASTIC SCATTERING (n, n') FROM A SINGLE LEVEL

The values of v_i and $\cos\psi_i$ given by

$$v_i = \frac{v}{A+1}(1 + R^2 + 2R\cos\theta)^{1/2}$$

$$\cos\psi_i = (1 + R\cos\theta)/(1 + R^2 + 2R\cos\theta)^{1/2}$$

where

$$R^2 = A^2\left(1 - \frac{A+1}{A}\frac{Q}{v^2}\right).$$

The quantity Q/v^2 is the ratio of the energy of the level to the energy of the incident neutron, and $\cos\theta$ is the cosine of the scattering angle in the center-of-mass system. $\cos\theta$ can be determined by either of the two methods used for elastic scattering.

C. INELASTIC SCATTERING USING THE STATISTICAL MODEL

The values of v_i and $\cos\psi_i$ are given by

$$v_i = \text{least of } \begin{cases}(T\,|\ln u_1 u_2\,|)^{1/2}\\ v\end{cases}.$$

where $T = Kv$ and K is a constant which depends on the isotope and $\cos\psi_i = 2u - 1$.

This process is equivalent to sampling from the distribution,

$$f(E_i) = \begin{cases} \dfrac{E_i e^{-E_i/T}}{\int_1^E E_i e^{-E_i/T} \, dE} & 0 \leqslant E_i \leqslant E, \\ 0 & \text{elsewhere} \end{cases}$$

where E is the energy of the incident neutron, and E_i is the energy of the neutron resulting from the collision.

D. FISSION

The fission spectrum is approximated by a histogram with 72 equal-area intervals.

$$\cos \psi_i = 2u_1 - 1$$
$$v_i = v_m + (v_{m+1} - v_m) \Delta,$$

where

$$m = \text{Grint } 72u_2$$
$$\Delta = 72u_2 - m.$$

VI. Examples of Calculations

Two examples of calculations are given. The first is the calculation of the critical size of an ideal spherical system. In this system the neutrons are assumed to travel with a single velocity in one material and the scattering is isotropic. The second is the calculation of the critical size and leakage spectrum of the Godiva (Frye et al., 1954; Mills, 1959) critical assembly.

A. ONE-VELOCITY PROBLEM

The purpose of presenting the one-velocity problem is that results can be compared with S_n calculations as well as exact solutions. Also this problem provides an illustration of the sources of statistical errors in the Monte Carlo calculation.

In the problem, distances are measured in units of mean free path and time is measured in units of average collision time. Since λ is the average collision distance and the average collision time is λ/v, this is equivalent to setting $v = 1$ and $\lambda = 1$. The problem is then completely specified by the parameter ν, the expected number of neutrons resulting from a collision. ν is taken to be equal to 1.02, 1.2, and 2 in three different problems.

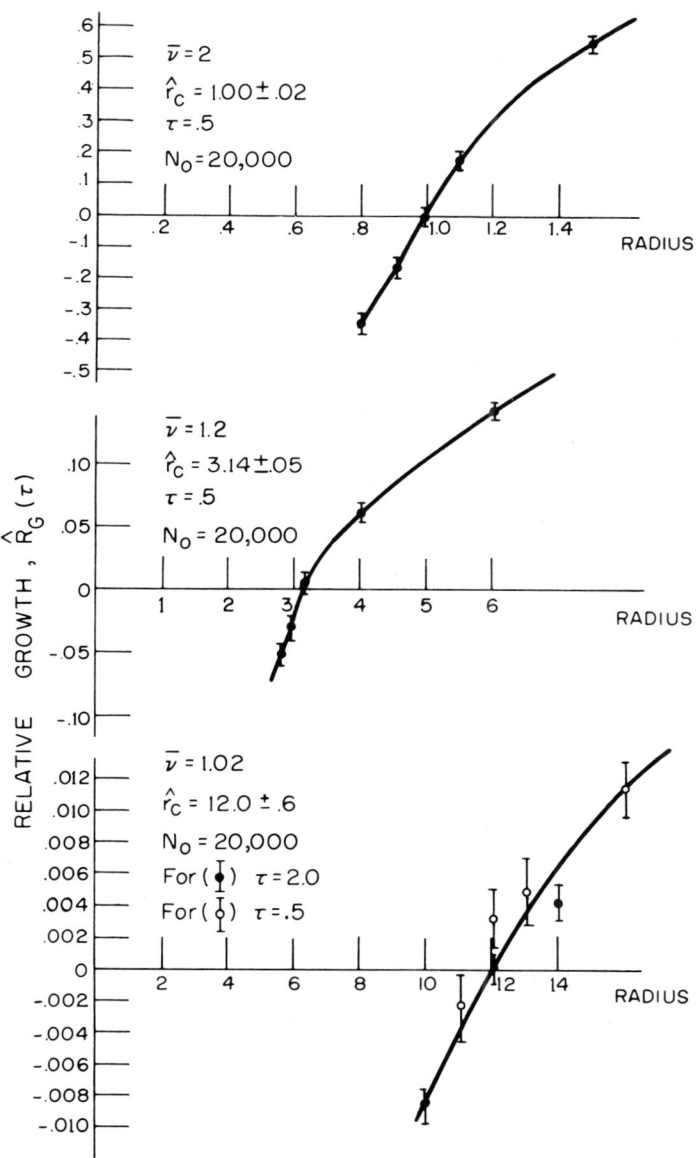

FIG. 2. Critical size for one-velocity problem.

As already explained, each Monte Carlo particle represents one neutron for convenience in particle following. Since the expected number of neutrons resulting from a collision is usually not an integer, the number of particles resulting from a collision is determined by the distribution

$$P_{N_1} = N_2 - \nu \equiv a_1$$
$$P_{N_2} = \nu - N_1 \equiv a_2,$$

where N_1 is the largest integer smaller than ν and $N_2 = N_1 + 1$. Notice that

$$a_1 \geqslant 0 \qquad a_2 \geqslant 0$$
$$a_1 + a_2 = 1$$
$$\nu = N_1 a_1 + N_2 a_2.$$

The critical radius is determined by plotting the relative growth R_G of the populations in a time t against the radius. The radius for which $R_G = 0$ is then the critical radius. Figure 2 shows R_G for the three problems. The indicated errors are from the estimated standard deviations The Monte Carlo and S_n values of the critical radii are given in Table I. The Monte Carlo calculations required $\frac{1}{2}$ to 4 min on the IBM 7090.

TABLE I

CRITICAL RADIUS FOR SPHERES

		Calculation method		
ν	Exact	S_4	S_8	Monte Carlo
1.02	12.027	12.028	12.028	12.0 ± 0.6
1.2	3.172	3.157	3.168	3.14 ± 0.05
2.0	0.9906	0.9790	0.9877	1.00 ± 0.02

The infinite radius case is included because the error analysis can be made. The analysis is as follows:

$$P(N) = \sum_{n=0}^{\infty} p_n P_{N|n},$$

$$E(N) = \sum_{N=1}^{\infty} NP(N) = \sum_{N=1}^{\infty} N \sum_{n=0}^{\infty} p_n P_{N|n} = \sum_{n=0}^{\infty} p_n E(N \mid n). \qquad (2)$$

$$Var(N) = \sum_{N=1}^{\infty} [N - E(N)]^2 P(N) = \sum_{N=1}^{\infty} [N - E(N)]^2 \sum_{n=0}^{\infty} p_n P_{N|n}$$

$$= \sum_{n=0}^{\infty} p_n \, Var(N \mid n) + \sum_{n=0}^{\infty} p_n [E(N \mid n) - E(N)]^2,$$

where $P(N)$ is the probability of one particle having N ancestors in time t, $P(N \mid n)$ is the probability of one particle having N ancestors after exactly n collisions, $E(N)$ is the expected number of ancestors in time t, Var (N) is the variance of N, and p_n is the probability of having n collisions in time t. t is numerically equal to the expected number of collisions because $\lambda = v = 1$. Therefore,

$$p_n = \frac{e^{-t}t^n}{n!}. \tag{3}$$

The probability generating function for $P(N \mid n)$ is given by the following recursion relation

$$G_{N\mid n}(s) = a_1 G_{N\mid n-1}(s) + a_2 [G_{N\mid n-1}(s)]^2,$$

where

$$G_{N\mid 0}(s) = s.$$

Then

$$E(N \mid n) = \left.\frac{\partial G_{N\mid n}(s)}{\partial s}\right|_{s=1} = (a_1 + 2a_2)^n = m^n \tag{4}$$

and

$$\mathrm{Var}\,(N \mid n) = \left.\frac{\partial^2 G(s)}{\partial s^2}\right|_{s=1} + \left.\frac{\partial G_{N\mid n}}{\partial s}\right|_{s=1} - \left(\left.\frac{\partial G_{N\mid n}}{\partial s}\right|_{s=1}\right)^2$$
$$= \frac{(2a_2 + m - m^2)}{1-m} m^{n+1}(1-m^n). \tag{5}$$

The preceding is for the cases: $1 \leqslant \bar{v} \leqslant 2$, and

$$a_1 = 2 - \bar{v}$$
$$a_2 = \bar{v} - 1.$$

Then

$$m = \bar{v}. \tag{6}$$

From Eqs. (2-5)

$$E(N) = \sum_{n=0}^{\infty} \frac{e^{-t}t^n}{n!} m^n = e^{t(m-1)} \tag{7}$$

and

$$\mathrm{Var}\,(N) = \sum_{n=0}^{\infty} \frac{e^{-t}t^n}{n!} \frac{(2a_2 + m - m^2)}{(1-m)} m^{n+1}(1-m^n) + \sum_{n=0}^{\infty} \frac{e^{-t}t^n}{n!} [m^n - e^{t(m-1)}]^2$$
$$= \frac{(2a_2 + m - m^2)}{m(1-m)} e^{-t} e^{tm} - e^{tm^2} + [e^{(m^2-1)t} - e^{2(m-1)t}]. \tag{8}$$

Equation (8) illustrates two of the sources of statistical error in the problem. The first term is the variance introduced by forcing the particles to represent an integral number of neutrons. This term is zero if $\bar{\nu} = 1$ or 2. The second term is the variance introduced by sampling the number of collisions. In the general problem other sources of error are introduced by having several materials, by particles escaping, and by the particles having different velocities.

The relative growth R_G is defined as

$$R_G = E\left(\frac{N-1}{t}\right).$$

For the infinite radius problem

$$R_{G\infty} = \frac{e^{t(m-1)} - 1}{t}.$$

The Monte Carlo estimate $\hat{R}_{G\infty}$ for the relative growth rate $R_{G\infty}$ is

$$\hat{R}_{G\infty} = \frac{1}{N_0} \sum_{i=1}^{N_0} \left(\frac{N_i - 1}{t}\right),$$

where N_0 is the number of samples, and N_i is the value of the number of ancestors from the ith particle. From the central limit theorem $\hat{R}_{G\infty}$ is approximately a normally distributed random variable. Its expected value is

$$E(\hat{R}_{G\infty}) = R_{G\infty}$$

and its variance is

$$\mathrm{Var}\,(\hat{R}_{G\infty}) = \frac{\mathrm{Var}\,(N)}{N_0 t^2} = \frac{2}{m N_0 t^2}\left[e^{(m^2-1)t} - e^{(m-1)t}\right].$$

The Monte Carlo estimates for $R_{G\infty}$ and standard deviations as well as the theoretical values are given in Table II.

TABLE II

RELATIVE POPULATION GROWTHS FOR INFINITE RADIUS

$m = \nu$	t	N_0	$\hat{R}_{G\infty}$	$\hat{\sigma}_{R_{G\infty}}$	$R_{G\infty}$	$\sigma_{\hat{R}_{G\infty}}$	α
1.02	0.5	20,000	0.0207	0.0015	0.0202	0.00198	0.02
1.20	0.5	20,000	0.2104	0.0047	0.2104	0.00686	0.20
2.00	0.5	20,000	1.282	0.018	1.2974	0.0238	1.00
1.02	2.0	20,000	—	—	0.0204	0.0103	0.02

The case for $m = 1.02$ and $t = 2$ is given to illustrate that accuracy is approximately proportional to the square root of calculation time. The calculation time is approximately proportional to the expected number of collisions which in the examples is equal to t. The ratio of the t's is 4; the ratio of the accuracy is 1.92.

B. Godiva Critical Assembly

The Godiva Critical Assembly was chosen for an illustration because it is both simple and a well-documented (Frye et al., 1954) experiment. The calculations were made using an atom density of 0.04801×10^{24} atoms/cm^3. The fraction of U^{235} was 0.935 and the fraction of U^{238} was 0.065. Cross sections used were supplied by R. J. Howerton (1958). In the calculation delayed neutrons were neglected. The experimental fast-critical radius is 8.73 cm.

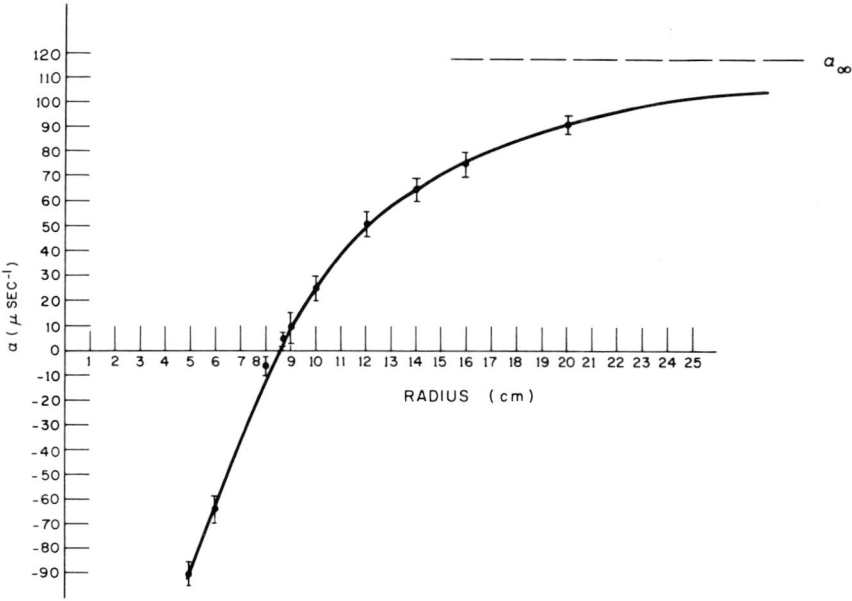

FIG. 3. α vs radius for sphere of Godiva composition.

The results are presented in Figs. 3 and 4. Figure 3 shows the α vs radius curve for the Godiva composition. The standard deviation for all except the 8.73-cm-radius point was approximately 3 μsec^{-1}. The computation time on the IBM 7090 was about 2.2 min for all except

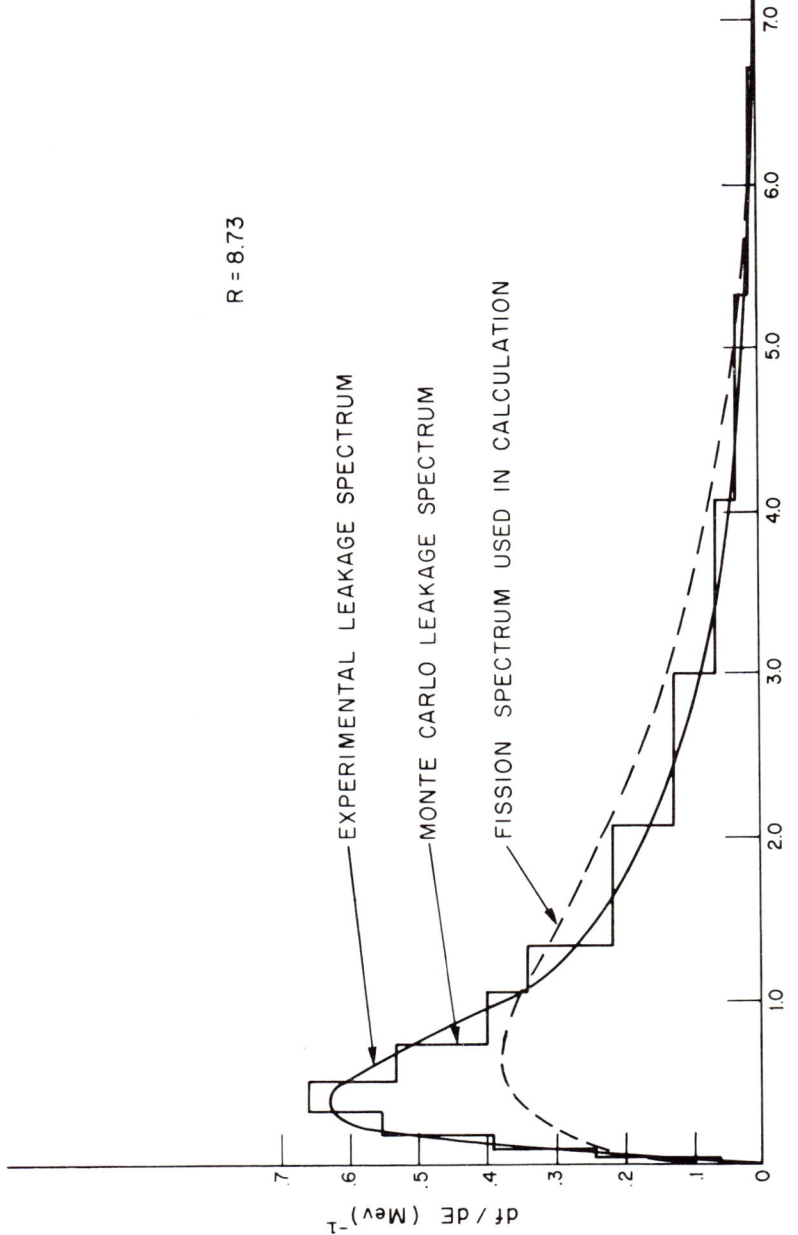

Fig. 4. Godiva leakage spectrum. Fraction of neutrons per unit energy vs energy.

the 8.73-cm-radius point. The time for this point was 8.5 min. Figure 4 shows the calculated and experimental leakage spectrum from Godiva and the fission spectrum used in the calculation.

The α vs radius curve provides a solution not just for the system described but for a set of dimensionally equivalent systems. These equivalent systems are characterized by the following:

$$\Sigma r = \text{constant}$$
$$\alpha/\Sigma v = \text{constant}$$
$$t\Sigma v = \text{constant}$$

where Σ is a macroscopic cross section, r is a linear dimension, t is a time between two events, v is a neutron velocity, and α is the multiplication rate. These relations must hold for all velocities and all cross section for equivalent systems. If the neutron velocity is unchanged between the new and old systems, the relationships become

$$\Sigma r = \text{constant}$$
$$\alpha/\Sigma = \text{constant}$$
$$t\Sigma = \text{constant}$$
$$v_{\text{new}} = v_{\text{old}}.$$

These relationships are useful in examining the gross effect of cross-section errors on the calculations of critical radii.

From Fig. 3 the calculated α for the measured critical radius of 8.73 cm is 5.09 μsec^{-1} with a standard deviation of 1.5 μsec^{-1}. This discrepancy cannot reasonably be accounted for by a statistical error. However, using the calculated critical radius of 8.5 cm and the equation $\Sigma_1 r_1 = \Sigma_2 r_2$, then

$$\Sigma_1/\Sigma_2 = r_1/r_2 = 8.5/8.73 = 0.974\ .$$

The error could therefore be accounted for by cross-section errors of about 4%. Since cross-section errors of 10% are not unusual, the error can reasonably be assigned to this source.

Acknowledgments

Much of the credit for the calculation belongs to Mark Z. Brown who did the programming for the problem, to Donald L. Reeves who prepared the cross-section data for the calculation, and to R. J. Howerton who collected and prepared the basic cross-section data.

References

DAVISON, B. (1957). "Neutron Transport Theory." Oxford Univ. Press, London and New York.

CARLSON, B. G. (1955). Solution of the Transport Equation by S_n Approximations. Los Alamos Scientific Laboratory Report No. LA-1891,

CARLSON, B. G., and BELL, G. I. (1958). Solution of the Transport Equation by the S_n Method. *Proc. 2nd Intern. Conf. on the Peaceful Uses of Atomic Energy*, P/2386. (1959). Geneva.

CARLSON, B. G. (1959). Numerical Solution of Transient and Steady-State Neutron Transport Problems. Los Alamos Scientific Laboratory Report No. LA-2260.

DAVIS, D. H. (1959). A Monte Carlo Calculation of Equilibrium Neutron Distributions and Multiplication Rates. Univ. of California Radiation Laboratory Report No. UCRL-5530. (Report classified Secret R. D., Title unclassified.)

KAPLAN, E. L. (1958). Monte Carlo Methods for Equilibrium Solution in Neutron Multiplication. Univ. of California Radiation Laboratory Report No. UCRL-5275-T, 1958.

GOAD, W., and JOHNSTON, R. (1959). A Monte Carlo Method for Criticality Problems. *Nuclear Sci. and Eng.* 5, 371.

MORTON, K. W. (1956). Criticality Calculations by Monte Carlo Methods. Harwell Atomic Energy Research Establishment Report No. AERE T/R 1903.

FRYE, G. M., GAMMEL, J. H., and ROSEN, L. (1954). Energy Spectrum of Neutrons from Thermal Neutron Fission of U^{235} and from an Untamped Multiplying Assembly of U^{235}. Los Alamos Scientific Laboratory Report No. LA-1670.

MILLS, C. B. (1959). Neutron Cross Sections for Fast and Intermediate Nuclear Reactors. LAMS-2255.

HOWERTON, R. J. (1958). Semi-Empirical Neutron Cross Sections. Univ. of California Radiation Laboratory Report No. UCRL-5351.

A Monte Carlo Calculation of the Response of Gamma-Ray Scintillation Counters

CLAYTON D. ZERBY

OAK RIDGE NATIONAL LABORATORY[*]
OAK RIDGE, TENNESSEE

I. Introduction	90
II. Gamma-Ray Scintillation Counters	91
A. Description of the Counter	91
B. Response of the Counter	92
III. Idealizations and Approximations	96
A. Idealization of the Counter	96
B. Approximating the Gamma-Ray Interactions	97
C. Approximating the Transport of Electrons and Positrons	99
D. Idealization of the Counter Response	102
IV. Sampling Procedures and Auxiliary Programs	103
A. Random Numbers	103
B. Basic Selection Techniques	103
C. Coordinate Systems	105
D. Sampling the Source	106
E. Cross Sections	106
F. Selecting Distances	107
G. Selecting a Uniformly Distributed Azimuthal Angle	108
H. Selecting a Random Isotropic Direction	109
I. Compton Scattering	110
J. Rotation of Coordinates	113
K. Pair Production	113
L. Annihilation Radiation	115
M. Bremsstrahlung Radiation	115
V. Details of the Monte Carlo Procedure	117
A. Estimating the Intrinsic Efficiency	118
B. Estimating the Analytic Zero Intercept	119
C. Details of the Transport Calculation and Estimation of the Response Spectrum	120
D. Calculating the Peak-to-Total Ratio	124
E. Broadening the Response Spectrum	125
F. The Computing Machine Program	125
VI. Results of the Calculations	125
Acknowledgments	133
References	133

[*] Operated by Union Carbide Corporation for the U. S. Atomic Energy Commission.

I. Introduction

THE PROBLEM OF DETERMINING the response of a scintillation counter to gamma rays is basically one of calculating the transport of that radiation through the scintillation detector. The gamma radiation from a source enters the detector, diffuses through it, and is either captured in the detector or escapes from it through one of the bounding surfaces. Secondary radiations created directly or indirectly by the source radiation in the process of interacting with the detector contribute to the transport problem and also are captured or escape. The quantity that must be calculated in order to determine the response of the counter is the total energy deposited in the detector; therefore, the amount of energy carried off by source radiation and secondary radiations that escape has to be determined.

The only practical way of obtaining a solution to this complicated transport problem is with the use of Monte Carlo methods programmed for calculation on high-speed, automatic computing machines. Solving the transport equation (or more accurately, the coupled transport equations, since the secondary radiations are included) by using standard numerical techniques is still impractical because modern high-speed computers are still incapable of the speed necessary to economically perform the calculations. Fortunately the counter problem is not particularly difficult when Monte Carlo methods are used. Various techniques that have been used and tested in a variety of other Monte Carlo transport problems are applicable in this case and can be used with an almost certainty that they will lead to the statistical accuracy required.

The application of Monte Carlo methods to the calculation of the response of gamma-ray scintillation counters is not new. Monte Carlo methods were employed by Berger and Doggett in 1956, by Davisson and Beach in 1959, and more recently by Miller and Snow in 1960. All of these calculations used analogue Monte Carlo techniques. The calculation of Miller and Snow differed from the earlier ones in that secondary bremsstrahlung and annihilation radiations were included in the transport problem. This is essential for source energies above approximately 1 Mev.

A calculation of the response of gamma-ray scintillation counters which also uses Monte Carlo methods but differs from the earlier calculations is described in the following sections. Like the Miller-Snow calculation, this calculation includes secondary radiation in the transport problem in order to make possible the consideration of higher source energies; however, the manner in which the secondary radiation is treated effectively restricts the calculation to energies below 10 Mev.

The present calculation differs from all the previous calculations in that analog procedures are not used for the complete problem.

Because of the variety of Monte Carlo techniques and methods used in this study, such as rejection techniques, systematic sampling, use of survival probabilities, weighting, and statistical estimation, its presentation serves as an introduction to many of the elementary Monte Carlo techniques, as well as to general methods employed in the solution of transport problems. The presentation begins in Section II with a detailed description of the gamma-ray scintillation counter problem itself in order to establish the necessity for considering various physical processes in the transport problem. The approximations and idealizations made in the problem are discussed in Section III. Following a brief description of basic Monte Carlo selection techniques, the details of the various selection techniques and auxiliary programs used in the calculation are presented in Section IV. In several instances alternate techniques are given if the procedure is of general interest. Section V contains a complete description of how the Monte Carlo calculation was carried out. Typical results of calculations for a 3-in.-diam. by 3-in.-high NaI(Tl) scintillation counter which have been compared with experimental results are presented in Section VI. References to several papers which report results calculated with the machine code that was produced for this calculation are also given in Section VI.

II. Gamma-Ray Scintillation Counters

It is appropriate here to describe the basic scintillation counter and to review its response characteristics. This is done to point out the necessity of considering various physical processes in the transport problem and to establish the validity of the simplifying assumptions or idealizations that are required to make the problem tractable. Fortunately the problem can be simplified considerably without loss of general applicability.

A. Description of the Counter

The components of a scintillation counter consist of two basic elements: (1) a scintillation detector which produces flashes of light when exposed to ionizing radiation and (2) a photomultiplier tube which is optically coupled to the detector and converts the light flashes into electrical pulses. When the counter is used to detect gamma rays, the light flashes are caused by secondary ionizing radiation produced by the incident gamma rays, and the associated electrical pulses from the photomultiplier are analyzed according to magnitude (pulse height) to obtain a response spectrum. Since the electrical pulses are almost proportional to the intensity of the light flashes and the light flashes are almost proportional to the energy deposited in the detector, the scintillation counter is especially useful as a spectrometer, although there are some complicating features that will be discussed subsequently.

For purposes of calculating the response of a gamma-ray scintillation counter, the composition of the detector is not of particular importance, and it is sufficient to note here only that at present considerable experimental attention seems to be focused on two alkali iodide crystals, NaI and CsI, which have been activated with the heavy impurity element thallium (Tl) and on organic liquid detectors having xylene or toluene as the basic material.

A schematic drawing of a typical but simplified source-counter configuration is shown in Fig. 1. The relative positions and sizes of

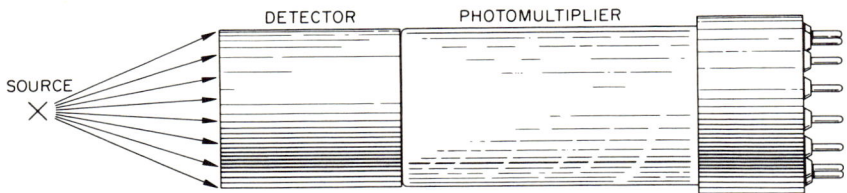

FIG. 1. Schematic drawing of the source-counter arrangement of a typical gamma-ray scintillation counter.

the components are typical of a small detector coupled to a photomultiplier tube and a small external source. If the detector were large, such as a large organic liquid detector, there would probably be several photomultiplier tubes coupled to it. In almost all cases the source-counter configuration has axial symmetry, or very nearly axial symmetry, which simplifies the spatial part of the transport problem.

A crystal is often modified to have beveled edges and/or an axial hole in the end nearest the source. In the latter case the source radiation is usually collimated to strike the detector within the hole. In some configurations the hole passes all the way through the detector and the source itself is located inside the detector.

B. RESPONSE OF THE COUNTER

It is obvious that if the detector were perfectly absorbing each incident photon would have an almost unique pulse-height response, depending on its energy. The pulse-height response spectrum from an arbitrary source could then be easily unfolded to obtain the energy spectrum of the source and one would have a very good spectrometer. It has been demonstrated, however, that the response of a scintillator to monoenergetic photons is not unique, even in the case of a hypothetical perfectly absorbing detector. This condition of nonuniqueness, which is the result of uncertainties in the intensity of light produced in the detector, the transmission of the light to the photomultiplier, and the conversion of the light into an electrical pulse by the photomultiplier, thus causes the response of a counter to a monoenergetic source to be spread out over a range of pulse heights.[1] This spreading, or so-called "line broadening," can be approximated by a Gaussian dependence given by

$$f(E, E')\, dE = (\pi\sigma^2)^{-1/2} \exp\left[-\frac{1}{2}(E-E')^2 \sigma^{-2}\right] dE, \tag{1}$$

where $f(E, E')\, dE$ is the probability that an energy E' deposited in the detector will give a response pulse with height E in dE.

The quantity σ in Eq. (1) has a dependence on E' which can be expressed empirically by the relation

$$\sigma = A\sqrt{E'} + BE', \tag{2}$$

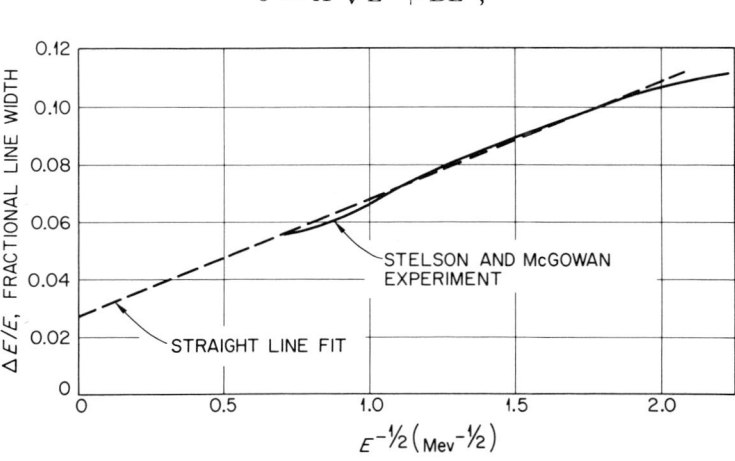

FIG. 2. Fractional linewidth, $\Delta E/E$, for a 3-in.-diam by 3-in.-high NaI(Tl) crystal as a function of $E^{-1/2}$. Here E is the energy absorbed by the crystal in units of Mev and ΔE is the linewidth at half maximum. The experimental data was obtained from Stelson and McGowan (1959).

[1] For a complete review of this subject see Mott and Sutton (1958), Breitenberger (1955), and Wright (1954).

where A and B are constants that differ from one counter to another. The plausibility of the empirical form of Eq. (2) can be observed by comparing it with experimental values of the fractional linewidth, $\Delta E/E$, of the total absorption line. A comparison with the experimental data of Stelson and McGowan (1959) for a 3-in.-diam. by 3-in.-high NaI(Tl) crystal is shown in Fig. 2, where ΔE is the linewidth at half maximum of a line at energy E and is related to σ by $\Delta E = (8 \ln 2)^{1/2}\sigma$. Although the empirical formula does not fit the experimental data perfectly, it gives a sufficiently accurate approximation to warrant its use. In this particular case the constants A and B are 0.01703 (Mev)$^{1/2}$ and 0.01147, respectively.

These two observations, that is, that the line broadening is fitted very well with the Gaussian distribution and that the empirical form of the energy-dependent parameter σ is fairly well established, are especially important since they indicate that in the calculation the

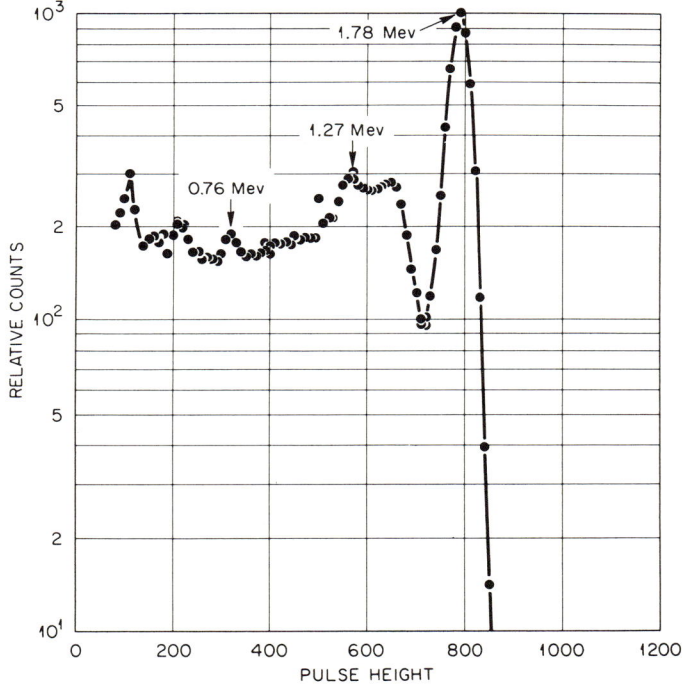

FIG. 3. Experimental Response Spectrum for a 3-in.-diam by 3-in.-high NaI(Tl) crystal exposed to a 1.78-Mev isotropic point source of gamma rays located on the centerline of the detector 36 cm from one end. The data was obtained from Lazar and Willard (1956).

counter can be assumed to give a unique response to gamma ray of the same energy and the broadening effect can be introduced after the main calculation is complete.

In addition to line broadening, however, the response of a real counter is complicated further by radiation which escapes from the detector, causing less than the full energy of the incident radiation to be deposited. Background effects can also cause distortions of the response spectrum. Examples of both these complicating features are apparent in Fig. 3, which is the experimental response spectrum observed by Lazar and Willard (1956) for a 3-in.-diam. by 3-in.-high NaI(Tl) counter exposed to 1.78-Mev gamma rays. The pulse height is given in energy units. It can be observed that the total absorption peak appears at 1.78 Mev. If the detector had been perfectly absorbing the entire response spectrum would consist of that peak alone since it corresponds to the case where all of the energy of the incident gamma rays is absorbed. At 1.27 and 0.76 Mev, however, there are two additional peaks which are accounted for by the loss of annihilation radiation. A typical sequence of events resulting in such a loss would be (1) the conversion of an incident photon into an electron-positron pair in a production event, (2) the energy degradation of the pair by inelastic collisions with bound atomic electrons, (3) the annihilation of the positron with an electron at rest to produce two photons of 0.511 Mev each, which in turn can travel through the detector leading to (4) the escape or absorption of either or both annihilation photons. The peak at 1.27 Mev corresponds to the case where only one annihilation photon escapes and no other energy is lost, while the peak at 0.76 Mev corresponds to the case where both photons are lost.

These two escape peaks are actually superimposed on an energy-loss spectrum for incident photons that inelastically scatter in the detector and then escape, taking with them part of their initial energy. As the source energy increases above the pair production threshold they become more prominent above the energy-loss spectrum and tend to increase in height relative to the total absorption peak.

There are several other energy-loss factors which cause distortions of the scintillation response spectrum. Typical of these are losses of energy by the escape of secondary bremsstrahlung radiation created by electrons which slow down in the detector or by loss of the ejected electrons themselves. Both of these effects become relatively more important as the source energy increases, the first because the average recoil energy of the electrons increases, resulting in more and higher energy photons being created in radiation collisions, and the second because a higher recoil energy means that the electron will have a longer residual path length before stopping, and, therefore, a greater chance of escaping. Of these two processes, the loss of energy by the escape of secondary bremsstrahlung radiation is far more important than by the loss of electrons for detectors of reasonable size and for sources below 10 Mev. This is because the path of the electron is usually quite devious as it slows down from energies below 10 Mev, and, hence, it migrates only a relatively short distance on the average from its point of origin before stopping.

The causes of background contributions to the response spectrum are so numerous that any discussion here must be limited to a few examples. Source radiation that initially misses the detector but is scattered into it by nearby apparatus, a wall, or even the air constitutes one source of background radiation. Another is the radiation that escapes from the detector and is scattered back into it. In most cases the background effects are observed in regions of low response of the counter and at low pulse heights. The erratic response and the increasing number of counts at low pulse heights shown in Fig. 3 are examples of the distorting effects of background contributions.

It should be pointed out that the difficulties discussed above have been in terms of a typical response spectrum for a monoenergetic source. Obviously, the response spectrum for a complex source would be much more complicated, consisting of a superposition of many such spectra, and unfolding these complex response spectra to obtain the incident gamma-ray spectra is very difficult. In general, the error in the unfolding process decreases as the ratio of the area under the total absorption peak to the total area of the response spectrum (the peak-to-total ratio) increases. Thus the peak-to-total ratio is an important parameter.

Much research has been devoted to understanding the response of scintillation counters in order to design better equipment in which the peak-to-total ratio is increased as much as possible. The hazard is making alterations that are costly with little gain in performance. Avoiding this situation is difficult without considerable data which is most easily obtained from calculations such as reported here.

III. Idealizations and Approximations

A. Idealization of the Counter

Since background effects are unique for any particular experimental arrangement, it is not practical to include this effect in a general calculation. Indeed, there is some advantage in being able to eliminate the background effects in order to estimate the true response of the counters. Background effects can then be estimated by subtracting the results of the calculation from experimental results.

Elimination of the background effects is particularly simple in the Monte Carlo calculation, since it is only necessary to assume that the source-detector configuration is suspended in an infinite vacuum. This idealization not only removes the background effects but also simplifies the calculation. The difference between calculation and experiment produced by this idealization is expected to be significant in regions of low pulse height and in regions of low response, such as in the valley between the total absorption peak and the energy loss spectra for low-energy sources.

B. Approximating the Gamma-Ray Interactions

There will be no attempt in this discussion to cover all the interactions which govern the behavior of a photon in a material, the subject having been covered previously by several authors (Heitler, 1954; Bethe and Ashkin, 1953). Here it is sufficient to discuss only the most important interactions in the energy below 10 Mev, since that is the maximum energy allowed in the calculation. (The reason for the energy limitation will be explained below in the discussion of how electrons are treated in the calculation.)

In the energy range below 10 Mev the most important interactions governing the transport behavior of a photon are the photoelectric effect, pair production, and Compton scattering. In the first two of these the photon disappears and charged particles are ejected from the atom; in the third process the photon changes direction and transmits part of its energy to an atomic electron.

The photoelectric effect is significant only at low photon energies where it increases in importance with decreasing energy and finally becomes the most probable reaction. In this interaction the photon transmits all of its energy to a bound atomic electron which is ejected with kinetic energy equal to the original photon energy less the electron binding energy. The binding energy was ignored in this calculation, however, since it is such a small fraction of the photon energy for high-energy photons and has little effect on the energy deposited in the detector for low-energy photons. The latter statement is based on the fact that a low-energy recoil electron usually deposits all of its energy in the detector because of its short range and also because the release of bremsstrahlung radiation is very improbable.

At energies above 1.02 Mev, pair production is possible and becomes increasingly important as the energy increases. In this process the photon disappears and part of its energy is converted into the formation of an electron-positron pair. The excess energy is shared between the pair as kinetic energy.

The cross sections for photoelectric events and pair production were obtained for the calculation from the paper by Grodstein (1957) which presents convenient tabulations of photon cross sections at different energies for different materials.

Compton scattering is competitive to both photoelectric events and pair production but has a cross section exceeding them only at intermediate energies, as can be seen in Fig. 4 where the photon cross sections for NaI are presented as an example.

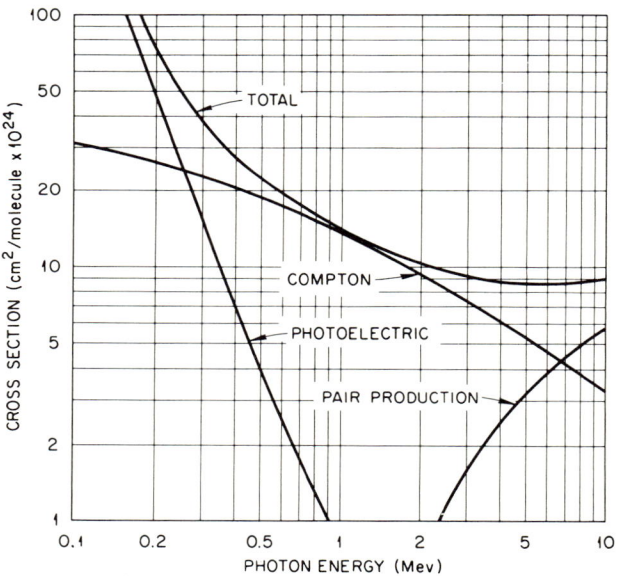

FIG. 4. Cross sections for NaI.

Polarization effects in Compton scattering were completely ignored in the calculation. It is true that the gamma-ray sources tend to generate unpolarized radiation; however, after several scatterings the radiation is no longer completely unpolarized and the penetration of the gamma rays through materials is increased. Thus the assumption of unpolarized radiation at every interaction is truly an approximation. The studies of this approximation by Spencer and Wolff (1953) indicate that the error in this estimated energy deposition in the present calculation should be negligible.

Rayleigh (coherent) scattering is another type of interaction which is very probable at low energies and should be noted here. In an event of this type, the photon is scattered by the atomic electron cloud and the entire atom recoils. The photon undergoes only a slight change in direction and suffers only an insignificant decrease in energy because of the small recoil energy taken up by the atom.

In problems dealing with the transport of photons it is convenient to ignore Rayleigh scattering completely by setting the cross section equal to zero as was done in this problem. This amounts to approximating the interaction by assuming straightahead scattering with no energy degradation. Such an approximation is reasonable under most conditions encountered in gamma-ray transport problems where photon energies in the low kilovolt range are not important.

In the course of the Monte Carlo calculation, it is convenient to choose an energy below which the photon is no longer of interest. This energy is usually chosen to be very low in comparison with the source energy, so that a negligible amount of energy is carried by the photon and serious biasing of the estimates of the energy deposited is not likely. A photon degraded below this cutoff energy is considered to be absorbed and all its energy deposited at that point of interaction. This is equivalent to the approximation that below the cutoff energy the photoelectric cross section is large and equal to the total cross section and all the energy of the recoil electron is deposited at the point of ejection from the atom. In this calculation the cutoff energy was selected as 5.11 kev.

C. Approximating the Transport of Electrons and Positrons

The two principal mechanisms by which electrons interact with matter and are degraded in energy are inelastic collisions with the electrons of the atom and radiative collisions with nuclei and electrons. The inelastic collisions dominate the degradation process at low energy, while the radiation collisions dominate at high energy. The approximate energy which separates these two regions is given by $E = 1600\ mc^2/Z$. This suggests that if the problem were restricted to energies below approximately 10 Mev, the region of primary interest in the present case, then inelastic events could be treated as the primary mechanism of energy degradation while radiation collisions could be treated as a perturbation. This, in fact, turns out to be the case and was used to calculate the bremsstrahlung spectrum as will be discussed below.

In the energy region below 10 Mev the residual path length of an electron is rather short, indicating the possibility of still further approxi-

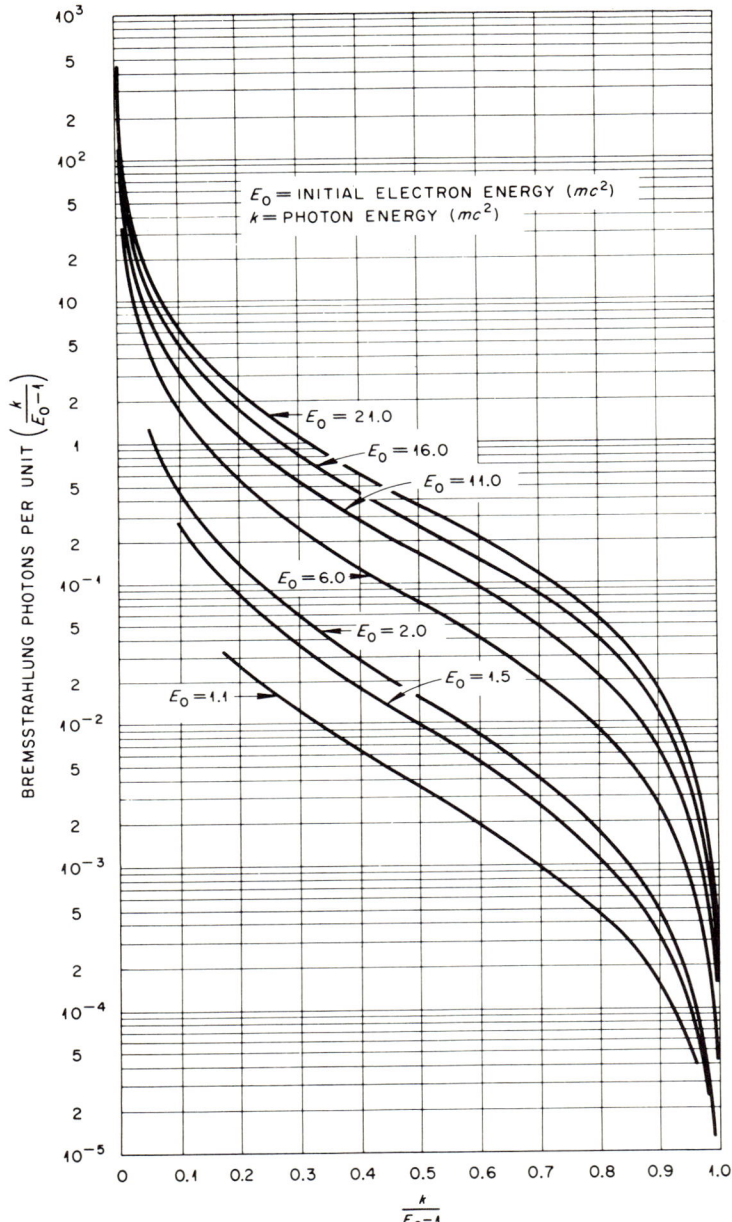

FIG. 5. Bremsstrahlung spectra from an electron slowing down in an infinite sodium iodide crystal. The curves are labeled with the initial total energy (kinetic plus rest mass energy) of the electron in rest mass units.

mations. As an example, in NaI(Tl) electrons of energies 1.0, 4.0, 7.0, and 10.0 Mev travel distances of 0.2, 0.9, 1.5, and 2.0 cm, respectively, before stopping. Of more importance, however, is the fact that the radial distance from the starting point to the point at which the electron stops is very much shorter than the residual path length because of the devious path taken by the electron. This last observation is made because it suggests that if the calculation were restricted to energies below 10 Mev it might be reasonable to let an electron stop at its point of origin, thus avoiding the tedious calculation of the electron transport. In this calculation this procedure was adopted.

The approximation that the electron stops at its point of origin also leads to simplification of the calculation of the spectrum of bremsstrahlung radiation given off by an electron. Since all the radiation will come from a point it is possible to use the average bremsstrahlung spectrum of photons created in the entire stopping process. The calculation of this average spectrum has been reported previously (Zerby and Moran, 1958) and will only be outlined briefly here. Born approximation cross sections for radiative collisions which include the effects of screening were used for bremsstrahlung events, and a continuous slowing-down model for inelastic (ionization) events was used. As pointed out above, the radiative collisions were treated as a perturbation. The results for NaI(Tl) are presented in Fig. 5. Each curve in Fig. 5, which is labeled with the total initial energy (kinetic plus rest mass energy) of the electron, presents the energy spectrum of photons with energies above $0.04\ mc^2$. The areas under these curves give the total number of photons released; the results are presented in Table I.

The direction of emission of the bremsstrahlung radiation can be assumed to be isotropic. This is because electrons at energies considered

TABLE I

TOTAL NUMBER OF BREMSSTRAHLUNG PHOTONS WITH ENERGY ABOVE $0.04\ mc^2$ RELEASED BY AN ELECTRON WHICH IS STOPPED IN NaI

Initial electron energy (Kinetic plus rest mass energy), mc^2	Number of photons
1.1	0.0009
1.5	0.0325
2.0	0.0972
6.0	0.8434
11.0	1.9770
16.0	3.1359
21.0	4.4144

here diffuse in an almost random direction after losing only a small fraction of their energy; hence, there should be almost no correlation between the original direction of the electron and the direction of the bremsstrahlung photons. An exception to this is the direction of the highest energy bremsstrahlung photons since they must be emitted before the electron has lost much energy by deflecting collisions. The relative probability of emitting these high-energy photons is very small, however, as can be seen in Fig. 5 and, therefore, little error should be introduced by assuming they are also isotropically emitted.

Positrons created in pair production events were assumed to yield the same bremsstrahlung spectrum as electrons although minor differences could be expected because of differences in ionization collisions. In addition to bremsstrahlung radiation, however, the positron is also the source of radiation when it is annihilated in a collision with an electron. As an approximation it was assumed that all the positrons came to rest before emission of the two annihilation photons so that each photon had an energy of 1.0 mc^2 and was emitted in a direction opposite to that of the other. Under these conditions there is no preferred orientation of the emitted radiation. Actually this procedure is a good approximation at energies below 10 Mev since over 90% of the positrons will be stopped before annihilation (Heitler, 1954). To be consistent with the previous approximations it was also assumed that the annihilation radiation emanated from the point where the positron was created.

D. Idealization of the Counter Response

As pointed out in Section II, A, the scintillation counter is especially useful as a spectrometer since the light flashes are almost proportional to the energy deposited in the detector and the electrical pulses from the photomultiplier are almost proportional to the intensity of light flashes. Much of the nonproportionality can be empirically accounted for in the broadening of the response spectrum as discussed in Section II, B; however, there remains still another effect which can be ignored but should be pointed out. In the experiments reported by Englkemeir (1956) and Managan (1959) it was shown that a counter gives a linear but not proportional response (other than broadening) for energy depositions in the detector greater than 70 kev. This nonproportionality is slight, however, and if neglected does not lead to significant error in the counter response.[2]

In this calculation the response spectrum was first determined by

---

[2] The nonproportionality also leads to an intrinsic line broadening as shown by Zerby et al. (1961), but this is taken into account in the final broadening of the response spectrum.

assuming a unique and proportional response of the counter to a given amount of energy deposited in the detector and was later broadened with the use of Eqs. (1) and (2).

IV. Sampling Procedures and Auxiliary Programs

This section is concerned with the construction of the distribution functions which characterize the various physical processes encountered in the transport problem and the development of the associated sampling routines. To introduce the procedures employed, some of the basic principles of sampling techniques will first be reviewed briefly. This will be followed by a description of each distribution function and sampling routine used in the calculation. In those cases where the procedure is of general interest, alternate sampling routines will be described.

No attempt will be made in this section to describe how all of the sampling routines logically connect to form the complete Monte Carlo calculation. (This will be done in Section V.) Instead, the procedures associated with each separate physical process encountered in the transport problem will be presented here as completely as possible. In some cases this will involve the description of data-handling routines or other auxiliary routines peculiar to the problem. In effect, this means that the major portion of the Monte Carlo program will be presented in this section, leaving only a few logical connecting steps to be described in Section V.

A. Random Numbers

Fundamental to all selection techniques used on high-speed computing machines is the use of pseudorandom numbers which are uniformly distributed in the interval [0,1]. These numbers, which are here designated as R_i, were generated as needed in the present problem for the IBM-704 computer by the congruence method (Taussky and Todd, 1954):

$$2^{35}R_{i+1} = 2^{35}R_i(5^{15}) \; [\text{Mod } 2^{35}],$$

where $2^{35}R_0 = 5^{15}$.

B. Basic Selection Techniques[3]

With the use of the pseudorandom numbers it is formally possible to obtain a random variable X_i from the probability distribution $f(x)$

---

[3] For a complete description of basic selection techniques see Kahn (1954).

by solving the equation

$$\int_{-\infty}^{X_i} f(x)\, dx = \int_0^{R_i} dy = R_i \tag{3}$$

for X_i. It is assumed in Eq. (3), as it will be throughout this discussion, that the probability distribution is properly normalized so that

$$\int_{-\infty}^{+\infty} f(x)\, dx = 1.$$

If the solution of Eq. (3) for X_i involves a complicated function of R_i, it is often possible to achieve a savings in computing time by rewriting the distribution function so that a "rejection technique" can be used. For example, if the solution of Eq. (3) for X_i involves relatively simple functions of R_i when $f(x)$ is replaced with the probability distribution $g(x)$, then the expression

$$f(x) = \frac{1}{E} \frac{E f(x)}{g(x)} g(x) \tag{4}$$

can be considered, where E is a constant selected so that the quantity $E f(x)/g(x)$ is always less than unity for all values of x.

A flow diagram of the rejection technique for selecting a random variable X_i from $f(x)$, based on Eq. (4), is presented in Fig. 6. An analysis of the flow diagram indicates that the probability of obtaining X_i in dx is $g(x)\, dx$, and the probability of accepting it is $E f(x)/g(x)$; therefore, the probability of accepting an X_i on any trial is

$$\int_{-\infty}^{+\infty} \frac{E f(x)}{g(x)} g(x)\, dx = E \int_{-\infty}^{+\infty} f(x)\, dx = E. \tag{5}$$

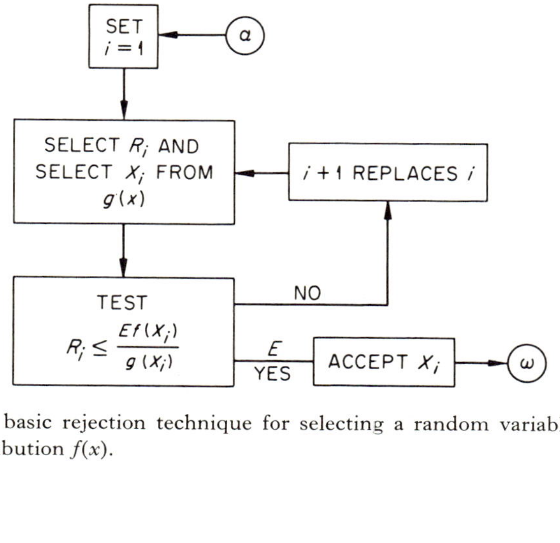

FIG. 6. The basic rejection technique for selecting a random variable X_i from the probability distribution $f(x)$.

The quantity E is appropriately called the efficiency of the rejection technique. The quantity E^{-1} is the average number of trials of the test in the rejection technique before an accepted value of X_i is obtained.

Another way to select random variables from a probability distribution is by systematic sampling. Use of this procedure instead of the method suggested by Eq. (3) yields a reduction in the variance, although it involves only a minor modification of that basic method. It does require, however, that the exact number of samples that are to be drawn be known from the beginning. If N samples are to be drawn from $f(x)$, then systematic sampling can be accomplished by replacing Eq. (3) by

$$\int_{-\infty}^{+\infty} f(x)\, dx = \frac{i - R_i}{N}, \qquad i = 1, 2, ..., N, \tag{6}$$

where the right-hand side of the equation has been divided into intervals of width $1/N$ and, rather than introducing a bias by taking the same fixed location within each interval, the location has been selected at random by introducing the pseudorandom number R_i.

Although systematic sampling is not used in general because the number of samples to be drawn must be known from the beginning, it is possible to use it for sampling the source in most cases. This is because the number of source particles that are to be considered is usually fixed at the beginning of a calculation.

C. Coordinate Systems

A brief discussion of coordinate systems is required here because many of the sampling schemes discussed in the following sections are intimately connected with the system selected. Throughout it will be implicitly assumed that the words "coordinate system" not only imply a means of designating a position but also a direction.

Although the source-counter configuration has axial symmetry, it is *not* advantageous to use a cylindrical polar-coordinate system even though the problem is independent of the azimuthal angle. A study of the machine operation involved in a translation from one point to another along a given direction with the respecification of the coordinates and direction after the translation clearly indicates that rectangular Cartesian coordinates, together with the familiar direction cosines, should be used. This is true in almost every case, including spherically symmetric problems.

Thus, in the present case the rectangular Cartesian coordinates of a point are designated by x, y, and z, and the direction cosines of a vector with respect to the coordinate directions $\hat{x}$, $\hat{y}$, and $\hat{z}$ are given by α, β, and γ, respectively.

D. Sampling the Source

One of the sources considered in this problem was an isotropic point located on the centerline of a cylindrical detector as shown in Fig. 1. Samples from this source were selected by using a systematic sampling technique as described in Section IV, B. A coordinate system was selected for this procedure where the detector axis was along the z-axis and the origin was at the face of the detector nearest the source. The probability distribution for the direction cosine of the source radiation with respect to the z-axis, γ, is then given by

$$k(\gamma) = \frac{1}{1-\gamma_0}, \qquad \gamma_0 \leqslant \gamma \leqslant 1. \tag{7}$$

Here, γ_0 is a constant equal to the minimum cosine with respect to the z-axis subtended by the detector. The azimuthal angle of the source is assumed to be uniformly distributed.

When the probability distribution in Eq. (6) is replaced with $k(\gamma)$ given in Eq. (7), the random variables γ_i obtained by systematic sampling are given by

$$\gamma_i = (1-\gamma_0)(i-R_i) N^{-1} + \gamma_0, \qquad i = 1, 2, ..., N. \tag{8}$$

There is no need to obtain the azimuthal angle for the source sample because of the assumed azimuthal symmetry. In all cases it is sufficient to allow the source sample to intercept the detector along the x-axis at the position $x = D\gamma^{-1}(1-\gamma^2)^{1/2}$, $y = 0$, and $z = 0$, where D is the source-detector separation distance. The direction cosines in this case are $\alpha = (1-\gamma_i^2)^{1/2}$, $\beta = 0$, and $\gamma = \gamma_i$.

E. Cross Sections

Because of the approximations introduced into the counter problem as discussed in Section III, it is only necessary to consider photoelectric events, pair production, and Compton scattering in the calculation of the transport of the photons. The cross sections for these events were taken from the tabulation of Grodstein (1957) and in turn introduced into the calculation in the form of tables of cross sections as a function of energy. Values of the cross sections at intermediate energies were obtained by linear interpolation. The energy intervals for the tabulation were selected to make possible quick access to the entries and to maintain less than 1% error in the interpolated values.

The principal advantage of obtaining the cross section from a table is that it is very much faster than calculating empirical formulas which have been fit to the data. The disadvantage is that the tables usually occupy much more space in the fast memory of a computer than the code for the empirical formulas.

F. SELECTING DISTANCES

The distances traveled by photons between interactions were randomly selected from an exponential probability distribution. In part of the calculation the exponential distribution was truncated to restrict the distance to a finite range for reasons that will be discussed in Section V. The truncated distribution is given by

$$f(x) = \frac{\Sigma \exp(-\Sigma x)}{1 - \exp(-\Sigma x_e)}, \quad 0 \leqslant x \leqslant x_e, \tag{9}$$

where Σ is the total macroscopic cross section, which is obtained from tables as described in Section IV, E, and x_e is the maximum allowed value of x.

Random variables X_i can be obtained from the truncated exponential distribution by replacing the $f(x)$ in Eq. (3) by the expression given in Eq. (9). The random variables X_i are then easily shown to be related to the pseudo-random numbers by

$$X_i = -\Sigma^{-1} \ln\left\{1 - R_i\left[1 - \exp\left(-\Sigma x_e\right)\right]\right\}. \tag{10}$$

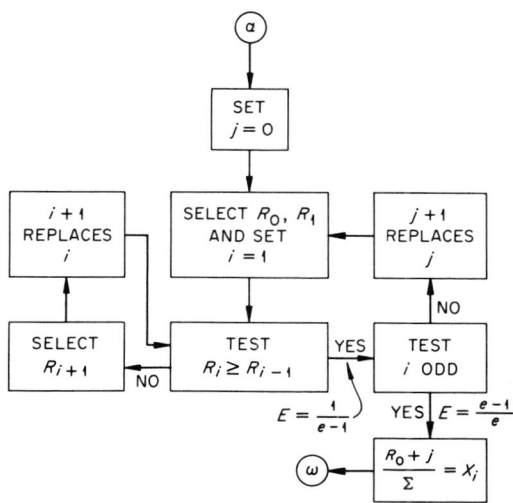

FIG. 7. Von Neumann's Method of Selecting random variables X_i from the Probability distribution $f(x) = \Sigma \exp(-\Sigma x)$, $0 \leqslant x < \infty$.

In another part of the calculation the range of the variable x in the exponential distribution was not restricted ($x_e = \infty$). Under these conditions it is practical to consider an alternate technique for selecting the random variable X_i. An ingeneous method for selecting the random variable X_i from the distribution $f(x) = \Sigma \exp(-\Sigma x)$ for $0 \leqslant x \leqslant \infty$ has been given by von Neumann (1951) and is presented in Fig. 7. From the efficiencies indicated on the tests in the routine (the efficiency is the probability of obtaining a "yes" answer in the test), it is easy to show that $e^2(e-1)^{-1} \simeq 4.3$ pseudorandom numbers are required on the average for each accepted random variable.[4] This means that

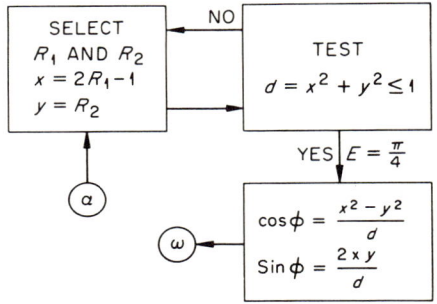

FIG. 8. Von Neumann's method of randomly selecting $\sin \phi$ and $\cos \phi$ where ϕ is uniformly distributed on the interval $[0, 2\pi]$.

at present the von Neumann routine is slightly faster than using Eq. (10) (with $x_e = \infty$) because of the logarithm that has to be calculated in that equation.

If in Eq. (10) the product Σx_e were always large, then it would be practical to select the random variable X_i from $f(x) = \Sigma \exp(-\Sigma x)$, where $0 \leqslant x \leqslant \infty$, and reject the selection if $X_i > x_e$. The efficiency of this procedure is $1 - \exp(-\Sigma x_e)$, which indicates the necessity for having Σx_e large.

G. SELECTING A UNIFORMLY DISTRIBUTED AZIMUTHAL ANGLE

The selection of a random azimuthal angle, ϕ_i, from a uniform distribution on the interval $[0, 2\pi]$ is quite common in Monte Carlo calculations. The construction of a selection technique for this distribution is trivial with $\phi_i = 2\pi R_i$ satisfying all the requirements.

[4] The average number of pseudorandom numbers required for an accepted random variable has been misquoted by von Neumann (1951) and Kahn (1954).

In the event that the sine and cosine of the azimuthal angle are required instead of the angle itself, as is usually the case, then calculation of the trigonometric functions becomes time consuming and the alternate method given by von Neumann (1951) should be considered. He circumvented the calculation of the trigonometric functions, and all other complicated functions as well, by the selection technique shown in Fig. 8. In the present calculation von Neumann's method was the one that was used.

H. Selecting a Random Isotropic Direction

It is appropriate that the preceding section be followed with a discussion of the equivalent problem in three dimensions. Actually there are several methods for randomly selecting the direction cosines of an isotropic vector where no complicated functions need to be calculated,[5] but this discussion will be restricted to the two most efficient techniques. The first one, invented by Coveyou (1958), is shown in Fig. 9. As

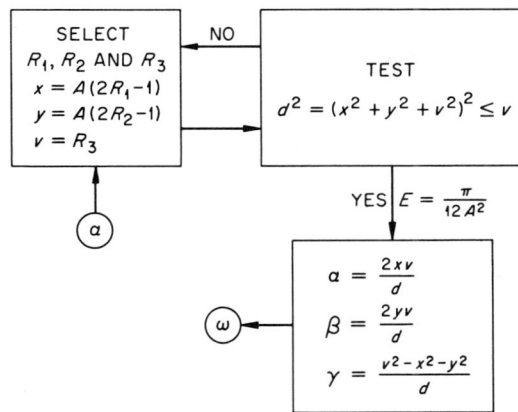

Fig. 9. Coveyou's method of randomly selecting the direction cosines of an isotropic vector. The value of the constant A is discussed in the text.

originally presented, the constant A was taken to be equal to unity; however, an analysis has shown that for a maximum efficiency A should be set equal to $3^{1/2}/4^{2/3}$. If this value of A is used, then approximately 5.42 random numbers are required on the average for an accepted set of direction cosines.

The second method is shown in Fig. 10, the first few operations being the same as von Neumann's method (Fig. 8). In this case approximately

[5] As an example, see Cook (1957).

5.15 random numbers are required for an accepted set of direction cosines, in contrast to 5.42 required in Coveyou's method.

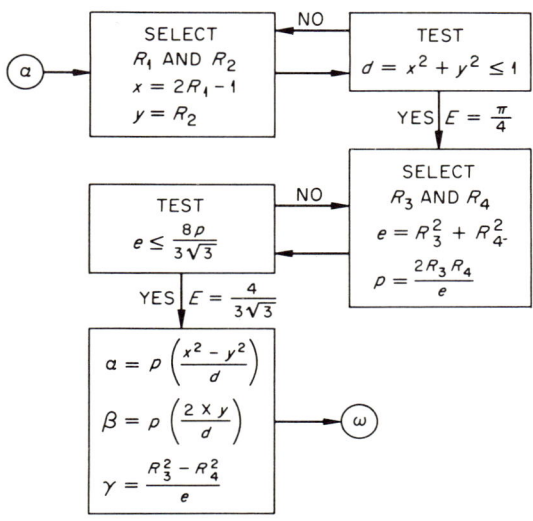

FIG. 10. A method of randomly selecting the direction cosines of an isotropic vector.

I. Compton Scattering

If it is assumed that only unpolarized photons are to be considered, as discussed in Section III, B, then the angular distribution of Compton-scattered photons is independent of the azimuthal angle. The differential cross section for this process is given by the Klein-Nishina formula,

$$\frac{d\sigma}{d\mu} = 2\pi r_e^2 \left(\frac{\alpha'}{\alpha}\right)^2 \left[\frac{\alpha}{\alpha'} + \frac{\alpha'}{\alpha} - 1 + \mu^2\right], \tag{11}$$

where r_e is the classical electron radius and μ is the cosine of the polar angle of scattering. If the energy of the photon is expressed in mc^2 units, then the initial energy of the photon, α, is related to the final energy, α', by the Compton relation

$$\alpha' = \frac{\alpha}{1 + \alpha(1 - \mu)}. \tag{12}$$

The integral of Eq. (11) over all values of μ yields the total Compton cross section,

$$\sigma = 2\pi r_e^2 \left\{ \frac{1+\alpha}{\alpha^3} \left[\frac{2\alpha(1+\alpha)}{1+2\alpha} - \ln(1+2\alpha)\right] + \frac{1}{2\alpha}\ln(1+2\alpha) - \frac{1+3\alpha}{(1+2\alpha)^2}\right\}, \tag{13}$$

which was tabulated for use in the calculation (see Section IV, E).

Figure 11 presents a flow diagram of a selection technique devised by Kahn (1954) for obtaining random variables from the probability distribution

$$p(\mu) = \frac{1}{\sigma}\left(\frac{d\sigma}{d\mu}\right). \tag{14}$$

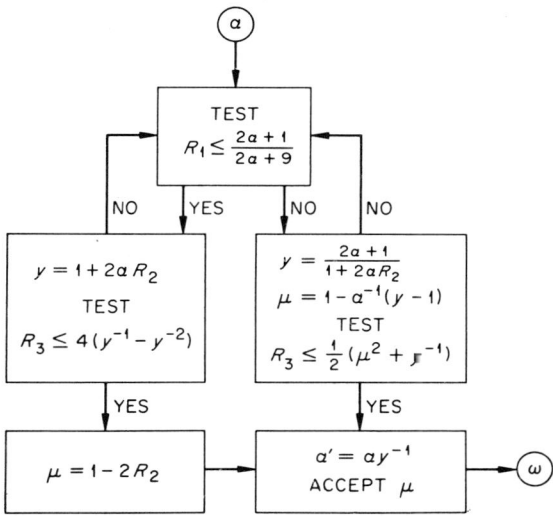

FIG. 11. Kahn's method of randomly selecting the cosine of the polar angle of scattering μ and the energy after scattering, α', of a compton-scattered photon. The over-all efficiency for the tests is given in the text.

The over-all efficiency of the rejection technique is approximately 0.65 at $\alpha = 2$ but drops steadily as the energy increases. At $\alpha = 20$ the efficiency is 0.34.

A sampling procedure which applies for the distribution $p(\mu)$ when $\alpha > 2.733$ and maintains a high efficiency at the higher energies is given in Fig. 12. In this selection technique it is assumed that the value of σ given in Eq. (13) has been calculated or obtained from a table previous to entering the routine.

The right-hand column in Fig. 12 corresponds to the selection of a random variable from the distribution $f(y) = [y \ln (1 + 2\alpha)]^{-1}$ given by Kahn (1954), and for $\alpha = 4$ the flow of the calculation favors this path. The probability of taking this path increases steadily as the energy increases. The over-all efficiency of both tests in that column varies from 0.75 to 0.93 with a trend toward the higher efficiency as the energy increases (Kahn, 1954).

It should be pointed out that the flow diagram for the right-hand column shown in Fig. 12 is not as complicated and time consuming as it appears. Modern floating point machines such as the IBM-704 can perform the indicated operations very simply and rapidly.

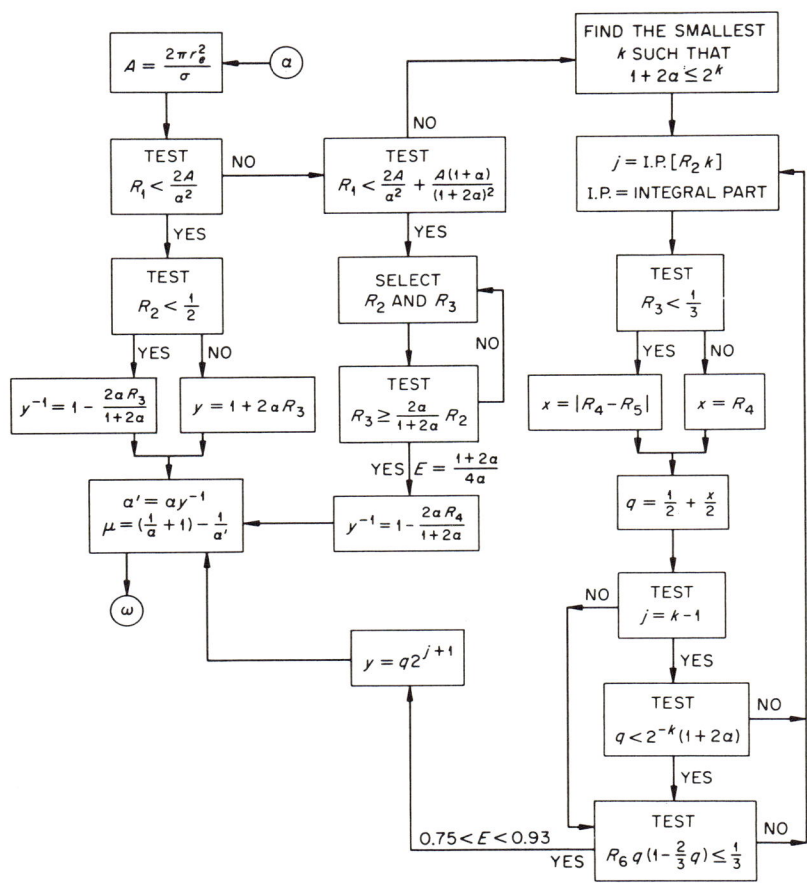

FIG. 12. A method of randomly selecting the cosine of the polar angle of scattering, μ, and the energy after scattering, α', of a compton-scattered photon. This method is valid only for $\alpha > 2.733$. The efficiency of the routine is discussed in the text.

To complete the specification of all angles involved in a Compton scattering, it is necessary to obtain the sine and cosine of a uniformly distributed azimuthal angle. This is done with the use of the von Neumann technique presented in Fig. 8. Thus, with the use of Fig. 11

or 12 together with Fig. 8 the cosine of the polar angle of scattering, μ, and the cosine and sine of the azimuthal angle, $\cos \phi$ and $\sin \phi$, all relative to the initial direction of the photon, are obtained. The direction cosines of the scattered photon are obtained by performing a rotation of coordinates.

J. Rotation of Coordinates

Assuming that the cosine of the polar angle of scattering is μ, the azimuthal angle is ϕ, and the direction cosines of the initial direction are α, β, and γ, then the direction cosines of the scattered photon, which are designated by α', β', and γ', are calculated as shown in Fig. 13.

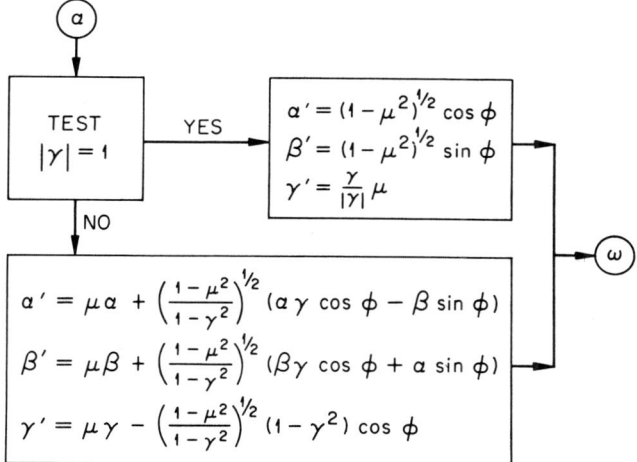

Fig. 13. Rotation of coordinates to obtain the direction cosines of a compton-scattered photon.

K. Pair Production

In pair production events the initial energies of the ejected electron and positron are required in order to determine the amount of bremsstrahlung radiation that is created by these particles in the slowing-down process. A selection technique was devised for this process with the use of an expression presented by Hough (1948) that approximates the Bethe-Heitler expression for pair production very well for photon energies below 10 Mev. Hough's formula can be written in the form of the probability distribution

$$h(x) = \frac{1}{1+M} h_1(x) + \frac{M}{1+M} h_2(x), \quad 0 \leqslant x \leqslant 1, \quad (15)$$

where $x = (E_- - 1)(\alpha - 2)^{-1}$. The quantities E_- and α are the electron and photon energies in mc^2 units, respectively,

$$h_1(x) = \frac{8}{\pi} x^{1/2}(1-x)^{1/2}, \tag{16}$$

and

$$h_2(x) = 30x(1-x)(1-2x)^2. \tag{17}$$

The quantity M in Eq. (15) is given by

$$M = \frac{P}{13.91\pi},$$

where

$$P = Q - 0.52 \quad \text{for} \quad \alpha > 4.2$$
$$= 0 \quad \text{for} \quad 2 < \alpha \leqslant 4.2$$

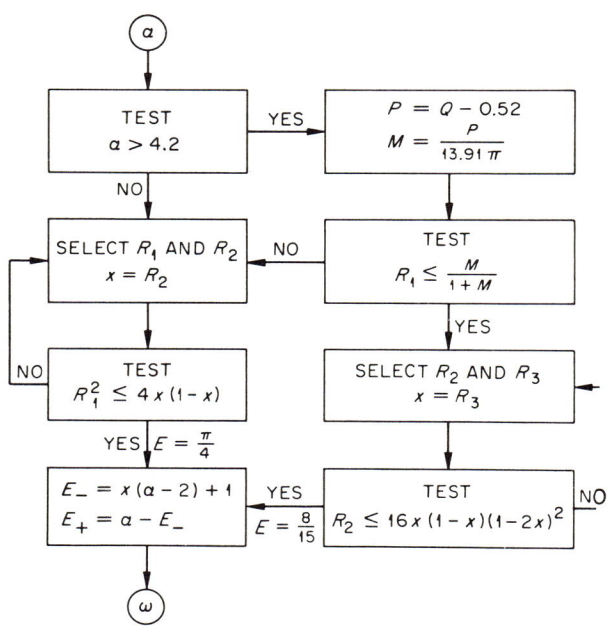

FIG. 14. A method of randomly selecting the initial energies of the electron and positron created in pair production by photons with energies below 10 Mev.

and

$$Q = (1 - 2\alpha^{-1})\left[\frac{4}{3}(1-\alpha^{-2})(L-1) - 4\alpha^{-2}F(F-1) - 16\alpha^{-4}F(L-F)\right] \tag{18}$$

with

$$L = 2\alpha^2(\alpha^2 - 4)^{-1} \ln\left(\frac{\alpha}{2}\right),$$

$$F = \alpha(\alpha^2 - 4)^{-1/2} \ln\left[\left(\frac{\alpha}{2}\right) + \left(\frac{\alpha^2}{4} - 1\right)^{1/2}\right].$$

The selection technique for the energies E_- and E_+ of the electron and positron, respectively, based on Eq. (15), is shown in Fig. 14. In this routine the function Q, which is given in Eq. (18) and is dependent only on the photon energy was contained as a table in the fast memory of the computing machine. Values of Q were obtained from the table by linear interpolation.

L. Annihilation Radiation

The annihilation of the positron created in pair production was assumed to take place at rest and at the position where the positron was created as discussed in Section III, C. The two photons that are released in this case each have an energy of 0.511 Mev. The first photon was given a random direction with the aid of the routine shown in Fig. 10, and the other was given a direction opposite to the first.

M. Bremsstrahlung Radiation

To make use of the bremsstrahlung spectra presented in Fig. 5, the data must be prepared in acceptable form for random sampling. This is a case where there are no analytic expressions for the spectra and numerical data must be used.

If the spectral data is represented by the function $k(E_0, \alpha)$, where E_0 is the initial energy of the electron and α is the energy of the photon in mc^2 units, then

$$g(E_0, \alpha) = \frac{k(E_0, \alpha)}{N(E_0)}, \qquad 0.04 \leq \alpha \leq E_0 - 1 \tag{19}$$

is a normalized probability distribution if

$$N(E_0) = \int_{0.04}^{E_0-1} k(E_0, \alpha)\, d\alpha.$$

The values of $N(E_0)$ are given in Table I.

A sampling procedure for $g(E_0, \alpha)$ is most easily obtained with the method suggested by Eq. (3). This procedure gives

$$\int_{0.04}^{\alpha_i} g(E_0, \alpha)\, d\alpha = B_i, \qquad (20)$$

which must be solved for α_i. (The manner in which B_i was obtained will be described below.) The results of Eq. (20) were prepared in the form of a table of α_i as a function of B_i since obtaining values of α_i from the table is most efficiently performed when the data is prepared in this manner. An example is shown in Fig. 15 where α_i is plotted as

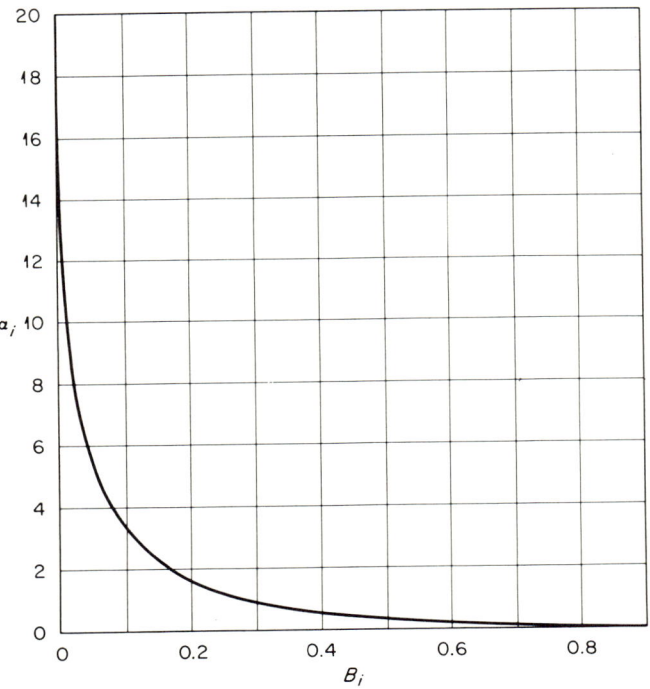

FIG. 15. A graph of the random bremsstrahlung photon energy, α_i, as a function of the Number B_i for an electron with initial energy 21 mc^2. These data were obtained from the solution of Eq. (20) in the text.

a function of B_i for an electron energy of 21 mc^2. This data, which is typical of the data at all energies, can now be used to select all the bremsstrahlung photons released by the 21 mc^2 electron in the process of slowing down. The number of photons released, $N(E_0)$, was obtained from a tabulation of the data presented in Table I.

Since the average number of bremsstrahlung photons released was not an integral number in general and varied considerably with the initial energy of the electrons, the procedure adopted was to always select two sample photons from the bremsstrahlung spectrum by systematic sampling. This was easily accomplished by letting

$$B_i = \frac{1 - R_i}{2}, \quad i = 1, 2,$$

and giving each sample photon the energy α_i which corresponded to its B_i value as indicated in Fig. 15. The fact that two samples were always selected was compensated for by giving each of them the weight $2^{-1} N(E_0)$, so that the total weight corresponded to the average number of photons released. In this case the weight may be thought of as the number of photons represented by the sample.

Tables of α_i as a function of B_i were prepared for each of the electron energies indicated on the bremsstrahlung spectra shown in Fig. 5. With this data a double linear interpolation scheme was used for selecting each of the two sample bremsstrahlung photons for electrons of intermediate energies. An interpolation using the variable B_i was performed first and followed by an interpolation using the electron energy.

A random isotropic direction was given to each of the bremsstrahlung photons with the use of the scheme presented in Fig. 10. The reason for this procedure is discussed in Section III, C.

V. Details of the Monte Carlo Procedure

The Monte Carlo procedures and techniques for solving particle transport problems are usually formulated on a microscopic basis where the possible events that can happen to a particle are considered in sequence. The simplest application of this approach is to analog problems in which random histories of individual particles are generated in a manner which corresponds to one's concept of reality. Since the individually calculated histories are intended to correspond to physical particle histories, and since there are necessarily few of these in order to be economical, the analog problem is actually the mathematical equivalent of a very low-intensity experiment. This means that estimates of events that are not very likely with analog procedures will in general have large statistical errors associated with them.

Mathematical modification of the analog problem to increase the accuracy of estimates of events which are improbable can also be formulated on a microscopic basis. Seldom is it necessary to refer to the Boltzmann transport equation to establish these procedures, although that is the equation that is to be solved. There are exceptions to this rule, of course, with the methods of the exponential transformation (Kahn, 1950) and conditional Monte Carlo (Drawbaugh, 1961; Penny and Zerby, 1961) serving as examples.

In the following discussion of the procedures used in the scintillation counter problem, the individual events will be considered in sequence in order to describe the statistical estimation and the weighting procedure used to estimate the pulse-height response function. This method of describing the problem also serves to logically connect the selection techniques and auxiliary programs in those cases where the connection has not already been established in Section IV.

In the description of the estimate of the intrinsic efficiency and analytic zero intercept, a somewhat different approach is more expedient. In these cases, which are described in the following two sections, the procedures can be best formulated from integrals. This approach serves as an elementary example of the use of Monte Carlo methods.

A. Estimating the Intrinsic Efficiency

The intrinsic efficiency, which is an important and useful quantity for estimating the over-all efficiency of detectors, is defined to be the fraction of incident photons which make at least one collision in the detector. The discussion of the calculation of this quantity will be restricted to the case where the source is external to the detector as shown in Fig. 1.

The intrinsic efficiency η is given by the expression

$$\eta = \int_{\gamma_0}^{1} \left\{ 1 - \exp\left[-\Sigma x(\gamma)\right] \right\} k(\gamma) \, d\gamma, \tag{21}$$

where $k(\gamma)$, given in Eq. (7), is the probability distribution of the direction cosine of the source radiation with respect to the z axis, Σ is the total cross section, and $x(\gamma)$ is the distance through the detector as a function of γ. The quantity in the braces is the probability that an incident particle with direction cosine γ will not penetrate the detector. The intrinsic efficiency η is, therefore, the average value of the nonpenetration probabilities.

In the analog procedure, a random variable γ_i would be selected from $k(\gamma)$ by systematic sampling according to Eq. (8), a random distance X_i would be selected from the exponential distribution by the method shown in Fig. 7, and a test would then be made to determine whether X_i was in the detector. The number of cases where $X_i \leq x(\gamma_i)$ divided by the total number of trials N would then be the estimate of η.

RESPONSE OF GAMMA-RAY SCINTILLATION COUNTERS 119

A considerable improvement in the above procedure is achieved if, instead of selecting a random distance, the distances $x(\gamma_i)$ are calculated and the nonpenetration factors averaged so that the estimate of η is

$$\hat{\eta} = \sum_{i=1}^{N} \eta_i \equiv \sum_{i=1}^{N} N^{-1} \left\{ 1 - \exp\left[-\Sigma x(\gamma_i)\right] \right\}. \tag{22}$$

This procedure, which is an example of statistical estimation, significantly reduces the statistical error. In fact, in this example if N is large, the estimate $\hat{\eta}$ is almost equivalent to the results of standard numerical integration since the source was systematically sampled. The only statistical uncertainty in this procedure is in the location of the grid points in each of the N intervals of the variable γ according to Eq. (8).

B. Estimating the Analytic Zero Intercept

The intercept of the pulse-height response spectrum with the zero pulse-height value was calculated as accurately as the intrinsic efficiency and in a very similar manner. Such a value is quite useful for terminating the curves through the statistical estimates of the response spectrum since the response in the low pulse-height region is usually small relative to the rest of the spectrum and has an associated large fractional statistical error.

The contribution to the response spectrum in the limit of zero pulse height can be shown to come entirely from those photons which scatter straight ahead only once before escaping the detector. If the response spectrum is calculated in units of counts per unit energy (pulse height), then the desired quantity is

$$\nu = \int_{\gamma_0}^{1} \left(\frac{d\Sigma}{d\alpha}\right)_0 x(\gamma) \exp\left[-\Sigma x(\gamma)\right] k(\gamma) \, d\gamma , \tag{23}$$

where $k(\gamma)$, $x(\gamma)$, and Σ are the same as defined for Eq. (21). The quantity $(d\Sigma/d\alpha)_0 = 2\pi r_e^2 N_0 mc^2 \alpha^{-2}$, where N_0 is the electron density, was obtained from the Klein-Nishina formula given in Eq. (11) in the limit of straightahead scattering.

Using systematic sampling of $k(\gamma)$ according to Eq. (8) and statistical estimation, the estimate of ν is

$$\hat{\nu} = \sum_{i=1}^{N} \nu_i \equiv \sum_{i=1}^{N} N_0^{-1} \left(\frac{d\Sigma}{d\alpha}\right)_0 x(\gamma_i) \exp\left[-\Sigma x(\gamma_i)\right], \tag{24}$$

which again corresponds closely to a numerical calculation when N is large.

C. Details of the Transport Calculation and Estimation of the Response Spectrum

The part of the Monte Carlo calculation which treated the transport of the radiation in the detector was divided into two distinct sections. In one part the histories of the bremsstrahlung and annihilation photons created indirectly by interactions of the source photons or other secondary photons were generated by an analog procedure, while in the other part the histories of the source photons were generated by using a scheme of weighting with survival probabilities.

In the discussion below, the analog procedure is described first and is followed by a description of the procedure using survival probabilities. This is followed by a description of the statistical estimation procedure for obtaining the response spectrum.

1. *The Analog Calculation*

In the analog calculation, weighting was used, but only because of the use of the bremsstrahlung selection technique discussed in Section IV, M. To be complete, therefore, the concept of weight is introduced here. For present purposes the definition of the weight is most conveniently given as the number of physical photons represented by the sample photon in the calculation.

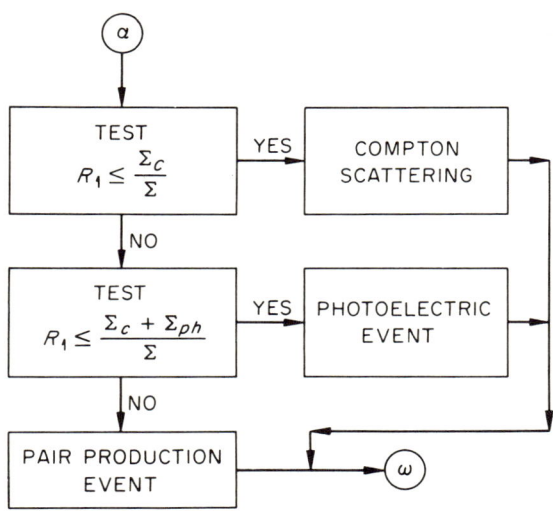

Fig. 16. The random method of selecting the type interaction a photon has in the analog monte carlo procedure.

Once the weight, energy, position, and direction of a secondary photon was determined, the analog calculation proceeded by selecting the distance to the next collision from the exponential probability distribution by the von Neumann technique shown in Fig. 7. If the selected distance carried the photon outside the detector, it was considered to have escaped and to have caused an energy loss equal to its current energy times its weight. If the collision was inside the detector, however, then a choice between the various interactions that are possible had to be made. The random procedure for selecting the type of interaction is shown in Fig. 16, where $\Sigma = \Sigma_c + \Sigma_{ph} + \Sigma_{pa}$ is the total cross section and Σ_c, Σ_{ph}, and Σ_{pa} are the Compton, photoelectric, and pair production cross sections, respectively.

If the secondary photon has a photoelectric event, its energy was converted into the kinetic energy of the ejected electron and bremsstrahlung photons were selected according to the scheme discussed in Section IV, M. The weight assigned to the bremsstrahlung photons was a product of the initial photon weight and the weight given to them by the selection routine. The history of each of these photons was generated in turn by the analog procedure.

If, on the other hand, a pair production event took place, the following sequence of events occurred: (1) the energy of each member of the electron-positron pair was selected according to the scheme shown in Fig. 14, (2) the bremsstrahlung photons from each charged particle were selected by the methods described in Section IV, M, and given weights equal to the product of the initial photon weight and the weight assigned to them in the selection technique, and (3) the annihilation photons were selected according to the methods described in Section IV, L, and given weights equal to the initial photon weight. The histories of the four sample photons created in this manner were then generated by the analog technique.

If a Compton scattering event had taken place, then the parameters describing the scattered photon were determined as discussed in Sections IV, I and IV, J. The kinetic energy of the ejected electron was taken as the difference between the initial and final photon energies, and the bremsstrahlung photons were selected according to the methods described in Section IV, M. The weights assigned to the bremsstrahlung photons were the product of the initial photon weight and the weights assigned to them by the selection technique, as before. The histories of the three sample photons that result from a Compton event were then generated by the analog technique.

It is easy to visualize the cascade that could take place in the calculation as the secondaries produced secondaries which produced secondaries, etc. The extent of the cascade was restricted, however, by terminating the branches when (1) the photon escaped from the detector, (2) the photon was degraded in energy to below 5.11 kev and absorbed (see Section III, B), or (3) the weight of the photon decreased below 0.001 of the weight of the photon that initiated the cascade. The third method was chosen for economy reasons and because it was not likely to introduce a bias in the estimations.

2. *Source Photon Histories and Use of Survival Probabilities*

Once a photon was selected from the source as discussed in Section IV, D, the distance through the detector $x(\gamma_i)$ was calculated and used in the estimation of the intrinsic efficiency and analytic zero intercept as described in Sections V, A and V, B. The distance to the first collision point was then selected from the truncated exponential distribution according to Eq. (10) where $x_e = x(\gamma_i)$. This forced the collision to be inside the detector and prevented the escape of the source radiation. To adjust for this unnatural behavior, a weight was attached to the sample photon which was equivalent to the probability that it survived or remained in the detector (and in the calculation). Thus the weight of the photon, which was initially set equal to N^{-1} to automatically take care of averages, was adjusted by multiplying by the factor $1 - \exp[-\Sigma x(\gamma_i)]$.

At the position selected for the next interaction, the photon was only allowed to experience a Compton scattering event. This again forced the photon to survive, so that its weight again had to be adjusted, this time by the factor Σ_c/Σ, which is the probability of scattering (probability of surviving). The energy and direction of the photon after scattering were selected by the methods described in Sections IV, I and IV, J.

The distance y from the last point of interaction through the detector to the exit surface along the direction after scattering was then calculated. The distance to the next point of interaction was then selected from the truncated exponential distribution according to Eq. (10) as before, but with $x_e = y$, and the weight was adjusted by the factor $1 - \exp[-\Sigma y]$. The cross sections, of course, were evaluated at the current energy of the photon.

The process of selecting distances and then scattering the photon as described above was repeated sequentially until either the photon was degraded in energy to below 5.11 kev or its weight was below 0.001 N^{-1}; thus the process was terminated in a manner similar to the termination of the analog histories.

3. *Estimation of the Response Spectrum*

After each Compton scattering of a source photon, but before the distance to the next collision was selected, the energy of the ejected electron was calculated and the bremsstrahlung photons produced by it were selected as described in Section IV, M. The histories of each of these photons and all cascade photons derived from them were then generated by the analog calculation to determine the total energy lost from the detector. The energy lost in each of the Compton events of the source radiation by this process was accumulated to be used in the estimation of the response spectrum.

With the procedure as described above it was possible to calculate a contribution to the response spectrum after each Compton scattering of the source photons by statistical estimation. This was done by multiplying the current weight of the photon (after it had been adjusted by the factor Σ_c/Σ) by the probability of escape, $\exp(-\Sigma y)$, and adding the product into the pulse-height interval which included the total energy deposited up to that particular collision. In this case the energy deposited is the source energy minus the current energy of the source photon minus all energy lost from the detector by the escape of secondary photons produced in all the Compton events.

The additional contributions required to completely determine the response spectrum were obtained by first treating each collision of the source radiation as a photoelectric event and then as a pair production event (if the energy was above threshold) before allowing the Compton scattering to take place. In the case of a photoelectric event, the bremsstrahlung photons from the photoelectron were selected as described previously and the energy lost from the detector that could be attributed to these was determined in the analog calculation. The weight of the source photon before the Compton scattering was then multiplied by the probability of having a photoelectric event, Σ_{ph}/Σ, and the product was added into the pulse-height interval corresponding to the energy deposited. The energy deposited in this case is the source energy minus the energy lost by the escape of secondary radiation from all previous Compton scattering events of the source photon minus the energy lost by the escape of secondary radiations produced in the photoelectric event. The contribution from the pair production event was calculated in a similar manner except that the loss of annihilation radiation had to be accounted for too.

In those cases where the history was terminated by the source photon being degraded in energy to below 5.11 kev or by it being decreased in weight to below 0.001 N^{-1}, the contribution to the response spectrum was calculated as if a photoelectric event had taken place but with no secondary radiation being lost from the detector. The total weight of the source sample was deposited in the response spectrum in this case rather than the product of the weight and the factor Σ_{ph}/Σ.

D. CALCULATING THE PEAK-TO-TOTAL RATIO

In the calculation of the peak-to-total ratio only those contributions (counts) for which the energy deposition was within 5.11 kev of the source energy were included in the total absorption peak. The ratio of the contributions in the total absorption peak to all contributions to the response spectrum was taken as the peak-to-total ratio.

It should be noted that a slightly different value of the ratio would be obtained at the higher source energies if the spectrum were broadened first and then the total absorption peak peeled off from the overlapping energy-loss spectrum. This occurs because the buildup of counts near the total absorption peak (as defined here) caused by small energy losses by bremsstrahlung radiation are not easily separable from the peak and are usually included in it. The peak-to-total ratios from experimental response spectra are obtained by the peeling off process and should deviate from the estimates in this calculation at high source energies because of the above reason. In addition, however, there should be some difference between calculation and experiment because background contributions in the experimental data tend to decrease the peak-to-total ratio from the estimated value.

The standard deviation of the peak-to-total ratio was also estimated in the course of the calculation although it is not quoted in the results presented in this paper. Calculation of the standard deviation was performed by first totaling the contributions to the total absorption peak from each separate source photon. The final estimate of the contribution to the total absorption peak was then calculated as

$$T = N^{-1} \sum_{i=1}^{N} T_i, \qquad (25)$$

where T_i is the total for the ith photon. The estimate of the standard deviation associated with T was calculated as

$$\sigma_T = \left\{ N^{-1} \left[N^{-1} \sum_{i=1}^{N} T_i^2 - T^2 \right] \right\}^{1/2}. \qquad (26)$$

The estimate of the peak-to-total ratio P and its standard deviation σ_P was obtained by dividing the quantities given in Eqs. (25) and (26) by the intrinsic efficiency, $\hat{\eta}$, which was estimated as discussed in Section V, A. Thus the estimate of the peak-to-total ratio and its standard deviation was

$$P = \hat{\eta}^{-1}T \tag{27}$$

and

$$\sigma_P = \hat{\eta}^{-1}\sigma_T, \tag{28}$$

respectively.

E. Broadening the Response Spectrum

From the previous discussion of how the response spectrum was estimated, it should be noted that no account was taken of the line broadening. This was done in an auxiliary calculation where the broadened spectrum $g(E)$ was generated from the unbroadened spectrum $k(E')$ by numerically integrating

$$g(E) = \int_0^{E_0} k(E')F(E, E')\, dE'. \tag{29}$$

The function $F(E, E')$ is given in Eq. (1), and the quantity E_0 is the source energy.

F. The Computing Machine Program

The complete Monte Carlo calculation that was described above was programmed for the IBM-704 computer. Actually three different codes resulted in order to treat detectors of NaI, CsI, and xylene. The principal difference between the codes was the difference in the cross sections, however, so that each code occupied about 6500 words of fast memory space. Several different source-detector configurations were allowed in the calculation as described in Section II, A.

Running times on the IBM-704 changed from one problem to another as the size of the detector or the source energy changed. The running time increased slightly with an increase of the detector size and rapidly with an increase in source energy. A typical set of times for a $3\frac{1}{2}$-in.-diam by 5-in.-high CsI detector for 1000 source photons was 9.9, 16.7, and 22.8 min for source energies of 0.662, 1.368, and 2.754 Mev, respectively. The running time is directly proportional to the number of source photons.

VI. Results of the Calculations

The experimental response spectra from scintillation counters with a 3-in.-diam by 3-in.-high NaI(Tl) detector are very useful for comparing

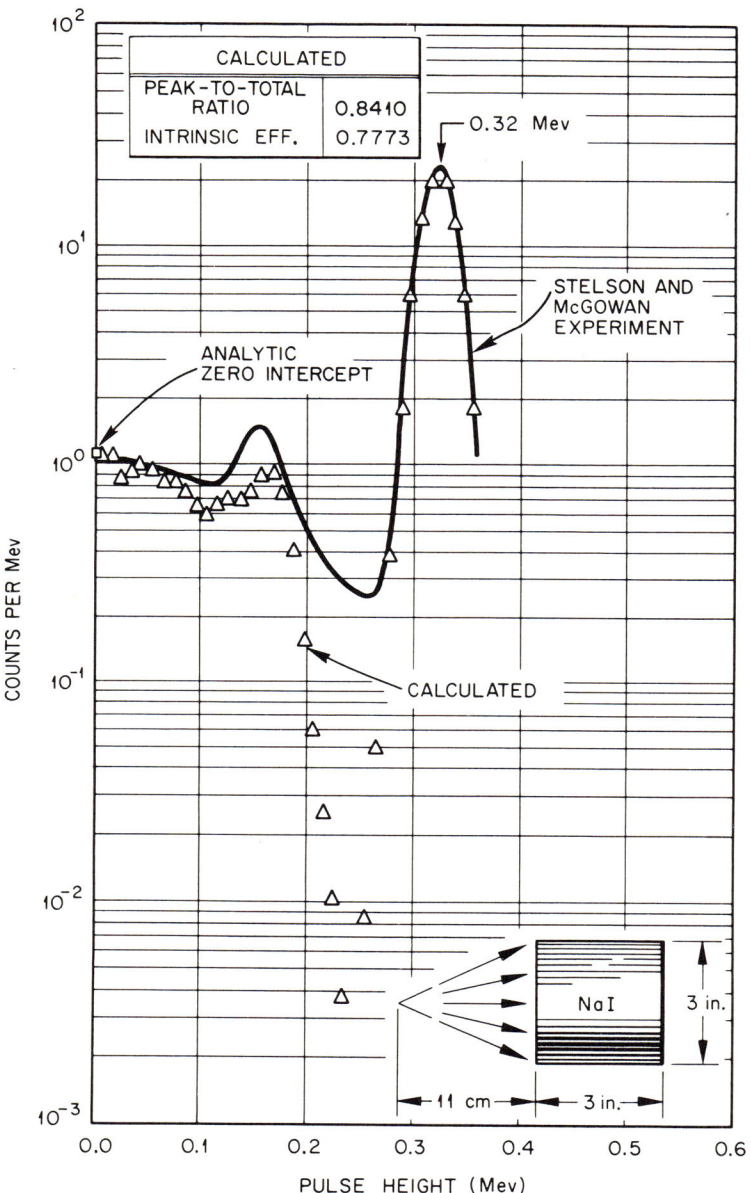

FIG. 17. Response Spectrum of a 3-in.-diam by 3-in.-high NaI(Tl) crystal detector exposed to 0.32-Mev incident gamma rays. The source was an isotropic point located on the centerline, 11 cm from the end of the crystal.

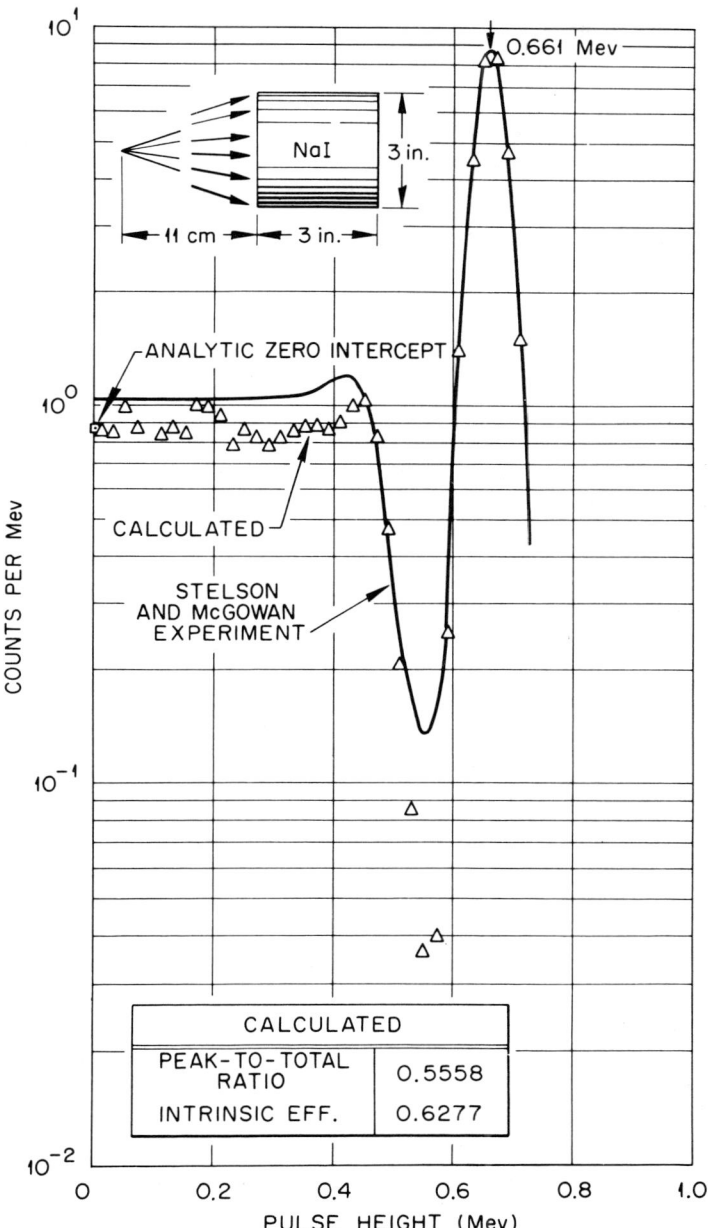

FIG. 18. Response spectrum of a 3-in.-diam by 3-in.-high NaI(Tl) crystal detector exposed to 0.661-Mev gamma rays. The source was an isotropic point located on the centerline, 11 cm from one end of the crystal.

with calculations since that detector has been used quite extensively and the reported response spectra are quite comprehensive.[6] Comparisons of the calculation with the response spectra from such a counter are reported here in an effort to demonstrate the general accuracy and validity of the Monte Carlo calculation.

Figures 17 through 21 show comparisons of the calculation with the experimental results for a counter with a 3-in.-diam by 3-in.-high NaI(Tl) detector for source energies from 0.32 to 7.48 Mev. In all these figures the pulse height is given in units of Mev and the calculated data are normalized to a total response of one count. The experimental data are normalized to the calculated data at the total absorption peak. The constants A and B used in Eq. (2) for broadening the spectrum (see Section V, E) were chosen to best fit the experimental data in each case. Each calculated case represents the results of averaging the estimates from 2000 source samples.

Figures 17 and 18 show the experimental results of Stelson and McGowan (1959). In their experiments the counter was suspended in the middle of a room by a minimum support so as to minimize background effects, and the resulting pulse-height spectrum was adjusted in an attempt to remove the remaining background contributions. The calculated data show a deeper valley just below the total absorption peak than the experimental data and tend to underestimate the experimental data at the smaller pulse heights. This deviation can be attributed to background effects that were not accounted for in adjusting the experimental data. At the higher source energy (Fig. 18) the deviation between calculation and experiment at low pulse heights becomes more obvious.

For source energies above 1 Mev the calculated data were compared with the experimental data of Lazar and Willard (1956) shown in Figs. 19 through 21. In these comparisons the calculation again gives an underestimate of the experimental data at small pulse-height values, although there is good agreement at large pulse heights.

The use of the analytic zero intercept for terminating the curves through the estimates of the response spectra should be obvious from Figs. 19 and 20. Since it was estimated with a high degree of accuracy (see Section V, B), it also is of significance because it shows that the less accurate estimates of the response spectrum are statistically distributed about the smooth curve and are not biased on the low side as might be expected from the comparison with experiment.

The deviation between the calculated and experimental data at low

---

[6] See, for example, Heath (1957).

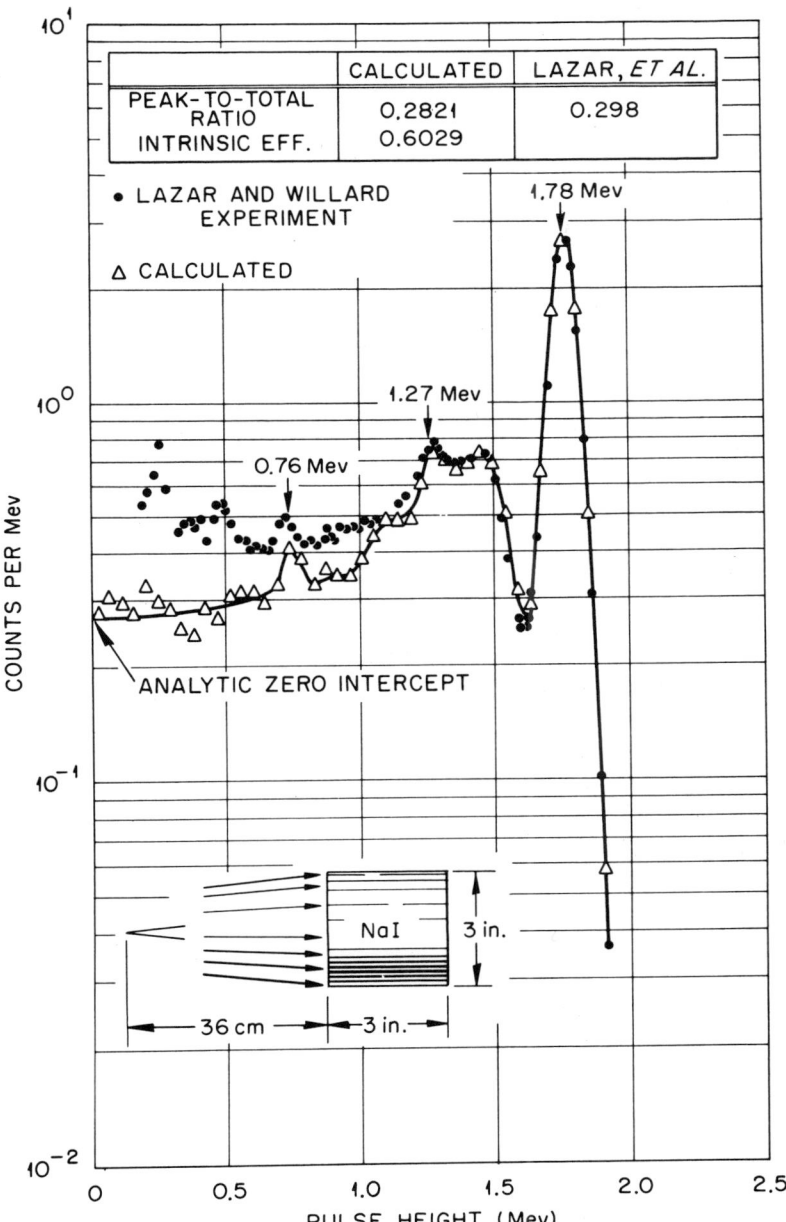

FIG. 19. Response spectrum of a 3-in.-diam by 3-in.-high NaI(Tl) crystal detector exposed to 1.78-Mev gamma rays. The source was an isotropic point located on the centerline, 36 cm from one end of the crystal.

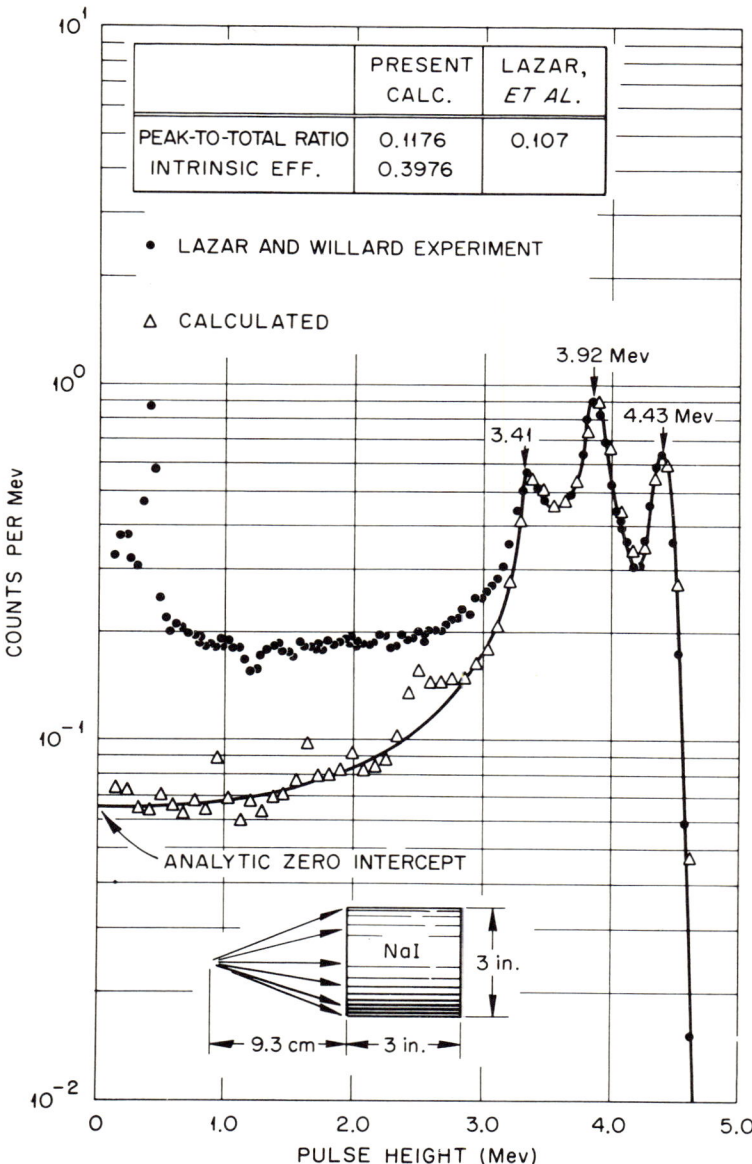

FIG. 20. Response spectrum of a 3-in.-diam by 3-in.-high NaI(Tl) crystal detector exposed to 4.43-Mev gamma rays. The source was an isotropic point located on the centerline, 9.3 cm from one end of the crystal.

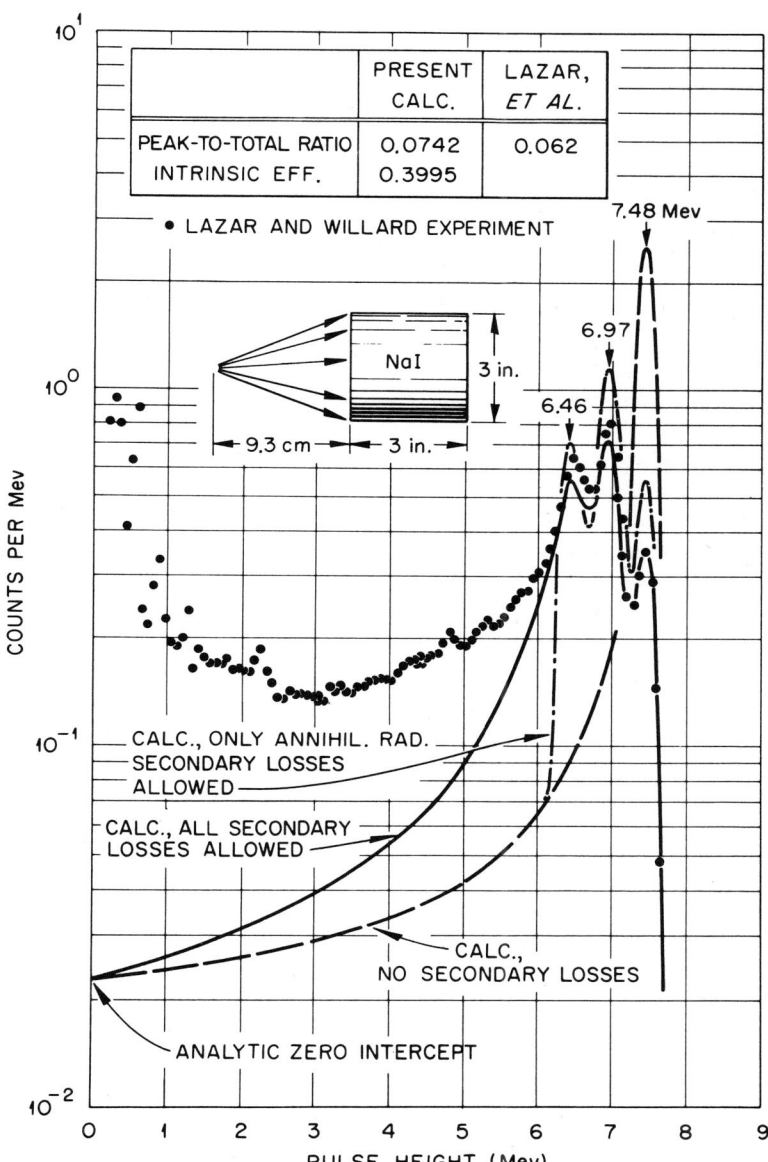

FIG. 21. Response spectrum of a 3-in.-diam by 3-in.-high NaI(Tl) crystal detector exposed to 7.48-Mev gamma rays. The source was an isotropic point located on the centerline 9.3 cm from one end of the crystal.

pulse heights must again be attributed to background contributions in the experiment for the cases shown in Figs. 19 through 21. This is based on the fact that the calculation is verified at both extremes of the spectrum: at the large pulse heights the calculation agrees with experiment and at the small pulse heights the statistical data approach the analytic zero intercept. A review of all the approximations embodied in the calculation indicates that there is no other valid explanation for the difference unless some obscure physical event happens in a scintillation counter that is not recognized at present.

It is rather interesting to observe how well the calculated data match the experimental data at high pulse-height values for the cases shown in Figs. 19 and 20. In both these cases the secondary escape peaks are quite prominent and there is some distortion of the spectrum from energy losses by bremsstrahlung radiation. The extent of these effects is indicated in Fig. 21 for the case of a 7.48-Mev source.

In Fig. 21 the calculated data are represented as a smooth curve for simplicity. The curve labeled "no secondary losses" was calculated by letting all secondary bremsstrahlung and annihilation radiation be absorbed at their point of origin. This meant that no histories needed to be generated in the analogue portion of the calculation. The curve labeled "only annihilation radiation losses allowed" was calculated by letting all bremsstrahlung radiation be absorbed at its point of origin and tracing the histories of the annihilation radiation by the analog calculation. The last curve labeled "all secondary losses allowed" followed the methods as described in this paper. As would be expected, the curve neglecting all secondary radiation losses does not agree with experiment at all. The curve including annihilation radiation losses agrees with experiment fairly well when the experimental data is normalized to it at the total absorption peak; however, the relative heights of the two escape peaks do not agree with experiment until all secondary losses are included. In the last case it can be observed that the two escape peaks are below the experimental data by approximately 10%. This does not necessarily indicate that the complete error should be attributed to statistics because of the apparently large background contribution appearing in this experiment.

On the basis of the apparently good comparisons of calculations with experiments, the codes have been used quite extensively by Zerby and Moran (1961) to survey the response of counters with 3-in.-diam by 3-in.-high NaI(Tl) detectors, by Moran (1960) to study response of counters with CsI(Tl) detectors, by Peelle (1960) to study the effects of the detector size and shape on the response spectrum from counters with large NaI(Tl) detectors, and by Van Dilla and Anderson (1961) to study the response of counters with very large liquid (xylene) detectors for whole body (human) counting. Many other applications of the codes have been made but have not been reported.

Acknowledgment

Above all the author wishes to express his appreciation to H. S. Moran for his care and precision in coding this calculation from sometimes very rough flow diagrams. The author is also indebted to P. R. Bell who originally suggested this problem and to R. W. Peelle and G. T. Chapman for their many helpful discussions. Appreciation is also extended to P. H. Stelson and F. K. McGowan for granting the use of unpublished experimental data and to R. R. Coveyou for granting the use of an unpublished selection technique.

References

Ashkin, J., and Bethe, H. A. (1953). In "Experimental Nuclear Physics" (E. Segré, ed.) Wiley, New York.
Berger, M. J., and Doggett, J. (1956). *J. Research Nat. Bur. Standards* **56**, 355.
Bethe, H. A., and Ashkin, J. (1953). In "Experimental Nuclear Physics" (E. Segré, ed.), Wiley, New York.
Breitenberger, E. (1955). *Progr. Nuclear Phys.* 4, 56.
Cook, J. M. (1957) *Math. Tables and Other Aids to Computation* 11, 81.
Coveyou, R. R. (1958). Oak Ridge National Laboratory, private communication.
Davisson, C. M., and Beach, L. A. (1959). A Study of Photons in Sodium Iodide Scintillation Crystals. U. S. Naval Research Laboratory Report NRL-5408.
Drawbaugh, D. W. (1961). *Nuclear Sci. and Eng.* 9, 185.
Engelkemeir, D. (1956). *Rev. Sci. Instr.* **27**, 589.
Grodstein, G. W. (1957). X-Ray Attenuation Coefficients from 10 kev to 100 Mev. National Bureau of Standards Report NBS-583.
Heath, R. L. (1957). Scintillation Spectrometry Gamma-Ray Spectrum Catalogue. AEC Research and Development Report IDO-16408.
Heitler, W. (1954). "The Quantum Theory of Radiation," 3rd ed. Oxford Univ. Press, London and New York.
Hough, P. V. C. (1948). *Phys. Rev.* **73**, 266.
Kahn, H. (1950). *Nucleonics* **6**, 27.
Kahn, H. (1954). Application of Monte Carlo. United States Atomic Energy Commission Report AECU-3259.
Lazar, N. H., and Willard, H. B. (1956). Physics Divison Semiannual Progress Report for Period Ending March 10, 1956. Oak Ridge National Laboratory Report ORNL-2076, p. 55.
Managan, W. W. (1959). *In* Proceedings of Sixth Tripartite Instrument Conference. AECL-805, Paper 5.10; see also "Applied Gamma-Ray Spectrometry" (C. E. Crouthamel, ed.), Chapter 2 Pergamon Press, New York.
McGowan, F. K., and Stelson, P. H. (1959). Oak Ridge National Laboratory, unpublished data.
Miller, W. F., and Snow, W. J. (1960). *Rev. Sci. Instr.* **31**, 39.

MORAN, H. S. (1960). Proceedings of the Total Absorption Gamma-Ray Spectrometry, Symposium. Report TID-7594, p. 113.
MOTT, W. E., and SUTTON, R. B. (1958). "Handbuch der Physik" (S. Flügge and E. Creutz, eds.), Vol. 45, p. 86. Springer, Berlin.
PEELLE, R. W. (1960). Proceedings of the Total Absorption Spectrometry Symposium. Report TID-7594, p. 89.
PENNY, S. K., and ZERBY, C. D. (1961). *Nuclear Sci. and Eng.* **10**, 75.
SPENCER, L. V., and Wolff, C. (1953). *Phys. Rev.* **90**, 510.
STELSON, P. H., and McGOWAN, F. K. (1959). Oak Ridge National Laboratory, unpublished data.
TAUSSKY, O., and TODD, J. (1954). *In* "Symposium on Monte Carlo Methods" (H. A. Meyer, ed.), Wiley, New York.
VAN DILLA, M. A., and ANDERSON, E. C. (1961). "Human Counters Using Liquid Scintillators," paper presented at the Symposium on Whole Body Counting, International Atomic Energy Agency, Vienna, Austria, June 1961 (to be published by IAEA).
VON NEUMANN, J. (1951). Monte Carlo Method. *Natl. Bur. Standards Appl. Math. Ser.* **12**, 36.
WRIGHT, G. T. (1954). *J. Sci. Instr.* **31**, 377.
ZERBY, C. D., and Moran, H.S. (1958). Bremsstrahlung Spectra in NaI and Air. Oak Ridge National Laboratory Report ORNL-2454.
ZERBY, C. D., and MORAN, H. S. (1961). Calculation of the Pulse-Height Response of NaI(Tl) Scintillation Counters. Oak Ridge National Laboratory Report ORNL-3169 and *Nucl. Instr. and Methods* **14**, 115, (1962).
ZERBY, C. D., MEYER, A., and MURRAY, R. B. (1961). *Nuclear Instr. and Methods* **12**, 115

Monte Carlo Calculation of the Penetration and Diffusion of Fast Charged Particles*

Martin J. Berger

NATIONAL BUREAU OF STANDARDS
WASHINGTON, D.C.

I. Introduction	135
II. General Description of the Monte Carlo Method	139
A. Relation to the Transport Equation	139
B. Detailed Case Histories	140
C. Condensed Case Histories	143
III. Particular Monte Carlo Schemes	144
A. Complete Grouping, Class I	144
B. Complete Grouping Class I'	152
C. Mixed Procedures, Class II	154
IV. Computational Aspects	157
A. Random Sampling	157
B. Flow and Arrangement of the Computations	159
V. Solution of Typical Problems	165
A. Backscattering of Electrons and Positrons	168
B. Transmission and Penetration of Electrons and Positrons	180
C. Energy Dissipation by Electrons in Bounded and Unbounded Media	188
D. Slowing-Down Spectrum and Pathlength Distribution of Electrons	191
E. Penetration of Protons	196
VI. Appendix: Single and Multiple Scattering Theories	202
A. Energy Loss	202
B. Angular Deflections	206
References	213

I. Introduction

THIS ARTICLE IS CONCERNED with the simulation, by random sampling, of the multiple Coulomb scattering of fast charged particles, for the purpose of solving electron and proton transport problems. Direct simulation of the physical scattering processes would be laborious because of the large number of Coulomb interactions that occur even in a short pathlength. An alternative approach is used instead, in which the diffusion process is imitated by letting the particles carry out an

---

* Work supported by the Office of Naval Research.

(artificially constructed) random walk, each step of which takes into account the combined effect of many collisions.

The state of the art of solving transport problems for fast charged particles can be indicated as follows. The mathematical complexities are considerable, because of the large number of variables (up to six) that enter into the transport equation, and because of the variety of interactions that may have to be considered jointly: elastic scattering by atomic nuclei, inelastic scattering by atomic electrons, production of secondary knock-on electrons, bremsstrahlung, and—possibly—nuclear interactions. When traversing even a thin layer of matter, an electron or proton will make an enormous number of collisions that result in small energy losses and deflections, and a relatively small number of "catastrophic" collisions in which they may lose a major fraction of their energy or may be turned through a large angle. The combined effect of all collisions is a complex process of diffusion and energy degradation whose realistic description requires an elaborate theory.

Ever since the end of the nineteenth century, when fast charged particles first became available to the experimenter, multiple scattering has received the attention of many prominent theorists, and a flourishing subbranch of mathematical physics has developed around the solution of the transport equation. The early phases of this work have been surveyed by Bothe (1933), and later work has been summarized by Rossi (1952), Fano (1953), and Birkhoff (1958), among others. The available multiple-scattering theories provide accurate predictions and have been well confirmed by experiments. Yet their applicability is limited because of the more or less severe restrictions that were necessary for an analytical treatment to be possible. For example, the theories of Williams (1939), Snyder and Scott (1949), and Molière (1948) describe angular distributions only, make the small-angle approximation, and consider energy loss either not at all or only with the disregard of statistical fluctuations. On the other hand, the theories of Landau (1944) and of Blunck and Leisegang (1950) deal with fluctuations in energy loss only, and assume the loss to be small compared with the initial energy of the multiply-scattered particle. The theory of Goudsmit and Saunderson (1940) places no limitation on the magnitude of the angular deflections, but disregards spatial deflections resulting from multiple scattering, a limitation that is shared by all the theories mentioned before. The moment-method of Lewis (1950) and Spencer (1955, 1959) is more complete in that it takes into account the spatial and angular aspects of the diffusion phenomenon, and has been extended to include fluctuations in energy loss (Spencer, private communication), but it is limited to applications where the medium is unbounded and homogeneous. This restriction also holds for the other theories mentioned above. Treatments assuming more realistic boundary conditions have been given by Fermi (see Rossi, 1952) who used a small-angle approximation, and by Bethe et al. (1938) whose work in turn was extended by Weymouth (1951), Roesch (1954), Meister (1958), and Archard (1961). This work is based on a simplified version of the transport equation (diffusion approximation) and is therefore restricted in its applicability. Finally, formal treatments such as those of Wentzel (1922), Wang and Guth (1951), and Breitenberger (1959) in principle yield general solutions, but in practice can be evaluated only with the use of drastic simplifications so that their generality is lost.

There is a large class of problems that arise in the context of experimental, technical or radiological physics for which presently available multiple scattering theories do not provide adequate answers. To mention a few typical examples: an experimenter may want to know the effect of the backscattering of electrons from components of a beta-ray spectrometer; he may be interested in the probability that an electron resulting from the decay of a stopped meson can reach a detector in a given experimental configuration; he may want to know how multiple scattering affects the response of radiation detectors such as ionization chambers, scintillation counters, and photographic plates. Beams of charged particles are often passed through thick layers of material to bring their energy to a desired value, and one would like to know the fluctuations in energy of the emerging particle beam. The radiological physicist may be interested in the transfer of energy from an X-ray or bremsstrahlung beam to secondary electrons which in turn distribute it to the medium. In all these problems, the boundary conditions imposed by the experimental configuration tend to be complicated, statistical fluctuations of energy losses and deflections may be of importance, and large losses and deflections cannot be disregarded.

Modern computers are a powerful new tool for the solution of such problems. There is a temptation to jettison analytical methods altogether, and to rely entirely on numerical methods. This could be done either by numerical integration of the transport equation, or by random sampling. The former approach, which—to the author's knowledge—has not yet been attempted, would be a formidable undertaking, in part because of the large number of the variables in the transport equation. However, a calculation patterned, for example, after the S_n method developed by Carlson (1955) for neutrons might very well be feasible. Random sampling by a direct analog Monte Carlo procedure would be quite costly, because of the enormous number of collisions that must be sampled. It should be mentioned, however, that MacCallum (1960) has reported a calculation of electron backscattering by this method.

In the present article we shall employ an approach in which numerical computation does not have an exclusive role but serves to combine several multiple-scattering theories into a coherent scheme. Each of the component theories covers some aspect of the diffusion phenomenon with particular accuracy, and the combination of theories is characterized by increased flexibility and applicability. Such an approach for charged particles was first used by Hebbard and Wilson (1955) and later by Sidei et al. (1957), Leiss et al. (1957), and Berger (1960). The essential feature is the grouping of many steps of the actual physical random walk into a single step of a "condensed" random walk. The transition probabilities for the condensed random walk are given by the appropriate multiple-scattering theories, and the number of steps in a walk are kept small enough (not more than, say, 100) so that a large number of walks can be sampled in a reasonable amount of time. Once the necessary grouping has been decided upon, the remaining Monte Carlo problem is similar to those encountered in conventional random-sampling treatments of neutron or gamma-ray transport problems. Thus the computations are relatively easy to set up, even if the boundary conditions and configuration of the medium are complex, but they are time-consuming and require a vast amount of predigested information to be stored in the computer memory. One could almost say that the problem is one of data processing rather than of analysis.

The multiple-scattering theories introduced into the Monte Carlo scheme are not quite complete. This, together with the effects of grouping, introduces some arbitrariness, and leads to a systematic error which is superimposed on the statistical error associated with random sampling. One of the purposes of the present work is to present a fairly substantial body of results by means of which this systematic error can be estimated by internal evidence and through comparison with experiments of independent calculations. For this reason, problems have been chosen which are typical but simple, involving one space variable only. Thus many of the comparisons will have to do with the reflection from, transmission through, and energy dissipation in thick foils. Relatively little will be said about bremsstrahlung, and even less about nuclear interactions, because of the scarcity of relevant calculational experience.

The computations to be described, insofar as carried out by the author, were accomplished with an IBM 704 computer. They are discussed from the standpoint of a physicist who does his own programming in a simplified coding language (FORTRAN), and who is intent on minimizing not only the machine running time but also the coding effort. Thus little attention is given to ways in which the computations could be speeded up through refined programming techniques in basic machine language, and efficiency is sought primarily through a suitable layout of the flow of calculation.

II. General Description of the Monte Carlo Method

A. Relation to Transport Equation

Although no direct use will be made of the transport equation in the sequel, we shall write it down briefly, in order to indicate the mathematical problems to be solved implicitly by the Monte Carlo method. It is a linear integro-differential equation of the form

$$\frac{1}{v}\frac{\partial F}{\partial t} + \mathbf{u} \cdot \nabla F + \mu F = \int_E^\infty dE' \int_{4\pi} d\mathbf{u}' F(E', \mathbf{u}', \mathbf{r}, t)\, \psi(E', \mathbf{u}'; E, \mathbf{u}), \quad (1)$$

where

t = time
v = particle velocity
$F(E, \mathbf{u}, \mathbf{r}, t)\, dE\, d\mathbf{u}$ = flux of particles, at time t and position $\mathbf{r}$, with energies in the interval $(E, E + dE)$ and directions in the interval $(\mathbf{u}, \mathbf{u} + d\mathbf{u})$. (Flux is defined as the number of particles that cross per unit time through a unit area of a surface perpendicular to $\mathbf{u}$.)
$\mu(E)$ = probability, per unit pathlength, of an interaction of any type between the particle and the medium.
$\psi(E', \mathbf{u}'; E, \mathbf{u})\, dE\, d\mathbf{u}$ = probability, per unit pathlength, that a particle with energy E' and direction $\mathbf{u}'$ will, as the result of a collision, acquire an energy in the interval $(E, E + dE)$ and a direction in the interval $(\mathbf{u}, \mathbf{u} + d\mathbf{u})$. If the flux of secondary electrons is also of interest, ψ must take into account their production, and must then be interpreted as a production rate.

The pathlength traveled by the particle may be used as a "clock" to measure time; i.e., the variable t may be replaced by the pathlength

$$s = \int_0^t v(t')\, dt', \quad (2)$$

in which case the term $1/v\ \partial F/\partial t$ in the transport equation is replaced by $\partial F/\partial s$. In stationary problems, $\partial F/\partial s = 0$, so that the pathlength variable could be dropped. In Monte Carlo calculations we shall retain it nevertheless, because it provides a useful parameter for the grouping of collisions.

Anticipating later developments, we also note that s is often allowed to play the role of an energy parameter, in the so-called *continuous slowing-down approximation*. In this approximation, fluctuations of the energy loss are disregarded, and the energy of the particle is taken to be a deterministic function of the pathlength traveled:

$$E(s) = E_0 - \int_0^s \left| \frac{dE}{ds}(s') \right| ds' , \qquad (3)$$

where dE/ds is the mean energy loss per unit pathlength (stopping power).

Various assumptions enter into the derivation of the transport equation which we also take over for the Monte Carlo calculations:

(1) The scattering centers (atoms and electrons) are distributed at random, although not necessarily with uniform density; correlations between the positions of different atoms and electrons are not taken into account;

(2) The charged particle, in the course of traversing the medium, interacts with one scattering center at a time. This implies the neglect of quantum-mechanical interference (electron diffraction) resulting from the coherent scattering by several centers. The trajectory of the particle is thus idealized as a zig-zag path, consisting of free flights interrupted by sudden collisions in which the energy and direction of the particle is changed.

A description of the diffusion process in terms of the transport equation is analogous to the use of Eulerian coordinates in hydrodynamics, in that one asks about the flux at a given point in space. By contrast, the Monte Carlo method uses Lagrangian coordinates; one attaches a label to a particular bit of fluid, i.e., a diffusing particle, and follows its history.

B. Detailed Case Histories

The trajectory of a particle can be described by the array

$$E_0, E_1, E_2, ..., E_n, ...$$
$$\mathbf{u}_0, \mathbf{u}_1, \mathbf{u}_2, ..., \mathbf{u}_n, ...$$
$$\mathbf{r}_0, \mathbf{r}_1, \mathbf{r}_2, ..., \mathbf{r}_n, ... ,$$

where E_n is the energy, $\mathbf{u}_n$ the direction, and $\mathbf{r}_n$ the position immediately after the nth collision, and where the index zero refers to the initial state of the particle. Such a trajectory can be generated by random sampling. The probability distribution for transitions from one state to the next, i.e., from one column of the array to the next, is determined by the single-scattering probabilities ψ and μ. In case the particle makes inelastic collisions in which secondary electrons are set in motion, the history of the latter must be followed separately.

By sampling many histories, one is able—in principle—to solve any diffusion problem. Suppose, for example, that we want to calculate the reflection and transmission of electrons by a foil, which is assumed to be bounded by the planes $z = 0$ and $z = d$. (See Fig. 1.) The Monte

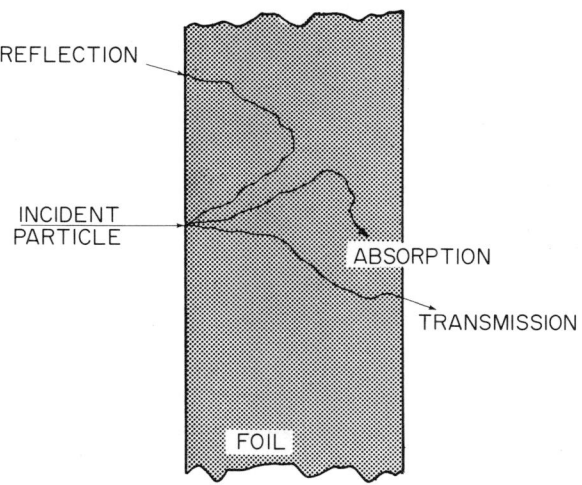

FIG. 1. Typical particle trajectories in foil.

Carlo procedure would consist of sampling many electron trajectories (usually called case histories), starting each one at $z = 0$ and following it until any one of three events happens for the first time: (1) z becomes negative (reflection), (2) z becomes greater than d (transmission), or (3) the residual range of the electron becomes so small that it can no longer escape from the foil (absorption). Dividing the scores thus obtained by the number of histories sampled, one obtains the reflection, transmission, and absorption coefficients for the foil. In order to determine these coefficients by using the transport equation, one would have to determine the reflected and transmitted currents,

$$J(z) = \int_0^{E_0} dE \int_{4\pi} d\mathbf{u} \, |\cos\theta| \, F , \qquad (4)$$

at $z = 0$ and $z = d$ (θ is the angle between **u** and the z-axis). The flux F would have to satisfy the boundary conditions that at the exit boundary ($z = d$) no electrons travel toward smaller z, and that at the entrance boundary ($z = 0$) no electrons travel toward larger z except those in the incident beam. The exact analytical solution of this boundary problem is not known.

Because of the strength and long range of the Coulomb interaction, the number of collisions in a typical charged-particle history is enormous. Some order-of-magnitude estimates for electrons are given in Table I.

TABLE I

ESTIMATED NUMBER OF COLLISIONS MADE BY ELECTRONS IN THE COURSE OF SLOWING DOWN.[a]

Energy interval (Mev)	Aluminum	Gold
0.5 − 0.25	2.9×10^4	1.7×10^5
0.25 − 0.125	3.4×10^4	1.7×10^5
0.125 − 0.0625	4.2×10^4	1.5×10^5

[a] Based on the Rutherford scattering cross section with screening correction.

They were obtained with an assumed Rutherford scattering cross section, modified to take into account screening of the nuclear charge by orbital electrons (see Appendix, B, 2). It can be seen that for pathlengths such that the electrons lose an appreciable fraction of their energy, the average number of collisions may be in the ten- or even hundredthousands. This is in great contrast to the behavior of gamma rays or neutrons. Only 20 to 30 Compton scatterings will reduce the energy of a photon from several Mev down to 50 kev. Similarly, 18 elastic collisions in hydrogen will reduce the energy of a neutron from 2 Mev to thermal energy, and even for heavy nuclei of atomic weight A approximately $9A + 6$ collisions are sufficient for the same energy reduction. Thus electron Monte Carlo histories might be up to several hundred times longer than neutron or gamma-ray histories. It is hard to see how biased sampling or importance sampling could remove this handicap. These techniques are designed to increase the likelihood, in the Monte Carlo calculation, of events which in the physical process are quite rare. They are not designed to increase the efficiency of Monte Carlo calculations in situations where the event of interest takes place only after a very long chain of intermediate collisions.

C. Condensed Case Histories

In order to reduce the required amount of computation we abandon the complete description of charged-particle histories and limit ourselves to snapshots taken at various times during the particle's history. A sequence of such snapshots provides a "moving picture" of the history which can be used for the solution of various diffusion problems. In order to select the times at which the snapshots are taken it is convenient, although not absolutely necessary, to use as clock the pathlength traveled by the particle. Thus we introduce condensed histories

$$0, s_1, s_2, ..., s_n, ...$$
$$E_0, E_1, E_2, ..., E_n, ...$$
$$\mathbf{u}_0, \mathbf{u}_1, \mathbf{u}_2, ..., \mathbf{u}_n, ...$$
$$\mathbf{r}_0, \mathbf{r}_1, \mathbf{r}_2, ..., \mathbf{r}_n, ... ,$$

where E_n is the energy, $\mathbf{u}_n$ the direction, and $\mathbf{r}_n$ the position of the particle when it has traveled a pathlength s_n from its starting point. A condensed history is sampled by letting the particle carry out a random walk in which each step, from state n to state $n+1$, takes into account the combined effect of many collisions. The transition probabilities for each step are determined by the appropriate multiple-scattering theories. There would be no question as to how this random walk should be set up, if a complete theory were available; but then the Monte Carlo calculation itself would become unnecessary. The important point is that even incomplete and partial theories, when suitably combined, can yield enough information so that condensed histories can be sampled with fair accuracy.

The size of the steps of the random walk, i.e., the pathlength intervals $\Delta s_n = s_{n+1} - s_n$, must be chosen with some care. On the one hand, the total *number of steps* should be kept as small as possible, because the length of the Monte Carlo calculation will be directly proportional to it. On the other hand, a small *step size* has the following advantages:

(1) In applications to boundary problems such as reflection and transmission by foils, most of the steps of the condensed history will lie entirely in the interior of the medium, so that multiple-scattering theories for unbounded media can be applied to them. Boundary affects need only be considered in the one section of the history in which the particle makes its escape from the medium. In this one step its state will not change much, so that even a very crude approximation may be adequate.

(2) The net angular deflection and energy loss in one step of the random walk are small, so that multiple-scattering theories with this restriction become applicable.

(3) Even though the correlation between deflections and energy losses within each step is disregarded by the multiple-scattering theory, this correlation is taken into account at least partially when the steps of the condensed random walk are combined.

III. Particular Monte Carlo Schemes

A scheme must provide, for each step of the random walk, a rule for selecting a pathlength $s_{n+1} - s_n$, an energy loss $E_n - E_{n+1}$, a change of direction from $\mathbf{u}_n$ to $\mathbf{u}_{n+1}$, and a spatial displacement $\mathbf{r}_{n+1} - \mathbf{r}_n$. A great variety of schemes are possible, which differ in regard to the theoretical input and the necessary amount of computation for random sampling. We shall list in this chapter some of these schemes, with emphasis on those which have actually been used. The execution of any one of these schemes takes a substantial amount of programming and computing effort, so that the list is neither as exhaustive nor as systematic as would be desirable. The multiple-scattering theories for each scheme are briefly indicated, but the detailed equations are relegated to the Appendix.

The schemes we shall discuss fall into two major categories. Class I, which is the simpler of the two, relies entirely on the grouping of collisions, and involves the use of a predetermined set of pathlengths. A variant, Class I', is based on a predetermined set of energy losses. Class II is based on a mixed procedure in which collisions with small energy losses and deflections are subject to grouping, but occasional "catastrophic" collisions, in which the loss or deflection are very large, are treated separately by conventional random sampling according to the single-scattering cross sections.

A. Complete Grouping, Class I

1. *Pathlength*

a. Logarithmic spacing. The pathlength is chosen such that, on the average, the energy of the particle is reduced by a constant factor k per step. Given E_n and s_n, s_{n+1} is then determined from the equation

$$1 - \frac{1}{E_n} \int_{s_n}^{s_{n+1}} \left| \frac{dE}{ds} \right| ds = k, \tag{5}$$

where dE/ds is the mean rate of energy loss per unit pathlength resulting from ionization [see Eqs. (A9-12) of the appendix].

Logarithmic spacing has the advantage that the average angular multiple-scattering deflection per step changes little from step to step. This can be shown by an approximate formula derived by Blanchard and Fano (1951) and Blanchard (1951) which holds at energies where bremsstrahlung losses can be disregarded. They find that the mean value of the cosine of the deflection angle, $\langle\cos\omega\rangle_{\mathrm{av}}$, can be estimated by the following rule of thumb:

$$\langle\cos\omega\rangle_{\mathrm{av}} \sim \left(\frac{E_{n+1}}{E_n} \frac{E_n + 2mc^2}{E_{n+1} + 2mc^2}\right)^{0.3Z}, \tag{6}$$

where Z is the atomic number of the medium and mc^2 the rest energy of the particle. When the kinetic energies E_n and E_{n+1} are smaller than $2\,mc^2$, the angular deflection depends only on E_{n+1}/E_n, but not on E_n and E_{n+1}, separately. Therefore, logarithmic spacing has the advantage that the distribution of angular deflections changes very slowly from step-to-step. It is often convenient to set the reduction factor $k = 2^{-1/m}$, where m is an integer, so that in the absence of energy loss fluctuations the particle would lose half its energy in m steps. Table II lists charac-

TABLE II

CHARACTERISTICS OF CONDENSED ELECTRON HISTORY IN ALUMINUM, MODEL I.

Step	Energy interval (Mev)		Δs (gm/cm²)	$\omega_{\max}{}^a$ (degrees)	$\langle\cos\omega\rangle_{\mathrm{av}}{}^b$
1	2.0	$-$ 1.9152	0.057476	11.4	0.945
17	1.0	$-$ 0.9576	0.028748	13.5	0.919
33	0.5	$-$ 0.4788	0.013163	15.2	0.893
49	0.25	$-$ 0.2394	0.005291	16.3	0.872
65	0.125	$-$ 0.1197	0.001749	16.5	0.857
81	0.0625	$-$ 0.0599	0.000603	15.8	0.848
97	0.03125	$-$ 0.0299	0.000184	14.7	0.842

[a] $\omega_{\max}$ = angle at which $A_{\mathrm{GS}}(\omega)\sin\omega$ peaks.
[b] $\langle\cos\omega\rangle_{\mathrm{av}}$ and $\omega_{\max}$ computed from Goudsmit-Saunderson theory, with Mott cross section.

teristic data for electrons in aluminum, for $m = 16$, $k = 0.9576$, which is a typical spacing that has been found advantageous. Results presented in Section V will indicate the effect of varying the magnitude of m.

b. *Mixed logarithmic spacing.* This is a procedure that has to be adapted to the diffusion problem under consideration. When the particle is in the interior of the medium, the logarithmic spacing described above is used with a fixed value of k (or m). But whenever the particle, in the course of its history, reaches a position such that the next step could carry it across a boundary of interest, that step is broken up into j steps with reduction factor $k' = k^{1/j}$ (or $m' = jm$). The energy and direction of the particle are well known only at the beginning and end of a step of a condensed random walk; their values at the time of the crossing of the boundary must be guessed by suitable interpolation. The error incurred thereby is reduced by forcing the crossing to occur in a very small step.

  c. *Uniform spacing.* The step size $s_{n+1} - s_n =$ constant. With this arrangement the angular deflection increases from step-to-step. The spacing constant may eventually have to be reduced toward the end of a long history, in order to limit the angular deflections to the desired small value.

2. *Energy Loss*

  a. *Continuous-slowing-down approximation.* The energy loss is determined from the equation

$$\Delta E_n = E_n - E_{n+1} = \int_{s_n}^{s_{n+1}} \left| \frac{dE}{ds} \right| ds . \tag{7}$$

If logarithmic pathlength intervals are used, $E_{n+1} = kE_n$.

  b. *Fluctuations of ionization loss.* E_n is selected from a distribution $W_I(\Delta E)$, which has been given by Landau (1944) and further refined by Blunck and Leisegang (1950) [see Eqs. (A17-21) of the appendix]. This distribution has been derived on the assumption that $\Delta E_n \ll E_n$.

  c. *Fluctuation of ionization and bremsstrahlung loss.* For electrons and positrons, at energies above one Mev in High-Z materials and several Mev in low-Z materials, radiative losses begin to make a significant contribution to the total energy loss. One must therefore select ΔE_n from a distribution taking both modes of energy loss into account. One such distribution has been derived by Blunck and Westphal (1951), on the assumption that $\Delta E_n \ll E_n$. They expressed it as a convolution of the ionization loss distribution $W_I(\Delta E)$ with a bremsstrahlung loss distribution $W_B(\Delta E)$,

$$W_{IB}(\Delta E_n) = \int_0^{\Delta E_n} W_I(\Delta E_r - \eta) W_B(\eta) d\eta . \tag{8}$$

Bremsstrahlung cross sections are complicated, and have been derived in a great variety of approximations applicable under different conditions (see Koch and Motz, 1959); the best formulation for Monte Carlo calculations needs further investigation.

3. Angular Deflection

Let $\mathbf{u}_n = (\theta_n, \varphi_n)$ and $\mathbf{u}_{n+1} = (\theta_{n+1}, \varphi_{n+1})$ denote the directions of the particle at the beginning and at the end of a step, and ω and $\Delta\varphi$ the polar and azimuthal multiple-scattering deflections in that step. It is understood that θ and φ are spherical coordinates in a system with the z-axis as polar axis, whereas ω and $\Delta\varphi$ are defined with respect to a spherical-coordinate system whose polar axis coincides with the direction of motion at the beginning of the step. We have then the well-known kinematic relations between change of direction and multiple-scattering deflections.

$$\cos\theta_{n+1} = \cos\theta_n \cos\omega + \sin\theta_n \sin\omega \cos\Delta\varphi, \tag{9}$$

$$\sin(\varphi_{n+1} - \varphi_n) = \frac{\sin\omega \sin\Delta\varphi}{\sin\theta_{n+1}}, \tag{10}$$

$$\cos(\varphi_{n+1} - \varphi_n) = \frac{\cos\omega - \cos\theta_n \cos\theta_{n+1}}{\sin\theta_n \sin\theta_{n+1}}. \tag{11}$$

The azimuthal deflection $\Delta\varphi$ is distributed uniformly between 0 and 2π, provided the medium is isotropic and polarization is disregarded. The deflection angle ω must be selected from one of many available multiple-scattering distributions. All existing theories allow the energy loss of the particle to be taken into account in the continuous-slowing-down approximation, but disregard the effect of fluctuations of energy loss.

a. Gaussian approximation. If the net angular multiple-scattering deflection is the result of the combined effect of many small individual deflections, each of the same order of magnitude, purely statistical considerations lead to a Gaussian distribution,

$$A_G(\omega)\,\omega\,d\omega = 2(\omega/\bar{\omega}^2)\exp(-\omega^2/\bar{\omega}^2)\,d\omega, \tag{12}$$

which is normalized to unity in the interval $(0, \infty)$. The mean square deflection $\bar{\omega}^2$ must be calculated from the appropriate single-scattering cross section, such as the Rutherford scattering law; there are ambiguities in this procedure because large individual deflections are not allowed to contribute to $\bar{\omega}^2$ as long as one wants to preserve the validity of the Gaussian approximation. Various prescriptions for the evaluation of ϖ^2 have been proposed (see Rossi, 1952). We shall not follow this up, because little use will be made of the Gaussian distribution in spite of its simplicity.

b. *Distribution of Molière* (1948). This theory takes into account the effect of occasional large individual deflection, neglected in the Gaussian approximation. It is formulated in terms of a "reduced scattering angle"

$$\vartheta = \omega/\chi_c \sqrt{B}, \qquad (13)$$

where χ_c and B are parameters which express the dependence on pathlength and energy [see Eqs. (A22-28) of the appendix]. The distribution has the form

$$A_M(\omega)\,\omega\,d\omega = \vartheta\,d\vartheta\left\{2\exp(-\vartheta^2) + \frac{f^{(1)}(\vartheta)}{B} + \frac{f^{(2)}(\vartheta)}{B^2} + ...\right\} \qquad (14)$$

where $f^{(1)}$ and $f^{(2)}$ are purely numerical functions tabulated by Molière (1948) and Bethe (1953).

The applicability of the Molière theory is subject to a number of limitations:

(1) It assumes a pathlength long enough for the occurrence of at least 20 collisions on the average; for our purposes, this restriction is not important.

(2) The net multiple-scattering deflection must be small (not greater than ~30 to 40 degrees).

(3) The parameters χ_c and B are evaluated on the basis of a single-scattering theory developed by Molière (1947) which is fairly exact, but does not distinguish between the scattering of positrons and electrons.

According to Bethe (1953), restriction (2) can be largely removed, and the applicability extended to large angles, through multiplication of the Molière distribution by a factor $\sqrt{\sin \omega/\omega}$. This is verified, in Table III, through a comparison—in a typical case—with a more exact theory. Restriction (3) has recently been removed by the work of Nigam *et al.* (1959) who fed into the Molière multiple-scattering formalism a single-scattering cross section based on a screened Coulomb potential and evaluated in the second Born approximation. Their theoretical distribution predicts differences in the multiple scattering of positrons and electrons (Nigam and Mathur, 1961). It has a structure similar to that of Molière, but with considerably more complicated numerical coefficients, not all of which have yet been evaluated (see also Fleischmann, 1960).

Table III

Comparison of Molière and Goudsmit-Saunderson Angular Multiple-Scattering Distributions.[a]

Interval (Degrees)	Molière[b]	Goudsmit-Saunderson		
		Rutherford	Mott electrons	Mott positrons
0–15	41473	40790	40736	41405
15–30	42333	42268	42601	42796
30–45	11770	12128	12321	11936
30–45	2766	2885	2854	2600
60–75	867	961	871	751
75–90	377	432	341	284
90–120	294	368	221	182
120–150	94	133	48	40
150–180	26	35	7	6

[a] Angular distribution of electrons and positrons slowing down in aluminum from 1.0 Mev to 0.9576 Mev. Pathlength is 0.0287 gm/cm² for electrons, 0.0294 gm/cm² for positrons. Distributions are normalized to 100,000 particles.

[b] Obtained from the Molière theory with the Bethe correction for large angles, $\sqrt{\sin \omega/\omega} \, A_M(\omega)$. Applies to electrons and positrons.

c. Theory of Goudsmit and Saunderson (1940). These authors derived the exact angular distribution of multiple-scattering deflections as a Legendre series,

$$A_{GS}(\omega) \sin \omega \, d\omega = \sum_{l=0}^{\infty} \left(l + \frac{1}{2}\right) \exp\left\{-\int_0^s G_l(s') \, ds'\right\} P_l(\cos \omega) \sin \omega \, d\omega, \quad (15)$$

where

$$G_l(s) = 2\pi N \int_0^\pi \sigma(\theta, s) \{1 - P_l(\cos \theta)\} \sin \theta \, d\theta. \quad (16)$$

N is the number of atoms per unit volume, s is the pathlength traversed by the particle, and $\sigma(\theta, s)$ is the single-scattering cross section, whose dependence on the energy is expressed, in the continuous-slowing-down approximation, through the dependence on the pathlength s.

The Goudsmit-Saunderson series has two great advantages. It applies to all angular deflections without restriction as to their magnitude, and it can be evaluated for any desired single-scattering cross section. For electrons and positrons we shall use it in conjunction with the Mott scattering cross section, modified to take into account the screening of the nuclear charge by the orbital electrons. This cross section, which includes relativistic and spin effects, differs considerably from the Rutherford cross section at large angles, and also predicts differences in the deflection of positrons and electrons.

In Monte Carlo applications, the pathlength s is usually small enough so that $A_{GS}(\omega) \sin \omega$ peaks at small values of ω which typically range from $10°$ to $25°$ and may be even smaller if very short steps are taken near a boundary of interest. The Legendre series then converges slowly, and twenty, forty, or even a larger number of terms may have to be included. This would be a difficult and tedious task, even with a high-speed computer, if Spencer (1955, 1959) had not indicated convenient recursion relations by means of which a large number of coefficients G_l can be computed easily and accurately. The details are indicated in the appendix, together with numerical examples of the Mott cross section, the angular multiple scattering distribution derived from it, and data illustrating the convergence of the Legendre series. We have found it possible to include as many as 100 terms in the Legendre series before encountering clearly recognizable round-off difficulties. However, such round-off error undoubtedly occurs, and should be further investigated with the use of double-precision arithmetic.

The Goudsmit-Saunderson theory is no more difficult to evaluate than the Molière theory, and furthermore allows us to treat large deflections with increased accuracy, which is important particularly for the investigation of backscattering. A large deflection may be very rare in any one step, but yet have a good chance of occurring somewhere in a long history consisting of 100 steps. Once it occurs it has a strong influence on the subsequent history of the particle; for example, a reversal of direction, at a point not too far from the entry into a foil, greatly increases the chance of eventual backscattering from the foil. A comparison of the angular distributions predicted by the Molière and Goudsmit-Saunderson theories, particularly at large angles, is given in Table III for conditions typical of those assumed in the Monte Carlo calculations.

4. *Spatial Displacement*

Let $\Delta \xi$, $\Delta \eta$ and $\Delta \zeta$ denote the spatial displacement of the particle in a single step of the random walk, in a Cartesian-coordinate system whose ζ-axis coincides with the direction of motion of the particle at the beginning of the step. As our later applications are all limited to one-dimensional problems, we shall be concerned here only with the change of the z-coordinate,

$$z_{n+1} - z_n = \sin \theta_n \cos \varphi_n \Delta \xi + \sin \theta_n \sin \varphi_n \Delta \eta + \cos \theta_n \Delta \zeta. \quad (17)$$

a. Inclusion of longitudinal and transverse displacement. The transverse displacements, $\Delta \xi$ and $\Delta \eta$, are correlated with the angular deflections ω and $\Delta \varphi$. This correlation has only been calculated in the Gaussian approximation (Rossi, 1952), with the result that (for small ω)

$$\Delta \xi = \tfrac{1}{2} \Delta s_n \left(\sin \omega \cos \Delta \varphi + k_x \sqrt{\frac{\overline{\omega^2}}{6}} \right) \quad (18)$$

$$\Delta \eta = \tfrac{1}{2} \Delta s_n \left(\sin \omega \sin \Delta \varphi + k_y \sqrt{\frac{\overline{\omega^2}}{6}} \right), \quad (19)$$

where k_x and k_y are random variables that are distributed independently according to a Gaussian distribution with mean zero and variance unity. When sampling on the basis of these formulas, one must of course exclude very large values of k_x and k_y for which $(\Delta \xi)^2 + (\Delta \eta)^2 + (\Delta \zeta)^2 > (\Delta s_n)^2$, but this has been found to be extremely unlikely. For $\overline{\omega^2}$ one can substitute $2(1 - \langle \cos \omega \rangle_{\text{Av}})$, with $\langle \cos \omega \rangle_{\text{Av}}$ evaluated from a more accurate theory than the Gaussian approximation. The distribution of $\Delta \zeta$ has been derived by Yang (1951) in the Gaussian small-angle approximation, and evaluated for two special situations ($\omega = 0°$, and an average over all values of ω). The evaluation of the distribution for arbitrary ω is possible but difficult. We have instead adopted a much simpler rule,

$$\Delta \zeta = \Delta s_n \frac{1 + \cos \omega}{2}. \quad (20)$$

This can be justified as follows: Clearly $\Delta \zeta$ cannot be greater than Δs_n, and an approximate lower bound is given by $\Delta s_n \cos \omega$. As a consequence of (15), for small Δs_n,

$$\langle \cos \omega(\Delta s_n) \rangle_{\text{av}} = \exp \left\{ - \int_0^{\Delta s_n} G_1(s') \, ds' \right\} \sim 1 - G_1(0) \Delta s_n. \quad (21)$$

General transport theory (Lewis, 1950) predicts that

$$\langle \Delta \zeta \rangle_{\text{av}} = \int_0^{\Delta s_n} \langle \cos (s') \rangle_{\text{av}} \, ds'$$

$$= \Delta s_n \frac{1 + \langle \cos \omega(\Delta s_n) \rangle_{\text{av}}}{2}. \quad (22)$$

This means that the rule (20) predicts the correct average displacement, and is an average of the upper and lower bounds of $\Delta \zeta$. Moreover, it is correct in the limit of very small ω, and remains plausible for large ω.

b. Inclusion of longitudinal displacements only. According to Eq. (17) the relative contribution of the transverse displacements $\Delta \xi$ and $\Delta \eta$ to $z_{n+1} - z_n$ can become important only when $\sin \theta_n$ is large, and even then will tend to be limited by azimuthal averaging. In problems such as reflection and transmission by foils, a particle incident normally ($\theta_0 = 0$) would have to spend a major part of its history traveling in a direction at right angles to its original direction in order for the transverse displacements to have a pronounced effect. But in this case the particle would be likely to get absorbed in the foil, so that one would expect the transverse displacements to have a small influence on reflection and transmission. It is difficult to put such considerations on a quantitative basis except by Monte Carlo calculations. Numerical experimentation, described in Section V, tends to confirm the above considerations. Therefore we have in many of our calculations set $\Delta \xi = \Delta \eta = 0$, a procedure which was also followed in the work of Hebbard and Wilson, Sidei *et al.*, and Leiss *et al.*

B. Complete Grouping, Class I′

Schemes in this class are variants of those in Class I, distinguished by the fact that the energy reached by the particle, rather than the pathlength traveled, serves as the clock for the timing of the snapshots of the particle history. Thus one preselects a set of energies, $E_0, E_1, E_2, ..., E_n, ...$, and determines the successive states of the particle as its energy drops to these values. The arbitrary preselection of energies is reasonable for protons, which can only lose a tiny fraction of their energy in an individual collision with an atomic electron (see Eq. (A13) of the appendix) and thus traverse a practically continuous range of energies. It would not be appropriate for electrons, which can lose a major fraction of their energy in a single collision so that they may jump over certain energy intervals altogether.

1. *Energy Loss*

   a. *Logarithmic spacing:* $E_{n+1} = kE_n$.

   b. *Uniform spacing:* $E_n - E_{n+1} =$ constant.

The advantages and disadvantages of these alternatives are the same as those of the uniform and logarithmic pathlength spacings in schemes of Class I.

2. Pathlength

a. Continuous slowing down approximation,

$$s_{n+1} = s_n + \int_{E_{n+1}}^{E_n} \left| \frac{dE}{ds} \right|^{-1} dE . \quad (23)$$

b. Consideration of pathlength fluctuations

$$s_{n+1} = s_n + \Delta s_n ,$$

where Δs_n is distributed normally with mean

$$\langle \Delta s_n \rangle_{\mathrm{av}} = \int_{E_{n+1}}^{E_n} \left| \frac{dE}{ds} \right|^{-1} dE \quad (24)$$

and variance $\langle (\Delta s_n)^2 \rangle_{\mathrm{av}} - \langle \Delta s_n \rangle^2_{\mathrm{av}}$ (see Eq. A15 of appendix.) Small corrections to this Gaussian distribution have been derived by Lewis (1952) but we have not used them.

TABLE IV

CHARACTERISTICS OF CONDENSED PROTON HISTORY IN LEAD, MODEL I'.

a. Approximately uniform spacing on linear energy scale.

Step	Energy interval (Mev)	$\langle \Delta_s \rangle_{\mathrm{av}}$ (gm/cm²)	p_s [a]	$\chi_c \sqrt{B}$ [b] (degrees)
1	338.5–330.0	4.931	0.080	1.79
6	285.0–270.0	7.672	0.055	2.84
12	195.0–180.0	6.308	0.048	3.48
18	105.0– 90.0	4.100	0.038	5.12
24	20.0– 15.0	0.397	0.036	7.13
30	3.0– 2.0	0.023	0.048	10.98

b. Uniform spacing on logarithmic energy scale.

Step	Energy interval (Mev)	$\langle \Delta_s \rangle_{\mathrm{av}}$ (gm/cm²)	p_s	$\chi_c \sqrt{B}$ (degrees)
1	338.5–285.3	29.795	0.032	5.04
6	143.9–121.3	7.623	0.036	5.34
12	51.6– 43.5	1.327	0.041	5.57
18	18.5– 15.6	0.226	0.048	5.80
24	6.6– 5.6	0.040	0.059	6.11
30	2.4– 2.0	0.008	0.079	6.71

[a] p_s = per cent pathlength straggling, defined by Eq. (A17) of Appendix.
[b] $\chi_c \sqrt{B}$ is the characteristic deflection angle of the Molière theory.

After the step size and pathlength have been determined, the selection of angular deflection and spatial displacement takes place exactly as in schemes of Class I. Thus the correlation between pathlength fluctuations and multiple-scattering deflections within a step of the random walk is again disregarded. Characteristics of a condensed proton history in a lead medium are shown, by way of example, in Table IV.

C. Mixed Procedures, Class II

In this class of schemes one excludes from grouping the individual collisions (denoted as catastrophic) in which the particle loses a large fraction ϵ of its energy, greater than, say, ϵ_c. The history of the particle is divided into sections, within which no catastrophic collisions occur and in which continuous-slowing-down is assumed. Each section is terminated by a catastrophic collision. This schematization has first been applied in electron Monte Carlo calculations by Schneider and Cormack (1959), and is illustrated by Fig. 2, adapted from their paper.

Let $(E_n^a, \mathbf{u}_n^a, \mathbf{r}_n^a)$ and $(E_{n+1}^b, \mathbf{u}_{n+1}^b, \mathbf{r}_{n+1}^b)$ indicate the state of the

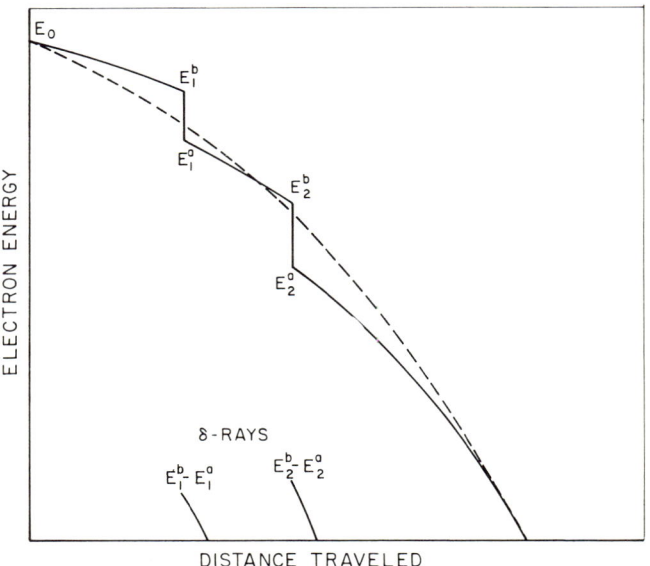

Fig. 2. Energy-pathlength plot of hypothetical electron case history. Solid curve corresponds to a Monte Carlo model of Class II with catastrophic collisions that lower the energy from E_1^b to E_1^a and from E_2^b to E_2^a and result in the occurrence of secondary knock-on electrons (delta rays). The dotted curve corresponds to the continuous-slowing-down approximation. Adapted from Schneider and Cormack (1959).

particle immediately after the nth, and immediately before the $n + 1$st catastrophic collision, respectively, and let s be the pathlength between these two collisions. The pathlength distribution function is

$$P(s)\, ds = \exp\left\{-\int_0^s \mu_c(s')\, ds'\right\} \mu_c(s)\, ds, \tag{25}$$

where μ_c is the probability, per unit pathlength, of a catastrophic collision with energy transfer $\epsilon > \epsilon_c$. To take the energy dependence of μ_c into account, one must change from a pathlength to an energy variable, using the stopping power $|\, dE/ds\,|_{\epsilon_c}$ evaluated for all collisions with fractional energy transfer $\epsilon < \epsilon_c$. Thus

$$P(s)\, ds = e^{-Q}\, dQ, \tag{26}$$

where

$$Q(s) = \int_0^s \mu_c[E(s')]\, ds' = \int_{E_{n+1}^b}^{E_n^a} \mu_c(E) \left|\frac{dE}{ds}\right|_{\epsilon_c}^{-1} dE. \tag{26a}$$

$Q(s)$ can be interpreted as the average number of catastrophic collisions in pathlength s. The evaluation of μ_c and $|\, dE/ds\,|_{\epsilon_c}$ is described in Appendix A.

The history of a particle is traced as follows:

1. Given E_n^a, $\mathbf{u}_n^a$, $\mathbf{r}_n^a$, Q is selected from an exponential distribution and used to determine s and E_{n+1}^b.

2. A Class I (or I′) scheme (with continuous-slowing-down approximation) is applied to determine the history of the particle until the instant before the next catastrophic collision (ending with direction u_{n+1}^b and in position $\mathbf{r}_{n+1}^b$).

3. The magnitude of the fractional energy loss ϵ in the $n + 1$st catastrophic collision is sampled from the appropriate single-scattering distribution $g(\epsilon; \epsilon_c)$ [see Eq. (A4) of the appendix].

4. The catastrophic collision leaves the position of the particle unchanged, i.e., $\mathbf{r}_{n+1}^a = \mathbf{r}_{n+1}^b$. The energy of the particle is reduced from E_{n+1}^b to

$$E_{n+1}^a = E_{n+1}^b(1 - \epsilon). \tag{27}$$

Associated with this energy loss is a deflection through an angle $\omega'(\epsilon)$ completely determined by conservation of momentum and energy in the collision [see Eq. (A3) of the appendix]. The new direction, $\mathbf{u}_{n+1}^a$ is determined by the customary trigonometric relations

$$\cos \theta_{n+1}^a = \cos \theta_{n+1}^b \cos \omega' + \sin \theta_{n+1}^b \sin \omega' \cos \Delta\varphi', \text{ etc.}, \tag{28}$$

where $\Delta\varphi'$ is an angle randomly distributed between zero and 2π.

5. The catastrophic collision leads to a secondary knock-on electron with energy $E_0 = E_{n+1}^b \epsilon$, whose direction is also determined by the kinematics of the collision, and whose history can be followed in turn.

A small set of corresponding E-, s- and Q-values for electrons in aluminum is shown in Table V. It illustrates the dependence of the mean

TABLE V

PROBABILITY OF ELECTRON-ELECTRON COLLISIONS WITH FRACTIONAL ENERGY LOSSES GREATER THAN ϵ_c, IN ALUMINUM MEDIUM.

	$\epsilon_c = 0.1$		$\epsilon_c = 0.03$		$\epsilon_c = 0.01$	
E (Mev)	s gm/cm²	$Q(s)$	s gm/cm²	$Q(s)$	s gm/cm²	$Q(s)$
1.0	0.0000	0.0000	0.0000	0.0000	0.0000	0.0000
0.5	0.3602	0.3391	0.3893	1.4610	0.4211	5.0390
0.25	0.5186	0.6920	0.5617	3.0509	0.6091	10.6461
0.125	0.5795	1.0677	0.6286	4.8000	0.6829	16.9308
0.0625	0.6004	1.4790	0.6518	6.7577	0.7088	24.0852

number of catastrophic collisions, Q, on the cutoff parameter ϵ_c (in the range $0.01 \leq \epsilon_c \leq 0.1$, which we have found useful in later calculations). In the limiting case $\epsilon_c = \frac{1}{2}$ (for electrons) one is led back to a Class I scheme, with complete grouping, whereas the other limiting case, $\epsilon_c = 0$, corresponds to a conventional Monte Carlo calculation in which all individual collisions are sampled. It is of course not necessary to use the same value of the cutoff parameter throughout the history of a particle; for example, one may require that those catastrophic collisions are singled out, in which an amount of energy greater than $\mathbf{E}_c$ is lost, so that $\epsilon_c = E_c/E$.

Among the advantages of Class II over Class I are the following: The initial state of secondary knock-on electrons is indicated unambiguously, angular deflections due to inelastic scattering can be calculated more accurately, and the correlation between energy-loss fluctuations and multiple scattering deflections is preserved more faithfully. On the other hand, the random occurrence of catastrophic collisions allows less storing of predigested transition probabilities for the condensed random walk, so that the computations become more laborious (see Section IV). It is possible to extend the Class II procedures to angular deflections, excepting from grouping all collisions associated with deflections greater than a cutoff angle ω_c. In problems where such individual collisions

IV. Computational Aspects

A. Random Sampling

Once a particular Monte Carlo scheme for condensed histories has been chosen, the required random sampling techniques are similar to those in conventional Monte Carlo calculations.

Many random numbers are needed, which must be uncorrelated and distributed uniformly between zero and one. We have used the well-known method of congruential multiplication, adapted to an IBM 704 computer, according to which so-called pseudorandom numbers ρ_n are generated by the following scheme:

$$j_0 = c$$
$$j_{n+1} = 5^d j_n \bmod 2^{35}, \qquad (29)$$
$$\rho_n = 2^{-35} j_n$$

where c and d are odd integers. Two types of sequences were employed. The main sequence, started with an arbitrary c and continued with a multiplicative factor 5^{11}, provided the starting number c for successive histories. In each history, the sequence thus started was continued with a multiplicative factor 5^{13}. This procedure was set up in order to facilitate "correlated" sampling. By this we mean comparison calculations, for example of the backscattering of electrons and positrons, in which two sets of histories are generated, one for electrons and the other for positions, each set using insofar as possible the same sequences of random numbers. Observed differences of backscattering are then only the result of differences in the scattering cross sections, and irrelevant statistical fluctuations are reduced. The use of main and secondary pseudorandom number sequences as described above has the advantage that one can produce sets of matched pair electron and positron histories such that each member of a pair, insofar as it is of the same length, will be based on exactly the same sequence of random numbers as its mate. This procedure can be applied not only to electron-positron differences, but also to the investigation of the different Monte Carlo models.

Many ingenious methods are available for sampling random variates x

from a distribution $f(x)$ with the use of pseudorandom numbers (see, e.g., H. Kahn, 1954). Many of them were developed when the memory capacity of computers was quite limited. The large capacity (typically 32,000 words) of present-day computers makes it preferable to use a less ingenious method which wastes memory space but can be carried out faster, by relying on the use of the cumulative probability distribution

$$F(x) = \int_{-\infty}^{x} f(x') \, dx' . \tag{30}$$

F is a random variable distributed according to

$$P(F) = f(x) \frac{dx}{dF} = 1 . \tag{31}$$

In other words, F is distributed uniformly between zero and one. One thus chooses a random number ρ, sets $F(x) = \rho$ and solves for $x = F^{-1}(\rho)$. This can easily be done with a table-look-up, provided one stores a large table of $F(x)$ in the computer memory. One procedure which has been found convenient is the following:

1. Compute $F(x_n)$ for a dense set of x-values spanning the range of $f(x)$.
2. By interpolation, find a set of x-values such that

$$F(\tilde{x}_m) = (m - \tfrac{1}{2})/M. \qquad m = 1, 2, \cdots, M$$

Store the $\tilde{x}_m$'s in the computer memory. They constitute, in good approximation, a set of "equally probable" variates from which one can make a selection.

3. To sample from $f(x)$, pick a random number ρ, and compute i = integral part of $M\rho$. The desired random variate is $\tilde{x} = x_i$. An equivalent but faster procedure, which cannot be done in FORTRAN language, is to use a section of the random number, in binary form to indicate the value of the address i. This requires M to be a power of 2.

Suppose we want so sample a random variate $\cos \Delta\varphi$, where $\Delta\varphi$ is an angle distributed with equal probability between 0 and π. If we are satisfied to specify $\Delta\varphi$ to the nearest degree, it is then sufficient to store in the computer memory 180 numbers ($\tilde{x}_m = \cos[(m - \tfrac{1}{2})\pi/180]$, $m = 1, 2, ..., 180$) and each selection requires the generation of one random number, one multiplication and a truncation of the product. This procedure can be compared with a more ingenious "rejection technique" which requires only a very few memory cells but more arithmetic.

1. Select a pair of random numbers, ρ_1 and ρ_2.
2. If $\rho_1^2 + \rho_2^2 \leq 1$, set $\cos \Delta\varphi = (\rho_1^2 - \rho_2^2)/(\rho_1^2 + \rho_2^2)$.

3. If $\rho_1{}^2 + \rho_2{}^2 > 1$, pick another pair of random numbers and try again. The probability of accepting a pair of random nubmers is $\pi/4$, so that on the average one must generate 2.55 random numbers, and perform 2.55 multiplications and one division, to select one value of $\Delta\varphi$.

The various statistical techniques such as biased sampling, importance sampling, Russian roulette, etc., which have been developed for application to gamma-ray or neutron transport problems also are applicable to condensed charged-particle histories. In other words, one could manipulate the rules of the game so as to increase the relative likelihood of interesting but ordinarily rare events such as penetration through a very thick foil.

In the present paper, the technique of correlated sampling, already touched upon in Section IV, A and further described in Section V, A, was used to obtain greater precision when comparing backscattering and transmission under slightly different conditions (e.g., for electrons and positrons). Particle histories were of course dropped as soon as they reached a stage where no further contribution to the problem at hand could be expected, for example, when electrons had penetrated so deep into a semiinfinite medium that their residual range was too small for reemergence, or when the residual electron displacement could not remove them from a layer of the medium in which energy dissipation was to be determined.

More elaborate variance-reducing techniques were not used in this paper, for a number of reasons. First, many of the problems under consideration, such as backscattering, transmission through foils of moderate thickness, and the spatial distribution of energy dissipation did not involve particularly rare events. Second, the Monte Carlo models were often not refined enough for the accurate calculation of extremely deep penetration for which statistical elaborations would have been most effective. Third, an increase of computing efficiency was often achieved through the use of the same set of histories for the simultaneous solution of several problems (i.e., reflection and transmission by a set of foils with different thicknesses) in which case the same biasing would not have been appropriate for all of them. However, it must be admitted that considerable computational economies in charged-particle Monte Carlo problems could be achieved through the use of more sophisticated sampling techniques.

B. Flow and Arrangement of the Computations

The remarks in this Section are based on the author's experience in programming a variety of exploratory calculations in FORTRAN language for an IBM 704 computer with a memory of 32768 words. Three main types of programs were needed: (1) A *Data Preparation Program* which pre-computes transition probabilities for condensed random walks; (2) A *Main Program* which generates Monte Carlo case histories and applies them to particular problems, and (3) A *Processing Program* which combines and analyzes the results obtained in various runs of the Main Program. Depending on the Monte Carlo model, different groups of such programs had to be written. Usually one Data Preparation Program was sufficient to provide input for a whole set of Main Programs. The Data Preparation Program was the most difficult and time-consuming to develop but had a running time of only a few minutes on the IBM 704. The Main Program was usually short and easy to code, but consumed large amounts of machine time, being largely repetitive. The Processing Program required relatively little programming or machine effort, and was needed to rescue the author from a flood of output data.

1. *Data Preparation Program*

The purpose is to make the random sampling in the Main Program as fast and painless as possible. This is achieved primarily through the tabulation of cumulative probability distributions, or related quantities, so that the sampling requires only a table-look-up but no further arithmetic. As an example, let us consider the generation of electron histories according to a model of Class I, assuming logarithmic spacing, the continuous-slowing-down approximation for energy loss, and the use of the Goudsmit-Saunderson multiple-scattering angular distribution together with the Mott scattering cross section. We assume further that the sampling of angular deflections is to be carried out with the use of a set of "equally probable" angles, as described in Section IV, A. The required program can be characterized by its input, function, and output. The *input* must specify:

(1) Properties of the medium (atomic number and weight, mean ionization potential, etc.).

(2) Details of the Monte Carlo model, including the spacing parameter k.

(3) The energy span to be covered, and the number of "equally probable angles" to be used for sampling.

The *function* of the program is to compute:

(1) The energies corresponding to each step.

(2) The mean energy loss rate by ionization.

(3) The pathlength for each step, from (1) and (2).

(4) The Mott scattering cross section at each energy.

(5) The cumulative Goudsmit-Saunderson multiple-scattering angular distribution, from (4).

(6) A table of "equally probable" scattering angles for each step, derived from (5).

The *output* of the programs consist of tables of items (1), (3), and (6) above. For example, if we require 96 steps which reduce the energy in steps of $2^{-1/16}$ from E_0 to $E_0 2^{-6}$, and 40 equally probable angles at each energy (these magnitudes are typical for some of our later calculations), then the output will comprise approximately 4000 numbers, which are loaded on tape or punched cards for later use, and also printed out to allow inspection and checking.

Now suppose that we drop the continuous-slowing-down approximation and decide to sample energy losses from a Landau distribution. We must then tabulate a set of "equally probable" values of the Landau's universal parameter λ [see Eq. (A17) of the appendix] from which the energy loss in each step can be obtained by simple arithmetic. A more severe complication arises from the fact that the Goudsmit-Saunderson angular distribution depends not only on the pathlength of the step but also on the energy of the particle at the beginning of the step which in the previous example was determined in advance but now is subject to statistical fluctuations. Thus the angular distribution at each step must be tabulated for at least a small set of initial energies, so that interpolation becomes possible. In Class II models, where even the pathlength per step is statistical rather than predetermined, more tabulations are necessary, and the amount of data per step may be several times larger than that required for the simplest model (Class I, continuous-slowing-down approximation). The author has no experience with data input greater than 12,000 words per problem. However, there is no doubt that with the magnetic tape input facilities of modern computers even much larger amounts of data can easily be fed into the Main Program, provided the latter is arranged to accept them in convenient form.

2. *Main Program*

We shall indicate the flow of the calculation for two types of Main Program, the first of which generates Monte Carlo case histories in series, one after the other, and is applicable to problems with a limited amount of data input whereas the second type generates many histories simultaneously, in parallel, and is appropriate when the data putin is very large. In the description we shall assume, for the sake of concreteness, that the problem to be solved is the transmission and reflection of particles by foils.

a. Series arrangement. The following breakdown indicates the logic of the program, but does not necessarily correspond in all details to the actual machine program.

(1) *Setup of Problem.* This is the main routine which links together the other subroutines listed below. It requires input parameters that specify the properties of the medium and of the diffusing particle, the characteristics of the Monte Carlo model, the number of histories to be sampled, the conditions for terminating a history, the configuration of the medium (i.e., the foil boundaries), etc.

(2) *Input Data.* If the required input data have previously been generated, they are read into the computer memory from magnetic tape. If not, the *Data Preparation Program* is put in operation, and the resulting tabulations are stored in the memory and also on magnetic tape for possible future use.

(3) *Start of History.* A history is begun by specifying the initial position, energy, and direction of the particle.

(4) *Advance.* The particle is allowed to make one step of the condensed random walk, selected with the use of the tabulated input data.

(5) *Scoring.* If the particle, as the result of the step just taken, crosses a foil boundary of interest, the energy and direction of escape, and the identifying label of the boundary are recorded.

(6) *Termination of History.* If the particle can no longer escape from the foil, or if some other condition for termination has been satisfied, one proceeds to subroutine (7). Otherwise, one proceeds to subroutine (4) and lets the particle take another step.

(7) *Termination of the Problem.* If the desired number of histories has been sampled, one proceeds to subroutine (8). Otherwise, one proceeds to subroutine (3) and starts another history.

(8) *Output.* Summary information is computed and printed out, such as reflection and transmission coefficients. This serves mainly as "quality control" indicating that the program has been running satisfactorily. The detailed information produced by the scoring subroutine is dumped on magnetic tape and provided with an identifying label so that it can be recovered easily for later use in a Processing Program.

b. Parallel arrangement. This requires a modification of the series program as follows: Initially, the input-data read into the computer memory are limited to a small energy range, extending from the highest particle energy to be considered down to an intermediate energy. Not just one but many histories are started and followed together until they are either terminated or go out of the energy range of the input data. When a history goes out of range, the position, energy, and direction of the particle are stored to allow later continuation. After the entire group of histories has been processed in this manner, another input-data set is read into the memory which covers a lower energy interval and replaces the old input data set in exactly the same memory location, which is permissible because the old set will not be needed again.

The histories which previously went out of range are now continued with the use of the new data. This procedure is repeated until the entire energy range of interest has been covered. Compared to a series program, some additional memory space is required to record the characteristics of histories that must be continued, but much less space is required for the storage of input data.

c. Computing time. The work for this paper was done with a great variety of experimental programs which were sometimes changed from run to run, so that it is not easy to arrive at exact time estimates. The preponderant amount of computing effort goes into the execution of the two subroutines *Advance* and *Scoring* which constitute the innermost loops of the Main Program. Time estimates can therefore be made in terms of the unit time required for doing one step of the random walk and analyzing its effects. This unit time must then be multiplied by the number of steps in a condensed history and the number of histories sampled. The total time per problem is equal to this product increased a few percent to allow for the execution of the other subroutines.

Depending on the complexity of the Monte Carlo scheme the unit time per step has been found to range from 5 to 12 milliseconds of IBM 704 time. With an IBM 7090 computer this time would be reduced by a factor between five and six.

3. *Processing Program*

It may, under some circumstances, be possible to achieve great computational economy by generating a set of Monte Carlo histories without regard to the conditions of any particular problem, storing them, and using them repeatedly later for various applications. (This would require a shift of the Scoring subroutine from the Main Program to the Processing Program.) Whether this is practical or not depends on the relative speed with which histories can be computed or read into and out of the computer memory. In a calculation with a desk-computer, the production of histories is quite laborious, whereas notebooks provide a large and easily accessible memory so that the reuse of histories is advantageous. This has been demonstrated by Sidei, *et al.* (1957) who efficiently reused different portions of a set of a few hundred Model-I histories to determine the transmission of electrons as a function of the incident energy and the albedo as a function of the source obliquity. When computers are used, the sample size tends to increase greatly and a vast amount of information must be transferred. For example, if 5000 histories of 50 steps are each generated, and three numbers (energy, direction, position) are recorded for each step, approximately 7.5×10^5 words have to be stored, which requires several hundred feet of magnetic tape. To calculate these numbers in the first place would take on the order of, say, 250 to 500 seconds, on an IBM 7090 computer. With the fasted tape equipment available for use with this computer, at most 10,000 words per second can be transferred from or to tape, so that at least 75 seconds would be required for recording the histories on tape. Actually, the time would be considerally longer, because the information would have to be organized in suitable blocks, separated by gaps on the magnetic tape, which would slow up the transfer of information. On the other hand, computers of very modern design, such as the IBM 7090, can simultaneously accept information from tape and perform arithmetic operations, which would make the tape read-in less of a burden. The relative merits of repeated use as recalculation of histories can only be decided with detailed knowledge of the characteristics of the computer and its associated input-output equipment, and is one which the author does not feel competent to answer. In any case, the error-free processing of such vast amounts of information would require an elaborate checking and indexing procedure involving a very substantial additional programming effort.

In the present work, histories were used in one run on the machine for the solution of several problems, which resulted in time savings up to 50% but led to the difficulty that there was not enough memory space for storing the various energy spectra, angular distributions, and other information to be printed out at the end of the run. This difficulty was overcome by recording, on binary cards or on magnetic tape, the details of a history whenever some event of interest occurred. For example, when the particle crossed one of many possible foil boundaries, the identifying number of the boundary, together with the energy and direction of the escaping particle, was compressed into one binary word, and when a sufficient number of these words had accumulated they were stored on tape or cards. If 5000 histories are to be applied to 5 problems, and if in each history one interesting event occurs for each problem (a considerable overestimate), only 25,000 words have to be stored, which presents no problem.

The function of the processing routine was then to combine the compressed information recorded in various runs of the Main Program, to normalize the results, compute reflection and transmission coefficients, evaluate their statistical accuracy, obtain energy spectra and angular distribution histograms by suitable classification, and so on. This took only a very short time, so that it was practical to repeat the processing whenever a new output format, or a new spectral classification, or some other additional information was desired.

V. Solution of Typical Problems

With one exception, the problems discussed in this section deal with charged particles whose energy is so low that radiative energy losses can be neglected. This reflects the state of the literature as well as the desire to confine the discussion to Monte Carlo calculations in which the use of condensed particle histories is essential. At extremely high energies, where bremsstrahlung and the resulting electron-photon cascade are of prime importance, multiple Coulomb scattering is a minor effect that is often treated approximately, without the more elaborate procedures required at lower energies. For example, very extensive Monte Carlo calculations of electron-photon cascades at energies up to 20 Bev have been made by Butcher and Messel (1960) in which the generation and transport of photons were treated by conventional sampling (without the grouping of collisions), and in which the Coulomb interactions of electrons were considered only as giving rise to a constant energy loss per unit pathlength. There is a need for further Monte Carlo calculations at intermediate energies, between, say, two Mev and several hundred Mev, in which both bremsstrahlung and charged-particle penetration are treated accurately.

In the review of previous work,[1] and of new calculations by the author, it will be necessary to specify the methods used for constructing

---

[1] The author has learned, after completion of this article, that J. F. Perkins has made extensive Monte Carlo calculations of electron backscattering and transmission, by schemes of Class I, based on the use of the Molière and Landau distributions. Unfortunately it was too late to include this material in the review.

TABLE VI

SUMMARY OF CLASS I MONTE CARLO MODELS OF CONDENSED HISTORY.

Procedure	Symbol
Pathlength	
a. Logarithmic spacing	PL
b. Mixed logarithmic spacing	PLM
c. Uniform spacing	PU
Energy loss	
a. Continuous-slowing-down approx.	EC
b. Fluctuations of ionization loss	EI
c. Fluctuations of ionization and bremsstrahlung loss	EIB
Angular deflection	
a. Gaussian approximation	AG
b. Molière theory	AM
c. Goudsmit-Saunderson theory	AGS
Displacement	
a. Longitudinal and transverse displacement	DLT
b. Longitudinal displacement only	DL

condensed particle histories. Table VI summarizes the procedures for schemes of Class I, the class which has found most frequent use, and assigns to each procedure a symbol. The symbols will be used in an abbreviated notation to identify the origin of results in figures and tables. For example, the notation

$$\text{Model } \{I, PL(16, 96), EC, AM, DLT\}$$

indicates the use of a Class I scheme; it means that the pathlengths for successive steps were selected with a logarithmic spacing such that $k = E_{n+1}/E_n = 2^{-1/16}$; that 96 steps were followed in each history (unless the history was terminated earlier through escape of the particle from the medium); that energy loss was calculated according to the continuous-slowing-down approximation; that angular deflections were sampled from the Molière distribution and that longitudinal as well as transverse multiple-scattering displacements were taken into account.

To take another example, the notation

$$\{\text{Model } I, PLM(16/8, 48), EI, AGS, DL\}$$

means that pathlengths were selected according to a mixed logarithmic scheme, with $k = 2^{-1/16}$ in the interior of the medium and with a reduced spacing $k = 2^{-1/16 \times 8} = 2^{-1/128}$ in the neighborhood of a boundary; that the history was followed through no more than 48 steps with the larger spacing; that fluctuations of energy loss by ionization were taken into account; that the Goudsmit-Saunderson angular distribution was used, and that only longitudinal displacements were considered. Unless the contrary is stated, the Goudsmit-Saunderson distribution was evaluated with the Mott scattering cross section.

The penetration of charged particles, under the assumption of continuous-slowing-down, is governed by an approximate scaling law. Let z denote the depth of penetration from the source, and

$$r_0 = \int_0^{E_0} \left| \frac{dE}{ds} \right|^{-1} dE \tag{32}$$

the mean range at the source energy E_0. When the spatial distribution of the diffusing particles is expressed in terms of the ratio z/r_0, the shape of the distribution is insensitive to the value of E_0, provided E_0 is smaller than twice the rest energy of the particles. This is a consequence of the scaling law for angular multiple-scattering deflections mentioned earlier [Eq. (6)], according to which the mean cosine of the deflection angle depends, to first order, only on the ratio of the energies of the particle at the end and the beginning of the path traversed, or —equivalently—on the ratio of the corresponding mean residual ranges. Taking advantage of the scaling law, we shall extend the generality of many of the Monte Carlo results by expressing them as functions of z/r_0. A small set of r_0-values for aluminum and gold, obtained with the use of Eqs. (A9-11), are given in Table VII. Many of the calculations to be discussed are for these materials which are representative of low-Z and high-Z materials.

TABLE VII

MEAN RESIDUAL RANGE, r_0, OF ELECTRONS AND POSITRONS (gm/cm²).

E_0 (Mev)	Aluminum, $I = 163$ ev		Gold, $I = 797$ ev	
	electrons	positrons	electrons	positrons
2.0	1.237	1.259	...	...
1.0	0.5568	0.5598	...	...
0.5	0.2258	0.2237	0.3451	0.3406
0.25	0.08196	0.07946	0.1288	0.1236
0.125	0.02706	0.02565	0.04372	0.04068
0.0625	0.008385	0.007778	0.01390	0.01256

In various tables the estimated errors of the results are shown. These are standard deviations indicating the statistical accuracy of the Monte Carlo calculations. Systematic errors will be discussed separately.

A. Backscattering of Electrons and Positrons

1. Albedo

The most important parameter characterizing backscattering is the total probability of reflection which is often called the *albedo*. Alternatively, the albedo may be defined as the ratio of the reflected to the incident current. The albedo of a foil increases with foil thickness until a saturation value is reached. In Table VIII some Monte Carlo

Table VIII
Dependence of Electron Backscattering on Foil Thickness[a].

Foil thickness	Ratio of foil albedo to semi-infinite-medium albedo	
z/r_0	aluminum[b]	gold[c]
0.025	...	0.16
0.05	0.07	0.43
0.10	0.23	0.83
0.15	0.43	0.97
0.20	0.71	1.00
0.25	0.89	...
0.30	0.98	...
0.35	1.00	...

[a] 0.5-Mev electrons incident perpendicularly on foil.
[b] Based on 5000 histories generated according to Model {I, PL(16, 48), EC, AGS, DL}.
[c] Based on 2000 histories generated according to Model {I, PL(32, 96), EC, AGS, DL}.

results are given which illustrate the approach to saturation backscattering for 0.5-Mev electrons incident perpendicularly on a foil. Complete saturation is reached with gold foils when $z/r_0 = 0.20$ and with aluminum foils when $z/r_0 = 0.35$. This implies that electrons are turned around sooner in gold than in aluminum, which corresponds to the fact that the mean angular deflection, over a pathlength with given fractional energy loss, increases with the atomic number of the medium. In the backscattering problems to be discussed from now on, the foil will be assumed to have saturation thickness, so that the albedo is a function of the source energy and direction of incidence only.

Figure 3 gives a comparison of the calculated and experimental values of the albedo, for perpendicular incidence on aluminum, as a function of the source energy. There is fair agreement, even through the Monte

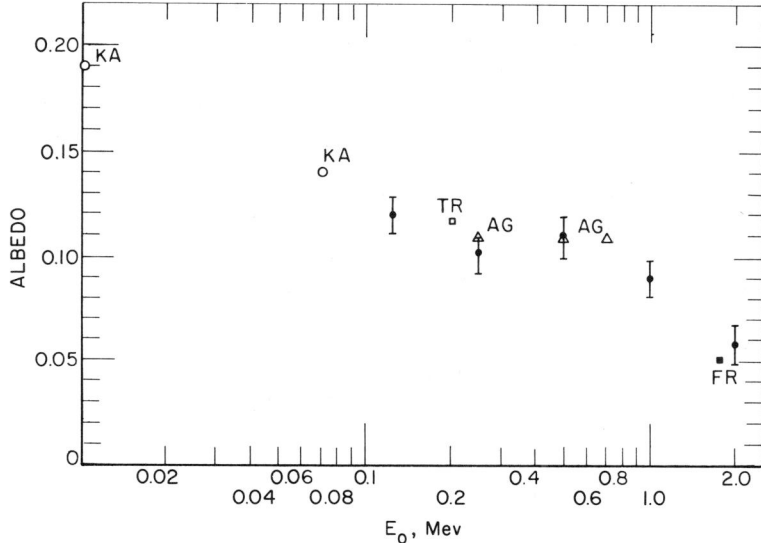

FIG. 3. Albedo as function of the source energy E_0, for electrons incident on a semi-infinite aluminum medium. Results of several experiments are given by: o KA, Kanter (1957); △ AG, Agu et al. (1958a); TR, Trump and Van de Graaf (1949); FR, Frank (1959). Monte Carlo, 1000 histories for each source energy E_0, generated according to Model {I, PL (16, 96), EC, AM, DLT}.

TABLE IX

ELECTRON ALBEDO OF SEMI INFINITE ALUMINUM MEDIUM.[a]

E_0 (Mev)	θ_0					
	0°	45°	60°	75°	90°	Isotropic
2.0	0.059	0.159	0.284	0.457	0.790	0.317
1.0	0.090	0.220	0.333	0.494	0.796	0.357
0.5	0.109	0.229	0.316	0.494	0.782	0.363
0.25	0.097	0.238	0.361	0.531	0.798	0.384
0.125	0.121	0.236	0.348	0.518	0.788	0.369

[a] Based on 1000 histories for each source energy E_0, calculated according to Model {I, PL(16, 96), EC, AM, DLT}. Standard deviation of albedo A is $\sqrt{A(1-A)/1000}$.

Carlo results are based on a relatively small sample of 1000 histories for each value of E_0, obtained by a very simple scheme of Class I. In Table IX is an extension of the calculated values to the case of oblique incidence, ranging from $\theta = 0°$ (perpendicular incidence) to $\theta = 90°$ (grazing incidence). The obliquity-dependence of the albedo is rather independent of the source energy, and is consistent with measurements by Kanter (1957) at energies between 10 and 70 kev. Table IX also gives albedo values for isotropic incidence (with $\cos \theta_0$ distributed uniformly between 0 and 1), which agree with measurements by Kanter (1957) with an Rb^{187} beta-ray source (max. energy 0.275 Mev), and with measurements by Suzor and Charpak (1952) with several different beta-ray sources (max. energies between 0.17 and 1.7 Mev), but are approximately 10% lower than results of Seliger (1952) with a P^{32} beta-ray (maximum energy 1.7 Mev).

2. Albedo Comparisons

a. Electron-positron albedo differences. To estimate such difference we have used the technique of correlated sampling. Pairs of electron and positron histories were sampled, both members of each pair being generated with the same sequence of random numbers insofar as possible.

To every electron history we assign a variable α_E which is equal to 1 if the electron is reflected and 0 otherwise, so that the electron albedo A_E is equal to the mean value $\overline{\alpha_E}$, taken over all sampled histories. A similar variable α_p is assigned to each positron history such that the positron albedo A_p is equal to $\overline{\alpha_p}$. The variances of α_E and α_p are

$$\sigma_E^2 = \overline{\alpha_E}(1 - \overline{\alpha_E}) \tag{33}$$

$$\sigma_P^2 = \overline{\alpha_P}(1 - \overline{\alpha_P}), \tag{34}$$

and their correlation coefficient (which must lie between -1 and $+1$) is

$$\rho(\alpha_E, \alpha_P) = \frac{\overline{\alpha_E \alpha_P} - \overline{\alpha_E} \cdot \overline{\alpha_P}}{\sigma_E \sigma_P}. \tag{35}$$

The quantity to be estimated is the ratio of the electron albedo to the positron albedo, A_E/A_p. The fractional standard deviation of the estimate of this ratio is

$$\delta = \frac{1}{\sqrt{N_0}} \left\{ \left(\frac{\sigma_E}{\overline{\alpha_E}} \right)^2 - 2\rho \frac{\sigma_E \sigma_P}{\overline{\alpha_E} \cdot \overline{\alpha_P}} + \left(\frac{\sigma_P}{\overline{\alpha_P}} \right)^2 \right\}, \tag{36}$$

where N_0 is the number of histories sampled. The larger the correlation coefficient, the smaller is the statistical error.

By way of example, Table X contains a classification of 10,000 pairs

TABLE X

ANALYSIS OF CORRELATED ELECTRON AND POSITRON HISTORIES.[a]

		Electrons		
		Reflected	Absorbed	Subtotal
Positrons	Reflected	891	46	937
	Absorbed	165	8898	9063
	Subtotal	1056	8944	10000
				Total

[a] Based on 10 000 electron and positron histories, generated according to Model {I, PLM (16/8, 48), EC, AGS, DL}, for a source energy of 0.5 Mev and perpendicular incidence on a semi infinite aluminum medium.

of correlated electron and positron histories, from which the following information can be obtained.

$$\overline{\alpha_E} = 0.1045;\ \overline{\alpha_P} = 0.0937;\ \overline{\alpha_E}/\overline{\alpha_P} = 1.127;$$
$$\sigma^2(\alpha_E) = 0.0944;\ \sigma^2(\alpha_P) = 0.0849; \tag{37}$$
$$\overline{\alpha_E \alpha_P} = 0.0891;\ \rho(\alpha_E, \alpha_P) = 0.885\ .$$

When these numbers are substituted into Eq. (36) one finds that the fractional standard deviation δ has the value 0.016. If the correlation coefficient were zero rather than 0.885, the value of δ would be 0.043. With the statistical error inversely proportional to $\sqrt{N_0}$, the use of correlated sampling in the example has therefore increased the effective

TABLE XI

COMPARISON OF ELECTRON AND POSITRON ALBEDO.

Medium	E_0 (Mev)	θ_0	Ratio of electron to positron albedo
Aluminum[a]	1.0	$0°$	1.08 ± 0.01
Aluminum[a]	1.0	Isotropic	1.01 ± 0.02
Aluminum[b]	0.5	$0°$	1.12 ± 0.02
Aluminum[b]	0.5	Isotropic	1.02 ± 0.01
Gold[c]	0.5	$0°$	1.20 ± 0.03
Gold[c]	0.5	Isotropic	1.09 ± 0.02

[a] 1000 histories, Model {I, PL (16, 96), EC, AGS, DLT}.
[b] 10,000 histories, $\theta_0 = 0°$ } Model {I, PLM (16/8, 48), EC, AGS, DL}.
   3,000 histories, isotropic source
[c] 5,000 histories, Model {I, PL.(32, 96), EC, AGS, DL}.

sample size by a factor $(0.043/0.016)^2 = 8.2$, without any increase in computation.

Table XI contains ratios of the electron albedo to the positron albedo, for aluminum and gold and various source energies. Electrons are backscattered more than positrons. The difference increases with the atomic number of the medium and is due mainly to the behavior of the Mott scattering cross section at large angles (see Table XIX) and of the corresponding multiple scattering angular distributions (see Tables III, XXII and XXIII).

The only experimental determination of the electron-positron backscattering difference appears to have been made by Seliger (1952) with isotropic sources (electrons from P^{32} with a maximum energy of 1.7 Mev, positrons from Na^{22} with a maximum energy of 0.58 Mev). As shown in Table IX, the albedo increases slowly with decreasing source energy. Seliger's results might thus be expected to result in an underestimate of the electron-positron albedo ratio because the average energy of his electrons was higher than that of his positrons. In fact, however, his ratios are 1.4 for aluminum and 1.3 for gold, and are much higher than the calculated ratios, particularly for aluminum. By allowing for the possibility of annihilation in flight, the calculated positron albedo would be reduced by approximately 1%, which is far too little to remove the discrepancy. It does not seem likely that further refinements of the Monte Carlo method will lead to significantly different results so that additional experiments would be desirable to clarify the situation.

An estimate of the electron-positron albedo ratio has also been made by Miller (1951) who evaluated a simple albedo theory of Bothe (1949) with the use of the Mott scattering cross section. Miller found a ratio 1.16 for 0.5-Mev electrons and positrons incident on mercury ($Z = 80$), for a source geometry intermediate between isotropic and perpendicular incidence. This is in good agreement with the results in Table XI for gold.

b. Model differences. Table XII illustrates how the calculated albedo depends on the type of multiple-scattering theory used to select angular deflections in the construction of condensed histories. The data in the table were obtained by correlated sampling, and the correlation coefficient ρ relating any two calculations was found to be of order 0.8. With a sample of 1,000 histories, the ratio of albedos for two different assumed angular distributions has a fractional standard deviation approximately equal to $0.1 \sqrt{2(1-\rho)} \sim 0.06$. Thus the albedo comparisons are not statistically conclusive, but they do show trends that conform to one's expectations. Possibly significant differences occur only for perpendicular

TABLE XII

DEPENDENCE OF CALCULATED ELECTRON ALBEDO ON MODEL OF CONDENSED RANDOM WALK[a]

Model Characteristics	Albedo	
	$\theta_0 = 0$	Isotropic source
b	0.086 ± 0.009	0.367 ± 0.015
c	0.100 ± 0.009	0.476 ± 0.015
d	0.090 ± 0.009	0.357 ± 0.015
e	0.088 ± 0.009	0.355 ± 0.015

[a] For 1-Mev electrons incident perpendicularly on a semiinfinite aluminum medium. Based on 1000 histories.
[b] Model {I, PL (16, 96), EC, AGS (Mott), DLT}.
[c] Model {I, PL (16, 96), EC, AGS (Rutherford), DLT}.
[d] Model {I, PL (16, 96), EC, AM (Molière), DLT (long. and transv.)}.
[e] Model {I, PL (16, 96), EC, AM (Molière), DL (long. only)}.

but not for isotropic incidence. This is plausible because the multiple-scattering theories differ mainly in their predictions concerning large angular deflections, and such deflections play an important role in promoting backscattering only when the direction of incidence is more or less perpendicular. The use of the Rutherford instead of the Mott cross section leads to an overestimate of backscattering that is expected from the small value of $\sigma(\text{Mott})/\sigma(\text{Ruth})$ at large angles. Finally, Table XII indicates that the neglect of the transverse multiple scattering displacements has no significant effect on the calculated value of the albedo.

Table XIII is concerned with the influence of stepsize on the calculated

TABLE XIII

DEPENDENCE OF CALCULATED ELECTRON ALBEDO ON STEP-SIZE OF CONDENSED RANDOM WALK[a]

Step-size and number of steps	Albedo
PL (4, 24)	0.083 ± 0.006
PL (8, 48)	0.092 ± 0.006
PL (16, 96)	0.108 ± 0.007
PL (32, 192)	0.119 ± 0.007
PLM (16/8, 96)	0.108 ± 0.006

[a] For 0.5-Mev electrons incident perpendicularly on a semiinfinite aluminum medium. Based on 5000 histories generated according to Model {I, PL, EC, AGS, DL}.

albedo. The latter increases slowly as the steps are made smaller and eventually appears to reach a limiting value, not far from the stepsize $k = 2^{-1/16}$ for aluminum which was used in many of the calculations. The underestimate of the albedo for large stepsize is presumably connected with the approximate treatment of spatial displacements in Monte Carlo schemes of Class I, the particles being allowed to penetrate too deep into the medium before being given a chance to turn around and reemerge. In this situation, the use of mixed logarithmic spacing is advisable, with fine steps near the boundary.

3. Angular Distribution of Backscattered Electrons

Whereas in a conventional Monte Carlo treatment, with assumed rectilinear propagation between collisions, there is no ambiguity regarding the direction with which a particle crosses a boundary of interest, this is not so for condensed histories. Suppose the crossing of the boundary $z = 0$ of a semiinfinite medium occurs in the nth step. One then merely knows the states (E_n, θ_n, z_n) and $(E_{n+1}, \theta_{n+1}, z_{n+1})$ at the beginning and end of the step, and must guess the direction θ_B at the intermediate position $z = 0$. Most authors have simply used the direction at the beginning of the step, setting $\theta_B = \theta_n$. However, when average multiple scattering deflections per step as large as 10 to 25° are allowed, a better approximation is desirable. In principle one should sample θ_B from an appropriate probability distribution, which is available only in the small-angle diffusion approximation (Rossi, 1952). We have used instead the following simple interpolation procedure:

(1) The pathlength s_B reached when the particle crosses the boundary $z = 0$ is taken to be

$$s_B = s_n + \frac{z_n}{z_n - z_{n+1}} (s_{n+1} - s_n). \tag{38}$$

This would be strictly correct if the propagation within the nth step were rectilinear.

(2) If ω is the multiple scattering deflection in the nth step, and ω_B the corresponding deflection in the part of the step in which the pathlength is increased from s_n to s_B, we let

$$\cos \omega_B = 1 - (1 - \cos \omega) + \left(\frac{s_B - s_n}{s_{n+1} - s_n}\right). \tag{39}$$

This relation, for small pathlength intervals $s_{n+1} - s_n$, would be exact if $\cos \omega$ and $\cos \omega_B$ were replaced by their mean values, and is also correct in the limiting $s_B = s_n$ and $s_B = s_{n+1}$. A small error arises from the assumption, implicit in Eq. (39), that the obliquity cosine is a monotone increasing function of the pathlength, so that the possibility of an S-shaped trajectory within the step, for example, is disregarded.

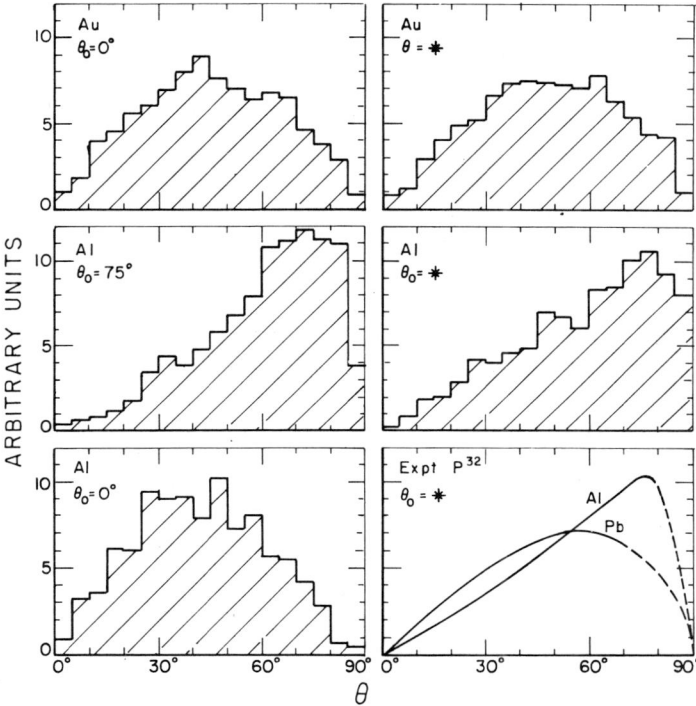

FIG. 4. Angular distribution of electrons reflected from a semiinfinite medium. The histograms represent Monte Carlo results for an incident energy of 0.5 Mev, obtained under the following conditions:

Medium	Direction of incidence	Number of histories	Monte Carlo model
Aluminum	0°	10,000	I, PLM (16/8, 48), EC, AGS, DL
Aluminum	75°	3,000	I, PLM (16/8, 48), EC, AGS, DL
Aluminum	Isotropic	3,000	I, PLM (16/8, 48), EC, AGS, DL
Gold	0°	5,000	I, PL (32, 96), EC, AGS, DL
Gold	Isotropic	5,000	I, PL (32, 96), EC, AGS, DL

The curves in the lower right-hand panel are from an experiment of Buys (1960) with an isotropic P^{32} beta ray source (maximum energy 1.7 Mev). Asterisk (*) indicates isotropic source.

(3) The deflection ω_B is inserted in Eq. (9) in place of ω, to obtain the direction of emergence θ_B.

Figure 4 shows angular histograms for electrons reflected from semiinfinite aluminum and gold media. To be consistent with our convention for describing histories, the angle θ should be taken between the direction of motion and the positive z-axis, and should thus lie between 90° and 180°. In fig. 4, we have instead taken it with respect to the negative z-axis, so that the angles of reflection range from $\theta = 0°$ (perpendicular to boundary) to $\theta = 90°$ (grazing). For perpendicular incidence, the angular distributions are diffuse and can be approximated by the function $\cos \theta \sin \theta \, d\theta$, i.e., by a cosine law, indicating that the electrons are reflected after so many collisions that they have "forgotten" their initial direction (a cosine-law for the emerging current corresponds to an isotropic flux near the boundary).

For isotropic incidence, electrons reflected from gold have a distribution not very different from a cosine-law but shifted slightly toward larger angles, whereas electrons reflected from aluminum emerge preferentially at very oblique angles. This can be understood qualitatively by a simple picture in which the reflected current is considered to consist of two components: the first the result of electrons incident more or less perpendicularly and reflected "diffusely," and the second the result of electrons incident obliquely which in turn are likely to emerge obliquely after few steps of the condensed random walk. For aluminum, the probability of diffuse backscattering is relatively small so that the obliquely reflected component dominates the angular distribution. This is not the case for gold, with its higher atomic number, in which diffuse backscattering has a much higher probability. Essentially the same argument has already been put forward by Seliger (1952) in the analysis of his experimental results, and is here confirmed by the Monte Carlo results. Further confirmation is provided by the experimental angular distributions, also shown in Fig. 4, obtained by Buys (1960) with isotropic beta-ray sources.

4. *Energy Spectrum of Reflected Electrons*

Figures 5 and 6 show energy spectra of electrons reflected from a semiinfinite aluminum medium. They were obtained by a Monte Carlo scheme of Class I (assuming the continuous-slowing-down approximation) and a subsequent correction to take into account energy loss fluctuations. The histograms labeled "b" and "a" represent the spectra obtained before and after application of this correction. The interval spacing of the histograms is logarithmic on an energy scale, and uniform in respect to the "lethargy" variable

$$u = m \log_2 (E_0/E) \tag{40}$$

that can also be interpreted as the number of steps (with energy reduction factor $k = 2^{-1/m}$) that are taken before the electron emerges from the medium. As expected, more steps are required on the average when the direction of incidence is perpendicular than when it is isotropic. In the latter case, one can clearly distinguish two components of the spectrum consisting of electrons emerging either after many or after few steps. This is the analog of the division of the angular distribution into a diffuse and oblique component discussed in the preceding subsection.

In the continuous-slowing-down approximation one obtains essentially a pathlength distribution $L(s)$ of the reflected electrons. In lowest approximation, the corresponding energy spectrum is

$$Y^{(0)}(E) = L[s(E)] \left| \frac{dE}{ds} \right|^{-1}. \tag{41}$$

Energy loss fluctuations can be treated, in the next approximation, by folding the pathlength distribution into an energy loss distribution,

$$Y^{(1)}(E) = \int_0^{r_0} L(s) \, W_I(E_0 - E; E_0, s) \, ds \tag{42}$$

where $W_I(\varDelta E, E_0, s) \, d(\varDelta E)$ is the probability that an electron with initial energy E_0 will have an energy loss between $\varDelta$ and $\varDelta E + d(\varDelta E)$ after traveling a pathlength s. Two small errors are made in this approximation. One disregards the correlation of the energy loss distribution with the pathlength distribution $L(s)$, and one evaluates W_I as if the medium were unbounded. Fully worked-out theoretical energy loss distributions are available only for the case $\varDelta E \ll E_0$. They can be extended, however, to the case of large E, which is important in the present application, through a Monte Carlo calculation according to a scheme of Class I, the condensed history specification being limited to the pathlength and energy variables. The energy loss in each step is sampled from the appropriate theoretical distribution, and the corresponding loss for long pathlengths is obtained by summation. By sampling in this manner one derives a correction matrix which is applied to the spectral histogram and converts it into a corrected histogram that takes into account energy loss fluctuations. Such a correction matrix depends only on the nature of the medium and on the pathlength intervals but not on the boundary conditions, and can be used for the correction of diverse spectra.

In a quick survey of the energy-loss fluctuation effect, we have evaluated the correction matrix with the use of the Landau energy loss distribution [Eq. (A17) of the appendix] instead of the more correct and complicated distribution of Blunck and Leisegang (1950). Figures 5

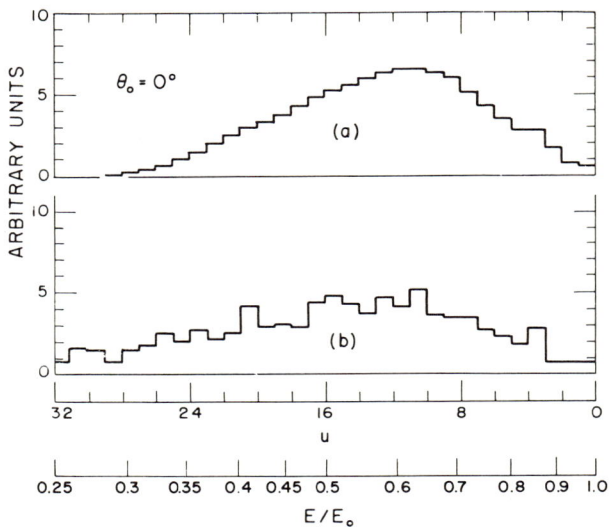

FIG. 5. Energy spectrum of electrons reflected from a semiinfinite aluminum medium. Source energy 0.5 Mev, perpendicular incidence. Histogram (b) represents results from 10,000 histories generated according to Model {I, PLM (16/8, 48), EC, AGS, DL}. Histogram (a) is the modified energy spectrum obtained from (b) by an approximate correction for energy loss fluctuations. Lethargy variable $u = 16 \log_2 (E_0/E)$.

TABLE XIV

TRANSMISSION OF ELECTRONS THROUGH ALUMINUM FOILS.[a]

a. Perpendicular incidence.

E_0, Mev	z/r_0			
	0.2	0.3	0.4	0.5
2.0	0.962 ± 0.006	0.847 ± 0.011	0.689 ± 0.015	0.469 ± 0.016
1.0	0.925 ± 0.008	0.785 ± 0.013	0.608 ± 0.015	0.373 ± 0.015
0.5	0.890 ± 0.010	0.751 ± 0.013	0.559 ± 0.016	0.361 ± 0.015
0.25	0.894 ± 0.010	0.738 ± 0.014	0.538 ± 0.016	0.306 ± 0.015
0.125	0.880 ± 0.010	0.731 ± 0.014	0.531 ± 0.016	0.321 ± 0.015

[a] Based on 1000 Monte Carlo histories for each source energy E_0, generated according to Model {I, PL (16, 96), EC, AM, DLT}.

b. Isotropic incidence.

E_0, Mev	z/r_0			
	0.2	0.3	0.4	0.5
2.0	0.635 ± 0.015	0.497 ± 0.016	0.339 ± 0.015	0.182 ± 0.012
1.0	0.585 ± 0.016	0.439 ± 0.016	0.298 ± 0.015	0.174 ± 0.012
0.5	0.590 ± 0.016	0.453 ± 0.016	0.306 ± 0.015	0.146 ± 0.011
0.25	0.577 ± 0.016	0.423 ± 0.016	0.268 ± 0.014	0.129 ± 0.011
0.125	0.558 ± 0.016	0.420 ± 0.016	0.264 ± 0.014	0.123 ± 0.011

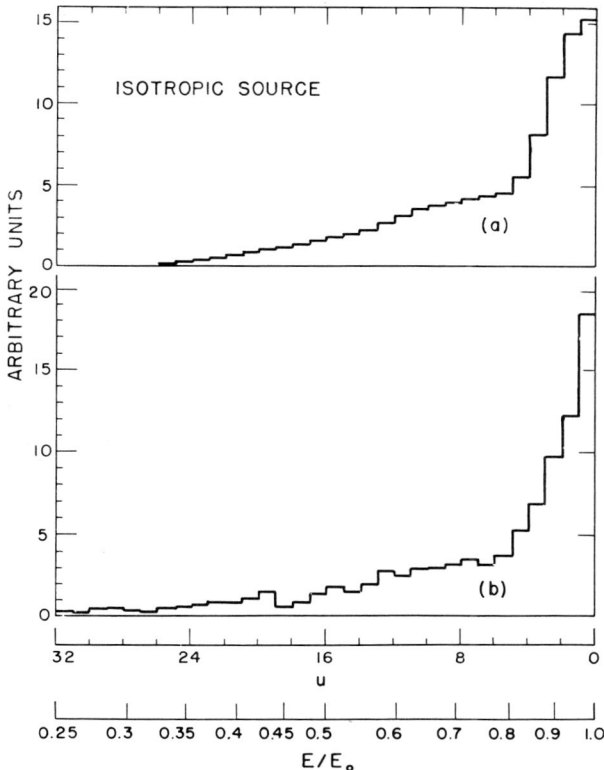

FIG. 6. Energy spectrum of electrons reflected from a semiinfinite aluminum medium. Source energy 0.5 Mev, isotropic source. Based on 3000 histories generated and utilized as described in caption of Fig. 5.

and 6 show that the correction is small for reflected electrons and results in a slight shift of the spectral distribution toward higher energies.

B. Transmission and Penetration of Electrons and Positrons

1. Transmission Coefficients

Table XIV contains Monte Carlo results on the transmission of electrons through aluminum foils, for various source energies and geometries. The outstanding feature is that the transmission coefficients, when expressed as functions of z/r_0, are insensitive to the value of the source energy (below 1 Mev), which is a confirmation of theoretical expectations.

In Fig. 7, Monte Carlo results are compared with recent precise experimental determinations of the transmission of electrons through aluminum and gold foils by Agu *et al.* (1958b). For small and intermediate foil thicknesses there is good agreement. For large thicknesses,

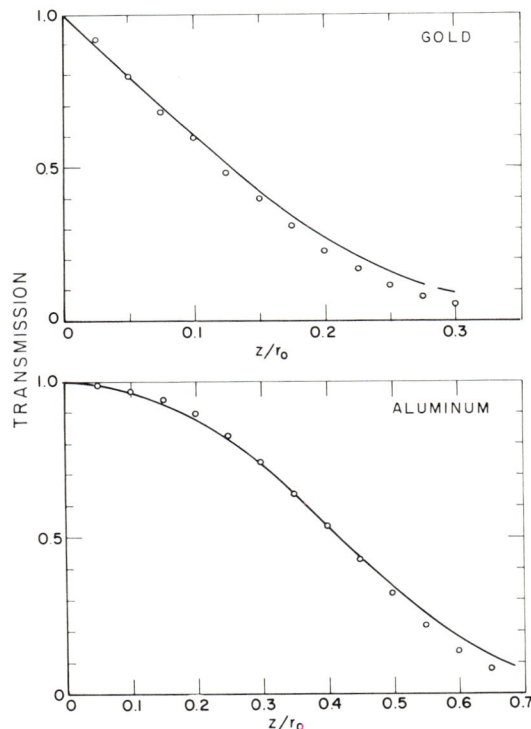

Fig. 7. Comparison of calculated and measured electron transmission through foils. The curves represent experimental results of Agu *et al.* (1958b) averaged for source energies between 0.25 and 0.75 Mev. The circles (o) represent Monte Carlo calculations for a source energy of 0.5 Mev, based on 5000 histories in aluminum, Model {I, PL(16.48), EC, AGS, DL}, and 2000 histories in gold, Model {I, PL (32,96), EC, AGS, DL}.

when the transmission becomes smaller than 30%, the experimental curves lie somewhat above the Monte Carlo results, which is due to the use of the continuous-slowing-down approximation. The disregard of energy-loss fluctuations of course greatly simplifies the computations, and Fig. 7 gives an indication of the rather wide limits within which this simplification will give reasonable results.

Just as positrons are backscattered less than electrons, their transmission is greater. Typical positron-electron transmission ratios are shown in Fig. 8. For gold they are in good agreement with experimental

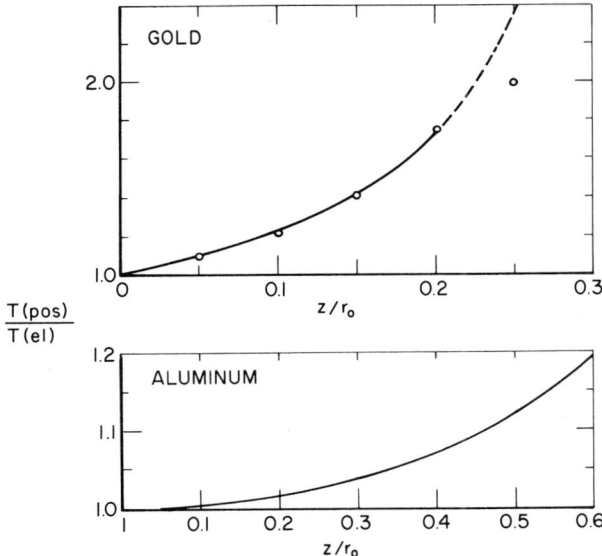

FIG. 8. Ratio of positron transmission to electron transmission through foils, for perpendicular incidence and a source energy of 0.5 Mev. Based on electron and positron histories generated as indicated in caption of Fig. 7. Circles (o) represent experimental results of Seliger (1955) with a 0.390-Mev source (r_0 = 0.245 gm/cm² gold for electrons, 0.240 gm/cm² gold for positrons).

results of Seliger (1955). Seliger's ratios for aluminum are greater than unity at 0.960 Mev but smaller than unity at energies below 0.336 Mev, so that a comparison with the calculated results at 0.5 Mev cannot readily be made.

The calculated ratios in Fig. 8 are actually somewhat too high because the positron transmission should be slightly reduced to take into account the possibility of annihilation in flight. From the theory of Bethe (1935) it can be estimated that the positron transmission for gold, for the conditions of Fig. 8, should be reduced by approximately one per cent.

TABLE XV

DEPENDENCE OF CALCULATED ELECTRON TRANSMISSION ON MODEL OF CONDENSED RANDOM WALK.[a]

z/r_0	b	c	d	e
0.1	0.985 ± 0.004	0.977 ± 0.005	0.983 ± 0.004	0.984 ± 0.004
0.2	0.920 ± 0.027	0.914 ± 0.028	0.925 ± 0.026	0.919 ± 0.027
0.3	0.795 ± 0.040	0.780 ± 0.042	0.785 ± 0.041	0.787 ± 0.041
0.4	0.609 ± 0.049	0.584 ± 0.049	0.608 ± 0.049	0.609 ± 0.049
0.5	0.389 ± 0.049	0.366 ± 0.048	0.373 ± 0.048	0.376 ± 0.049

[a] For 1-Mev electrons incident perpendicularly on an aluminum foil. Based on 1000 histories.
[b] Model {I, PL(16,96), EC, AGS(Mott), DLT}.
[c] Model {I, PL(16, 96), EC, AGS (Rutherford), DLT}.
[d] Model {I, PL(16, 96), EC, AM (Molière), DLT (long. and transv.)}.
[e] Model {I, PL(16,96), EC, AM (Molière), DL (long. only)}.

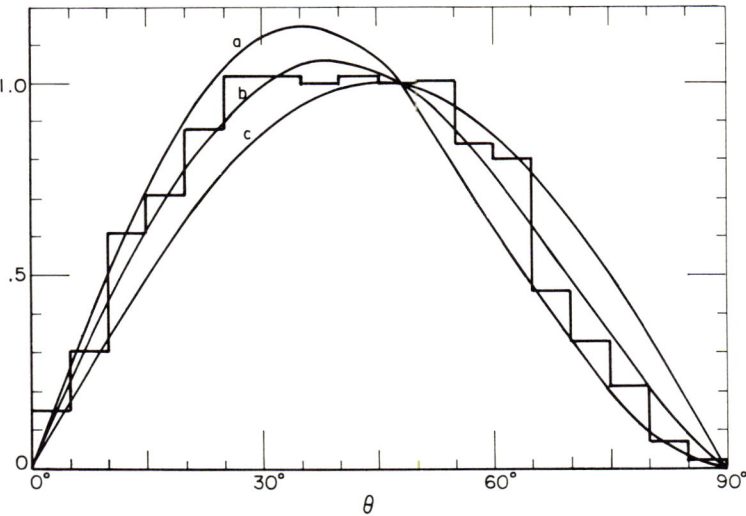

FIG. 9. Angular distribution of electrons transmitted through an aluminum foil 0.113 gm/cm² thick. Perpendicular incidence source energy 0.5 Mev. The histogram was obtained from the analysis of 4000 histories generated according to Model {I, PLM(16/8, 48), EC, AGS, DL}. Curves (a), (b) and (c) represent the functions $3\cos^2\theta\sin\theta$, $1.445 (0.717\cos\theta + \cos^2\theta)\sin\theta$, and $2\cos\theta\sin\theta$, respectively.

Table XV illustrates the dependence of the calculated transmission coefficients on the multiple-scattering angular distribution used in the Monte Carlo scheme. The differences are much less than in the comparison of backscattering in Table XII, and are hardly significant statistically. Again, the neglect of transverse spatial multiple-scattering deflections appears to be of no consequence.

2. *Angular Distribution and Energy Spectrum of Transmitted Electrons*

The direction of transmitted electrons as they cross the foil boundary was determined by the same kind of interpolation procedure used in the backscattering problem. Figure 9 shows a typical angular distribution, for 0.5-Mev electrons incident on a foil of thickness $z/r_0 = 0.5$. The distribution appears to lie between a cosine-law and a cosine-

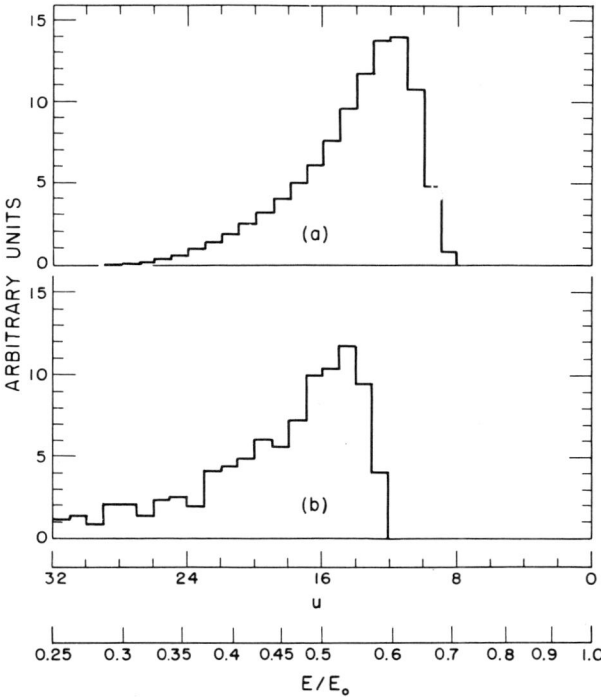

Fig. 10. Energy spectrum of electrons transmitted through an aluminum foil 0.113 gm/cm² thick. Perpendicular incidence, source energy 0.5 Mev. Histogram (*b*) is based on 4000 histories generated according to Model {I, PLM (16/8, 48), EC, AGS, DL}. Histogram (*a*) is the modified energy spectrum obtained from (*b*) through an approximate correction for energy loss fluctuations. Lethargy variable $u = 16 \log_2(E_0/E)$.

square law, and is actually described fairly well by the function $(0.717 \cos \theta + \cos^2 \theta) \sin \theta$, which is the leading term of a result derived by Bethe *et al.* (1938) in a treatment disregarding energy loss.

Figure 10 shows an energy spectrum of transmitted electrons, for the same conditions as in Fig. 9. Two histograms are shown, one obtained in the continuous-slowing-down approximation and the other corrected for energy loss fluctuations. The correction is similar to that made in the backscattering problem, but the shift of the corrected spectrum toward higher energies is much more pronounced.

Hebbard and Wilson (1955) have calculated the transmission of electrons through aluminum and gold foils of moderate thickness, taking energy loss fluctuations into account from the beginning rather than by a subsequent correction. Figure 11 shows their results for the

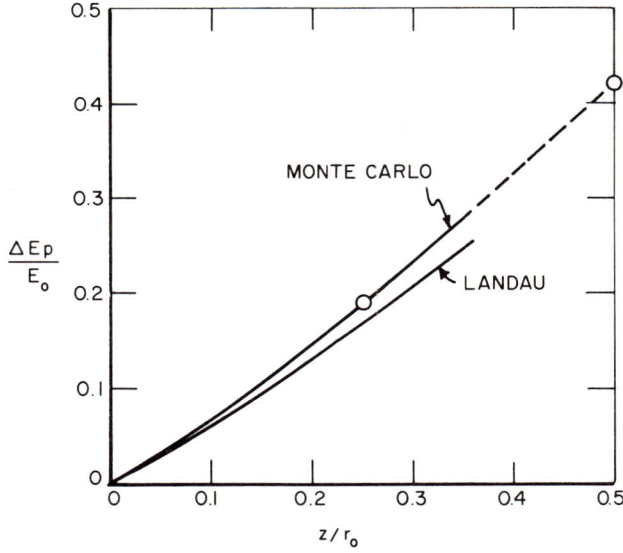

FIG. 11. Ratio of most probable energy loss ΔE_p, to the incident energy E_0, for 1-Mev electrons traversing an aluminum foil (Hebbard and Wilson, 1955). The Hebbard-Wilson curve labeled "Monte Carlo" is derived from 1000 histories generated according to Model {I, PU(Δs = 5.26 mg/cm^2), EI, AM, DL}. The circles (o) represent Monte Carlo results derived from a calculation similar to that described in Fig. 10.

most probable energy loss as a function of the foil thickness. Also shown are two Monte Carlo results derived from the data in Fig. 10 and similar data, which are in good agreement with the Hebbard-Wilson curve or its extrapolation, which indicates that our correction procedure has some validity. The deviation of the Hebbard-Wilson curve from the corresponding curve predicted by the energy loss theory of Landau exhibits the effect of multiple scattering detours.

3. *Emergence of Electrons from the Interior of a Medium*

We consider a plane isotropic source of electrons at $z = 0$, sandwiched by two foils of thickness z, and want to know the electron current that emerges through the exterior surfaces of the two foils. Such problems arise, for example, when foils are bombarded with neutrons and the induced radioactivity of the foils is measured by counting the number of beta particles emitted. We shall compare Monte Carlo results with a calculation by Meister (1958) based on the diffusion theory of Bethe et al. (1938). He assumed that after leaving the source the electrons travel a distance of s' in a straight line until their energy has fallen to some intermediate value E', and that thereafter their propagation can be described by a diffusion equation. The energy E' is so adjusted that the mean square displacement in the z-direction, as function of the electron energy, agrees as closely as possible with the result of an exact transport calculation.

In the age-diffusion approximation for electrons, continuous-slowing-down is assumed, and the flux $F_0(z, s)$, integrated over all electron directions, is calculated by solving the diffusion equation

$$\frac{\partial F_0}{\partial \tau} = \frac{\partial^2 F_0}{\partial z^2} \tag{43}$$

with the appropriate boundary conditions. The "age"

$$\tau = \int_{s'}^{s} \frac{ds''}{3G_1(s'')} = \int_{E}^{E'} \frac{dE''}{3G_1[s(E'')]} \left| \frac{dE}{ds} \right|^{-1} \tag{44}$$

is related to the scattering cross section through the function $G_1(s)$ previously defined in Eq. (16). The reciprocal of G_1 is called the transport mean free path. The associated current $F_1(z, s)$ is obtained from the flux by differentiation,

$$F_1(z, s) = -\frac{1}{3G_1(s)} \frac{\partial F_0}{\partial z}. \tag{45}$$

In order for the age-diffusion approximation to be applicable, the following condition, among others, must be satisfied,

$$\frac{d}{ds}[1/G_1(s)] \ll 1. \tag{46}$$

In the problem of a plane isotropic source sandwiched by two foils, the boundary conditions are that no electron current is incident on the foils from the outside. The current emerging from the foils, when normalized to unit source strength, can be interpreted as a transmission coefficient. Meister investigated this transmission coefficient by calculation and experimentally for a source emitting 0.312-Mev electrons (conversion electrons from an isomer of In^{115} produced by neutron irradiation). He found good agreement for gold and indium foils, but a discrepancy for aluminum, which he ascribed to the fact that Eq. (46) is satisfied only for media with atomic number $Z > 30$.

Monte Carlo calculations for aluminum are compared in Fig. 12 with

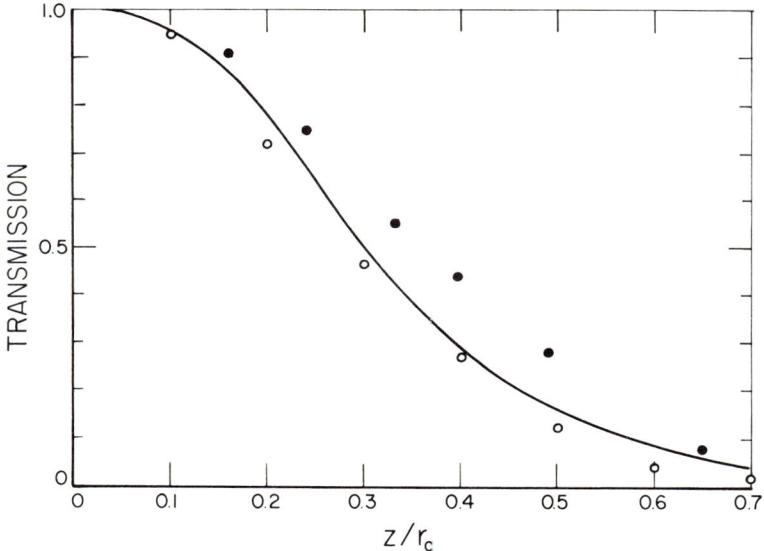

FIG. 12. Emergence of electrons from the interior of an aluminum medium. A plane isotropic source, emitting 0.312 Mev electrons (with mean range $r_0 = 0.114$ gm/cm²) is assumed to be sandwiched between two layers of aluminum of thickness z. The curve, derived from an age-diffusion calculation, and the solid points (●) (experimental) are from a paper by Meister (1958). The open circles (o) represent Monte Carlo results based on 4000 histories generated according to Model {I, PL(16,96), EC, AGS, DL}.

Meister's results. Surprisingly, they are in fair agreement with his age-diffusion calculation, but deviate even more than the latter from the experimental points. The fractional deviation increases with increasing foil thickness, except for the last experimental point at $z/r_0 = 0.65$. The possibility suggests itself that the deviation may be because of the neglect of energy loss fluctuations in both calculations. However, it is difficult to understand why the effect of these fluctuations is much greater in the present problem than for the transmission data described in Fig. 7.

4. Influence of Bremsstrahlung on the Penetration of Very Fast Electrons

The penetration of charged particles at high energies, where bremsstrahlung plays a paramount role, was first calculated by the Monte Carlo method by Wilson (1950, 1951). His work was concerned with the development of electron- and photon-initiated showers in lead at energies up to 300 Mev, and with the range and straggling of electrons in lead at energies up to 1000 Mev. In first approximation, he considered only energy losses by bremsstrahlung and ionization, and assumed the propagation of the radiation to be rectilinear. In a second approximation, he estimated the shortening of the electron mean range due to multiple-scattering detours. The calculations were done by hand, the

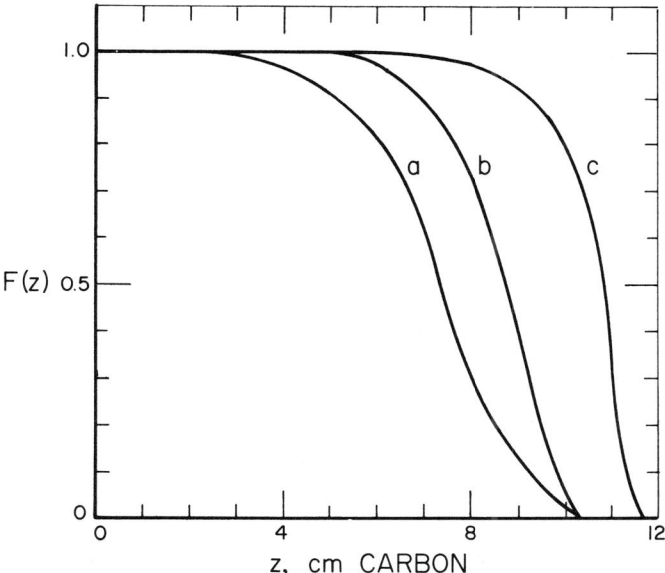

Fig. 13. Integral range-straggling distribution for 30-Mev electrons in carbon (Leiss et al., 1957). $F(z)$ is the fraction of the incident electrons that penetrate at least to depth z. Curve (c) takes into account only fluctuations of ionization losses, curve (b) in addition includes the effect of multiple scattering detours, and the final result, curve (a), also include effects of bremsstrahlung losses. Curve (a) was obtained from the analysis of 1000 histories generated according to Model {I, PU(1/4 cm carbon), EIB, AG, DL}.

random sampling being done with the use of a physical device (wheel of chance) rather than with the use of random numbers. The method involved the grouping of collisions, the step size of the condensed random walk being 0.2 of a radiation length (1.2 g/cm² of lead).

More elaborate calculations with a computer were done by Leiss *et al.* (1957) who determined the integral range straggling distribution $F(z)$, i.e., the fraction of the incident electron beam that penetrates at least to depth z, for a carbon medium and energies up to 55 Mev. They used a Class-I scheme with a uniform step size of 1/4 cm, compounding the energy loss per step from an ionization loss, sampled from the Landau distribution, and a bremsstrahlung loss, sampled from a distribution derived by Eyges (1949). The latter, for the conditions of the calculation of Leiss *et al.*, takes the form

$$W_B(\epsilon_B) \, d\epsilon_B = 0.01278(1 - \epsilon_B)^{1/4}[\log (1 - \epsilon_B)^{-1}]^{-0.9874} \, d\epsilon_B , \qquad (47)$$

where ϵ_B is the fractional energy loss as a result of bremsstrahlung. In a history generated in this fashion, very large energy losses are mainly due to bremsstrahlung. For example, the probability that a 20-Mev electron will, in a pathlength of 1/4 cm of carbon, lose and energy greater than 3 Mev is about 0.2% for ionization but 1.7% for bremsstrahlung, whereas the probability of a loss greater than 0.5 Mev is practically 100% for ionization, but only 3.6% for bremsstrahlung.

Figure 13 shows the integral range straggling distribution of Leiss *et al.* for 30-Mev electrons. Auxiliary curves are also shown which indicate the relative contributions to the shape of the distribution due to ionization loss fluctuations, bremsstrahlung loss fluctuations, and multiple-scattering detours. The corresponding mean range, in the continuous-slowing-down approximation, is 10 cm of carbon.

C. Energy Dissipation by Electrons in Bounded and Unbounded Media

In radiological applications it is of considerable interest to know the spatial pattern of energy dissipation by diffusing electrons. We have obtained some Monte Carlo results for plane sources in infinite and semiinfinite media, in the continuous-slowing-down approximation. The entire electron energy must be dissipated at points no farther from the source plane than the initial electron range r_0. Accordingly, the region $-r_0 \leqslant z \leqslant r_0$ was divided into forty equal subregions. When a step of a condensed history fell entirely within one of these subregions, the energy $E_n - E_{n+1}$ was assumed to be dissipated there. When the step straddled the boundary of two subregions, this energy was assumed to be deposited in both, in proportion to the pathlength in each. The step sizes were small enough so that only one boundary between subregions could be straddled at one time, and even this happened rarely. The same set of histories was used for the evaluation of energy deposition in infinite and semiinfinite media, which constituted another application of the technique of correlated sampling.

A very accurate calculation of energy deposition in infinite media has been carried out by Spencer (1955, 1959) who computed the spatial moments of the distribution function from a transport equation that was exact except for the use of the continuous-slowing-down approximation, and constructed the distribution from a knowledge of its moments and its asymptotic behavior for deep penetration. Figure 14, pertaining to a plane perpendicular source, shows a Monte Carlo histogram that is in good agreement with Spencer's result. Figure 15 contains a similar

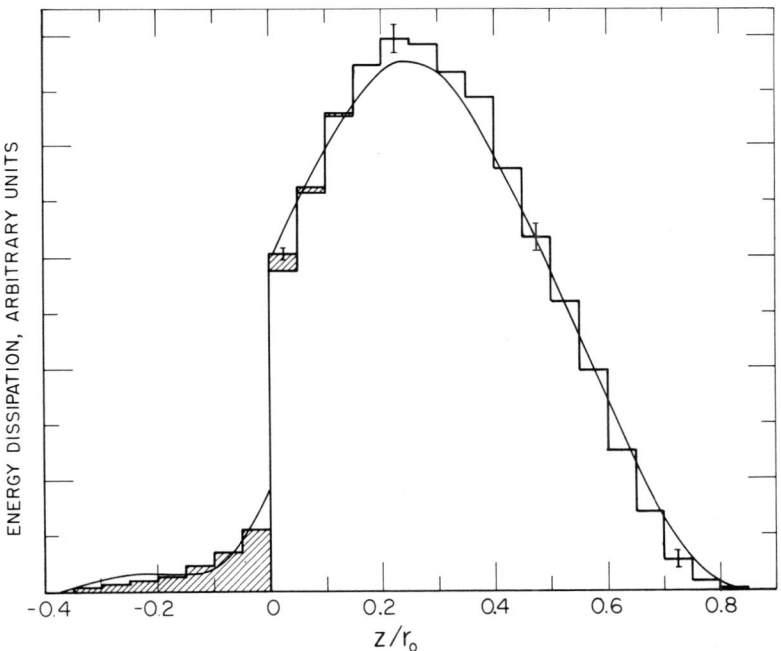

FIG. 14. Energy dissipation by electrons in aluminum. Electrons are assumed to originate from a 1-Mev plane perpendicular source at $z = 0$. The curve represents a calculation, by the moment-method, of Spencer (1959). The histogram is derived from 5000 histories generated according to Model {I, PL(16,48), EC, AGS, DL}. The shaded portions of the histogram indicate the reduction of energy dissipation, resulting from the escape of electrons, that would occur if the medium were semiinfinite ($0 < z < \infty$) rather than infinite.

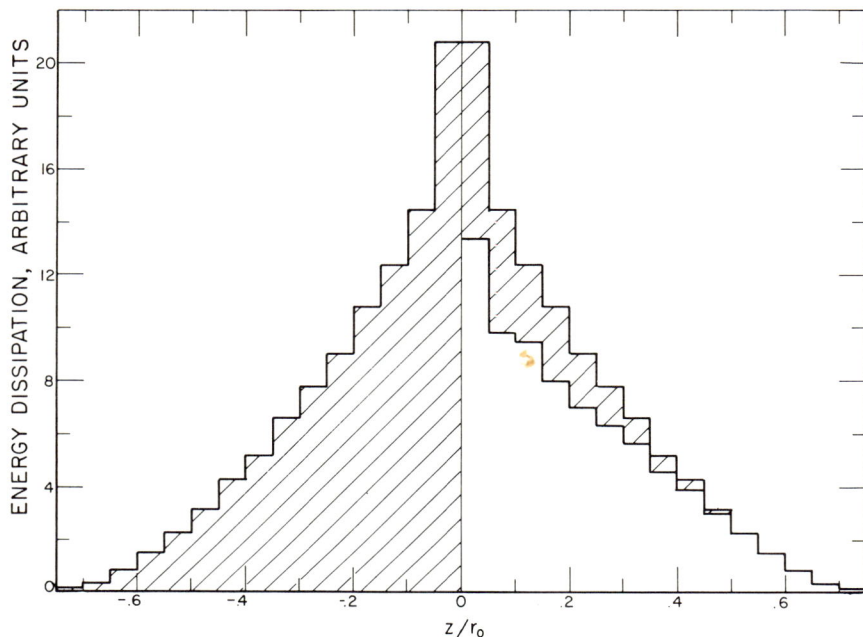

Fig. 15. Energy dissipation by electrons in aluminum. Based on 3000 histories generated as described in caption of Fig. 14, but for plane isotropic source at $z = 0$. Again, shaded portions of histogram indicate the reduction of energy dissipation due to electron escape from a semiinfinite medium.

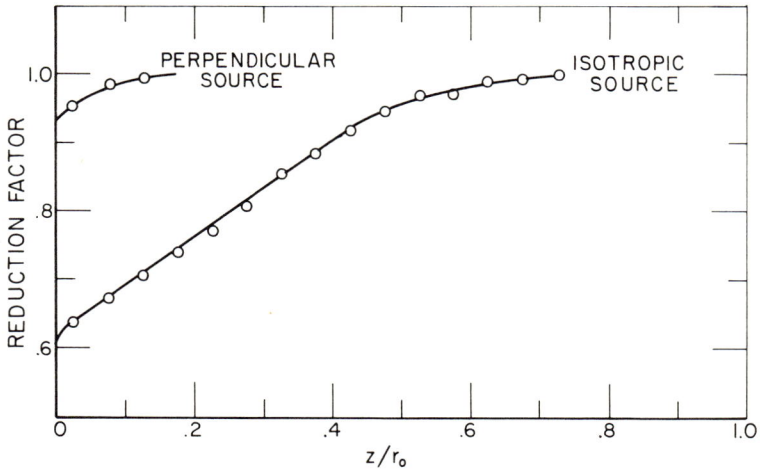

Fig. 16. Reduction of energy dissipation near the boundary of a semiinfinite medium. Based on Monte Carlo results for a 1-Mev source in aluminum shown in Figs. 14 and 15.

Monte Carlo histogram for a plane isotropic source. In both figures the histograms represent energy dissipation in an infinite medium, and their shaded portions indicate the reduction of energy dissipation that would occur if the medium were semiinfinite, because of the leakage of electrons through the boundary $z = 0$. The determination of such a boundary effect is a very trivial task for the Monte Carlo method but cannot be done by the moment-method. In Fig. 16, the boundary effect is plotted as a function of the distance from the boundary. The reduction of energy dissipation is seen to be small for the perpendicular source, whereas it is considerable and extends to great distances for the isotropic source. The latter situation is realized when a medium is covered with a thin layer of radioactive material emitting beta particles.

D. Slowing-Down Spectrum and Pathlength Straggling of Electrons

The problems discussed in this section, involving the energy and pathlength variables only, were first treated by Schneider and Cormack (1959) by a Monte Carlo scheme of Class II. Their computations were done by hand and were thus limited to a few hundred histories. The numerical results presented here are based on similar but more extensive machine calculations.

The first problem concerns the *slowing-down spectrum* which may be defined as the energy spectrum that is established as the result of multiple Coulomb scattering in a medium in which an electron source is distributed uniformly.

If the energy loss of the diffusing electrons were continuous, the slowing-down spectrum would be inversely proportional to the electron stopping power, which can be seen by the following argument. To each electron we may assign an age t, which is equal to the time it has spent in the medium since leaving the source. Under steady-state conditions, the number of electrons with ages between t and $t + dt$ is proportional to dt, but does not depend on t itself. The number of electrons with pathlengths between s and $s + ds$ is proportional to $dt/ds\, ds = ds/v$, and the flux of electrons with energies between E and $E + dE$ is

$$F(E)\, dE = Sv\left(\frac{ds}{v}\right) = S\frac{dE}{\left|\dfrac{dE}{ds}\right|}, \qquad (48)$$

where S is the source strength.

We recall that in Monte Carlo schemes of Class II the cumulative effect of collisions with small fractional energy losses $\epsilon < \epsilon_c$ is assumed to result in a continuous slowing down of the electrons, which is occasionally interrupted by catastrophic collisions with $\epsilon > \epsilon_c$ as the result of which the electron may jump across certain energy intervals (see Fig. 2). The expression for the slowing-down spectrum, in this schematization, must be modified to read

$$F(E)\, dE = SP(E, \epsilon_c) \frac{dE}{\left|\dfrac{dE}{ds}\right|_{\epsilon_c}}, \qquad (49)$$

where $P(E, \epsilon_c)$ is the probability that an electron will have an energy E at some stage of its history, and where $|dE/ds|_{\epsilon_c}$ is the stopping power evaluated with a cutoff ϵ_c. Both $P(E, \epsilon_c)$ and $|dE/ds|_{\epsilon_c}$ decreases as ϵ_c decreases. The applicability of Eq. (49) depends on the fact that their ratio tends toward a limiting value for small ϵ_c for which the Class-II schematization is still valid. (If ϵ_c were allowed to go to zero, there would be no more continuous-slowing-down.) $P(E, \epsilon_c)$ can easily be determined through inspection of a set of sampled Class-II histories.

TABLE XVI

SLOWING-DOWN SPECTRUM OF ELECTRONS IN ALUMINUM[a].

E (Mev)	$1 \left/ \left|\dfrac{dE}{ds}\right|\right.$ [b]	Spencer-Fano[c] theory	Monte Carlo[d]		
			$\epsilon_c = 0.1$	$\epsilon_c = 0.03$	$\epsilon_c = 0.01$
0.6450	0.647	(0.7025)	(0.710)	(0.748)	(0.750)
0.6400	0.646	(0.698)	0.703	0.737	0.740
0.6200	0.644	(0.681)	0.689	0.700	0.700
0.6000	0.640	(0.668)	0.678	0.674	0.690
0.5827	0.637	0.6581	0.667	0.661	0.657
0.5500	0.631	(0.641)	0.652	0.647	0.642
0.5191	0.625	0.6284	0.637	0.637	0.631
0.4625	0.610	0.6055	0.609	0.607	0.610
0.3270	0.555	0.5450	0.552	0.553	0.557
0.2312	0.487	0.4840	0.492	0.488	0.475
0.1635	0.414	0.4237	0.424	0.426	0.441
0.1156	0.342	0.3687	0.364	0.374	0.371
0.08176	0.275	0.3233	0.322	0.317	0.323
0.05779	0.219	0.2893	0.290	0.284	0.280

[a] The spectrum is given as differential track length and has dimensions g cm^{-2} Mev^{-1}. Source energy 0.6450 Mev. Values in parentheses are interpolated or extrapolated.
[b] Reciprocal of mean energy loss by ionization (evaluated with $\epsilon_c = 0.5$).
[c] As evaluated by McGinnies (1959).
[d] Based on 1000 Model-II histories for each value of ϵ_c.

The slowing-down spectrum has been studied by Spencer and Fano (1954) who developed a method for the accurate numerical integration of the appropriate transport equation. Table XVI gives a comparison of Monte Carlo results with a spectrum obtained by McGinnies (1959) according to the Spencer-Fano method. Both methods of computation give the same spectrum, to within the accuracy of the Monte Carlo results, and the latter are practically independent of the value of the cutoff parameter in the range $(0.01 \leqslant \epsilon_c \leqslant 0.1)$, at all spectral energies except very close to the source energy. Actually the spectrum diverges at the source energy and rises very steeply in its vicinity. The values in the top line of Table XVI are extrapolations indicative of the spectrum near the source energy. Energy loss straggling is seen to increase the slowing-down spectrum at high spectral energies compared to the value in the continuous-slowing-down approximation (inverse of the stopping power). This is similar to the shift of the spectra of reflected and transmitted electrons in Figs. 5, 6, and 10 toward higher energies. At low energies, the correct slowing-down spectrum again exceeds the reciprocal of the stopping power, because of the appearance of secondary knock-on electrons. The ratio of the total to the primary flux is given in Table XVII. Again, the Monte Carlo results are insensitive to the value of ϵ_c and in agreement with the predictions of the Spencer-Fano theory. Finally, Table XVIII shows the frequency distribution of the number of secondary knock-on electrons.

Within the schematization of Class II, the effect of energy loss fluctuations on the distance traveled by electrons in the course of slowing down is twofold. On the one hand, the pathlength is increased compared to its value in the continuous-slowing-down approximation,

TABLE XVII

RATIO OF TOTAL TO PRIMARY ELECTRON FLUX IN ALUMINUM.[a]

E (Mev)	Spencer-Fano[b] theory	Monte Carlo[c]		
		$\epsilon_c = 0.1$	$\epsilon_c = 0.03$	$\epsilon_c = 0.01$
0.2312	1.001	1.01	1.01	1.01
0.1635	1.031	1.03	1.04	1.05
0.1156	1.086	1.09	1.09	1.13
0.08176	1.180	1.20	1.18	1.21
0.05779	1.332	1.34	1.31	1.32

[a] Source energy 0.6450 Mev.
[b] Calculated by McGinnies (1959).
[c] Based on 1000 model-II histories for each value of ϵ_c.

TABLE XVIII

PRODUCTION OF SECONDARY ELECTRONS[a]

No. of secondary electrons	Monte Carlo[b]		
	$\epsilon_c = 0.1$	$\epsilon_c = 0.03$	$\epsilon_c = 0.01$
0	713	771	695
1	255	263	280
2	28	25	24
3	4	1	1

[a] Frequency with which a primary 0.6450-Mev electron gives rise to a cascade of 0, 1, 2, 3, ... secondary electrons with energies between 0.3225 Mev (the highest possible secondary energy) and 0.0578 Mev, in an aluminum medium.

[b] Based on the analysis of Model-II histories of 1000 primary electrons and their secondaries.

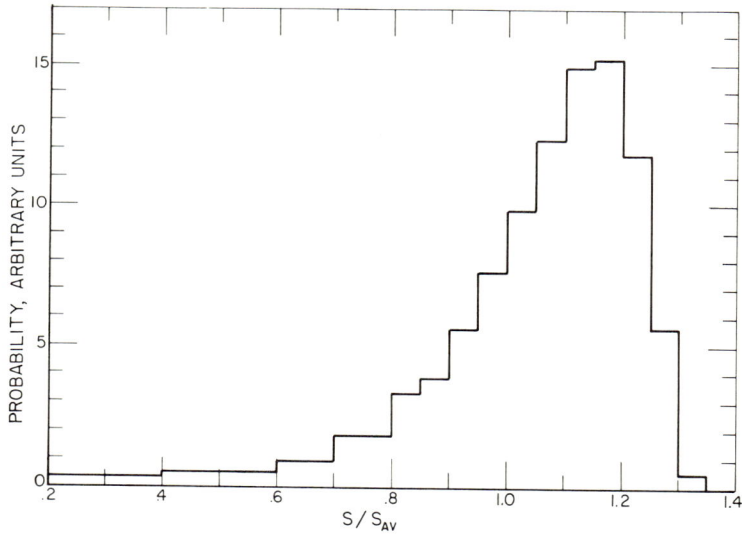

FIG. 17. Pathlength traveled by electrons in aluminum, in the course of slowing down from 0.5 to 0.25 Mev. The histogram was obtained from 7000 histories generated according to Model II with a cutoff $\epsilon_c = 0.01$. $s_{av} = 0.143$ gm/cm² is the mean pathlength calculated in the continuous-slowing-down approximation.

because of the energy transfer cutoff ϵ_c which reduces the mean energy loss per unit pathlength. On the other hand, the possibility of catastrophic collisions tends to reduce the pathlength required to reach a given energy. The interplay of the two effects leads to a pathlength distribution determined by the statistical occurrence of catastrophic collisions. Figure 17 shows a typical distribution for electrons slowing down in aluminum from 0.5 Mev to 0.25 Mev. It is strongly skewed such that the most probable pathlength exceeds the mean pathlength.

The mean pathlength given by the Monte Carlo calculation exceeds the value obtained in the continuous-slowing-down approximation from the expression

$$\langle s \rangle_{\text{av}} = \int_E^{E_0} \left| \frac{dE}{ds} \right|^{-1} dE, \qquad (50)$$

with $\left| dE/ds \right|^{-1}$ evaluated as the reciprocal of the mean of $\left| dE/ds \right|$. As shown in Table XIX, this discrepancy is greatest for very short pathlengths, when the electrons have lost only little energy, and eventually vanishes when they have been slowed down to very low energies. The reason for the discrepancy appears to be that it would actually be more correct to use in Eq. (50) the mean value of the reciprocal $\left| dE/ds \right|^{-1}$. Such an average is not readily available from stopping power theory. It is plausible, however, that the energy loss fluctuations will make the mean value of the reciprocal greater than the reciprocal of the mean value of $\left| dE/ds \right|$. The bearing of this discrepancy on the use of the continuous-slowing-down approximation in transport calculations should be examined.

TABLE XIX

PATHLENGTH OF ELECTRONS[a]

E (Mev)	$\langle s \rangle_{\text{av,MC}} / \langle s \rangle_{\text{av}}$	
	Aluminum	Gold
0.4788	1.24	1.28
0.4585	1.18	1.23
0.4391	1.15	1.18
0.4204	1.12	1.15
0.4026	1.10	1.13
0.3536	1.08	1.09
0.2973	1.05	1.06
0.2500	1.03	1.04
0.1768	1.01	1.01
0.1250	1.01	1.00
0.0625	1.00	1.00

[a] Pathlength of electrons slowing down from an initial energy of 0.5 Mev to energy E. $\langle s \rangle_{\text{av,MC}}$ is the average pathlength obtained from the analysis of 1000 Model-II histories and $\langle s \rangle_{\text{av}}$ is the corresponding average calculated on the assumption of continuous-slowing-down. Statistical error of the Monte Carlo results is ~ 0.02.

E. PENETRATION OF PROTONS

1. *Analysis of a Stopping Power Experiment*

Measurements of the penetration of protons through thick foils can be used to derive the value of the one parameter in the Bethe theory of stopping power that cannot yet be evaluated theoretically, namely, the mean ionization potential I (see Eqs. A9 and A12 of the appendix). We shall discuss Monte Carlo calculations pertinent to the analysis of a precise experiment by Mather and Segrè (1951) in which 340-Mev protons were incident on thick foils and the ionization due to transmitted photons was measured with an argon-filled ionization chamber.

If there were no multiple-scattering detours and energy loss fluctuations, the observed ionization as function of the foil thickness, $R(z)$, would vanish abruptly at a thickness z equal to the mean proton range r_0, and from this range one could readily determine I. Actually $R(z)$ has a rounded tail rather than an abrupt end, because of statistical fluctuations of the proton penetration, and a more elaborate analysis is needed. In Mather and Segrè's analysis, the statistical distribution of proton pathlengths (caused by energy loss fluctuations) was taken into account, but multiple-scattering detours were considered only insofar as they result in a *mean* difference between the depth of penetration and the pathlength. The shape of $R(z)$ thus derived was somewhat at variance with the observed shape. Mather and Segrè suggested that the value of I deduced by them was not likely to be affected by this discrepancy, and that nuclear interactions (disregarded in the analysis) were the most probable cause. The purpose of the Monte Carlo calculations was to find out to what extent a more thorough treatment of multiple scattering detours could remove the discrepancy, and to what extent this would alter the deduced value of I.

Expressed as a function of the proton flux $F(E, \theta, z)$, the observed ionization behind a foil of thickness z is

$$R(z) = \text{constant} \times \int_0^{E_0} dE \int_0^{\pi/2} \sin\theta \, d\theta F(E, \theta, z) \left| \frac{dE}{ds} \right|_A , \qquad (51)$$

where $|dE/ds|_A$ is the stopping power of the gas in the ionization chamber (argon). $R(z)$ was estimated from histories generated according to the Class-I' scheme, with the multiple-scattering angular deflections selected from a Molière distribution, the pathlengths in each step from a Gaussian distribution. Actually the pathlength distribution for short steps is not exactly Gaussian but is slightly skewed (Lewis, 1952). The skewness is similar to that for electrons (see, e.g., Fig. 17) but is much smaller for protons and decreases rapidly with the mean pathlength.

The error caused by the neglect of the skewness is expected to be unimportant because $R(z)$ was calculated only for large z such that the cumulative pathlength of the protons emerging from the foil was long enough for the Gaussian approximation to be very good.

The histories were terminated when the proton energy fell below 2 Mev, which was permissible because the remaining distance of travel (e.g., 50 mg/cm² of lead) was very small compared to the accuracy with which the foil thickness is known and compared to the scale on which $R(z)$ is expressed. Each proton emerging from the foil was given a "score"

$$\frac{1}{\cos \theta} \left| \frac{dE}{ds} \right|_A , \tag{52}$$

where the factor $1/\cos \theta$ provided the conversion from current to flux, and $| dE/ds |_A$ (evaluated with $I = 197$ ev for argon) the conversion from flux to ionization. The average of many scores yielded an estimate of $R(z)$.

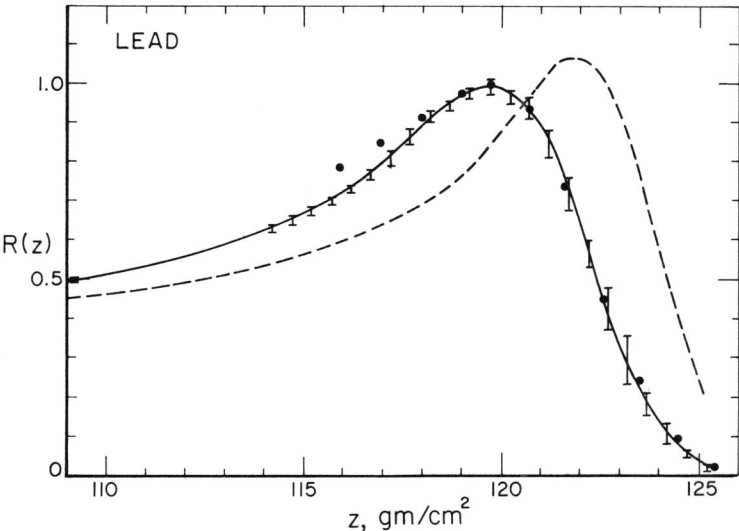

FIG. 18. Ionization $R(z)$ produced in an argon-filled ionization chamber by protons that have traversed a lead barrier of thickness z. Perpendicular incidences, source energy 338.5 Mev. The points (0) are from an experiment by Mather and Segrè (1951). The solid curve is obtained from 5000 histories generated according to Model I', with an assumed mean ionization potential $I = 802$ ev. The experimental and calculated values are both normalized so that the peak value of $R(z)$ is unity. The vertical bars represent limiting values of the Monte Carlo results, corresponding to different procedures for computing spatial displacements. The dotted curve is a Monte Carlo result obtained when multiple-scattering detours were disregarded.

In Figs. 18 and 19, calculated values of $R(z)$ for 338.5-Mev protons incident on lead and copper barriers are compared with the experimental results of Mather and Segrè (1951).[2] The adjustment of the

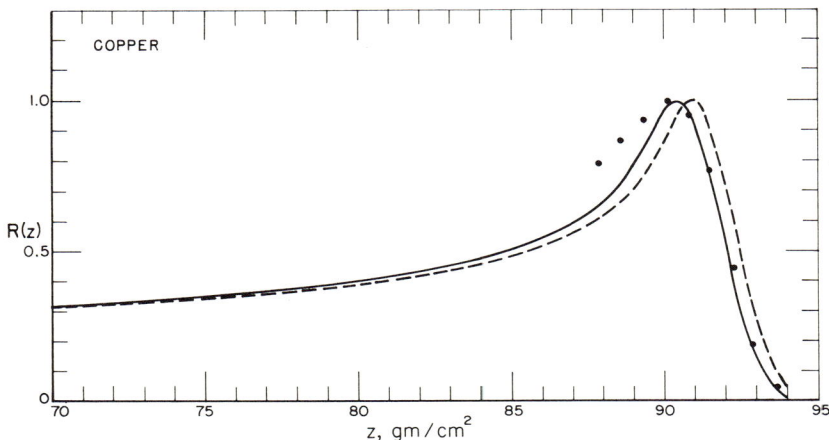

FIG. 19. Ionization $R(z)$, as described in caption of Fig. 18, but for a copper barrier. Assumed mean ionization potential in Monte Carlo calculation was $I = 305$ ev.

calculated data to the experimental data was based on the observation that the shape of $R(z)$ in the neighborhood of its peak remains practically unchanged when the assumed value of the mean excitation potential I is changed slightly, whereas the location of peak is shifted by an amount almost equal to the change of the mean range r_0. After some preliminary calculations with various I-values, a best fit in the immediate vicinity and to the right of the peak was obtained with $I = 802$ ev for lead and $I = 305$ ev for copper. These values are in good agreement with the values 792 ev, respectively, 304 ev, deduced by Mather and Segrè.

It should be noted that the uncertainty of the incident proton energy puts a limit on the accuracy with which one can determine I. For example, in lead at 340 Mev, $\Delta r_0/r_0 = 0.19 \ \Delta I/I$ at constant E_0, and $\Delta r_0/r_0 = 1.6 \ \Delta E_0/E_0$ at constant I, so that the same shift of the peak could be due to a 0.5% change of E_0 or a 5% change of I.

Inspection of Figs. 18 and 19 shows that to the left of the peak, the calculated results are slightly lower than the experimental results for lead, and significantly lower for copper. One reason for this discrepancy

[2] These data were used, but not explicitly described, in the publication of Mather and Segrè, and were put at the author's disposal through the courtesy of Dr. R. L. Mather.

could be the inhomogeneity of the energy of the incident proton beam. However, Mather and Segrè estimated this inhomogeneity to be no greater than 0.5%, which might be enough to remove the discrepancy in lead but not in copper. Thus, the more elaborate analysis of multiple-scattering detours tends to confirm Mather and Segrè's original conclusion that nuclear interactions may have to be taken into account in order to obtain a complete understanding of the observed $R(z)$ curves, and the indications are that such effects would be more important for copper than for lead.

As has been indicated in Section III, A, 4, the z-displacement of the proton in each step of a condensed history was related to the pathlength by an approximate formula instead of being sampled from the appropriate distribution. The error incurred thereby becomes important for very deep penetration, so that it was investigated by three correlated calculations, in which the spatial displacements were evaluated as follows:

Upper limit; $\quad\quad\quad\quad\quad\quad\quad \Delta\zeta = \Delta s_n$ $\quad\quad\quad\quad\quad$ (53)

Usual interpolation formula; $\quad \Delta\zeta = \Delta s_n \dfrac{1 + \cos\omega}{2}$ $\quad\quad\quad\quad\quad$ (54)

Approximate lower limit; $\quad\quad \Delta\zeta = \Delta s_n \cos\omega$. $\quad\quad\quad\quad\quad$ (55)

The solid curve in Fig. 18 was obtained with the use of Eq. (54) and the upper and lower ends of the vertical bars correspond to the results obtained with the use of Eqs. (53) and (55), respectively. Finally, Table XX shows that the value of $R(z)$ remains the same, within the limits of statistical accuracy, whether the step sizes for condensed histories are chosen uniform on a linear or a logarithmic energy scale.

TABLE XX

Dependence of Relative Ionization on Step-Size of Model-I' Histories.[a]

z (gm/cm²)	$R(z)$	
	Uniform spacing[b]	Logarithmic spacing[c]
115.8	0.704 ± 0.004	0.689 ± 0.004
117.8	0.863 ± 0.008	0.843 ± 0.008
119.8	1.000 ± 0.012	1.000 ± 0.011
121.8	0.707 ± 0.014	0.720 ± 0.015
123.8	0.181 ± 0.010	0.189 ± 0.011

[a] 338.5 Mev protons in lead. 5000 histories consisting of 30 steps each.
[b] Approximately uniform step-size as indicated by Table IV(a).
[c] Step-size uniformly spaced on logarithmic energy scale as indicated by Table IV(b), with energy reduction factor $k = (2.0/338.5)^{1/30} = 0.8428$.

3. Multiple-Scattering Detours

We next consider in more detail the difference between pathlength and depth of penetration,

$$s - z = \int_0^s [1 - \cos \theta(s')] \, ds' \qquad (56)$$

and compare Monte Carlo results with two theories.[3]

Bichsel and Uehling (1960) have put forward an approximate simple theory based on the assumption of the small-angle approximation and

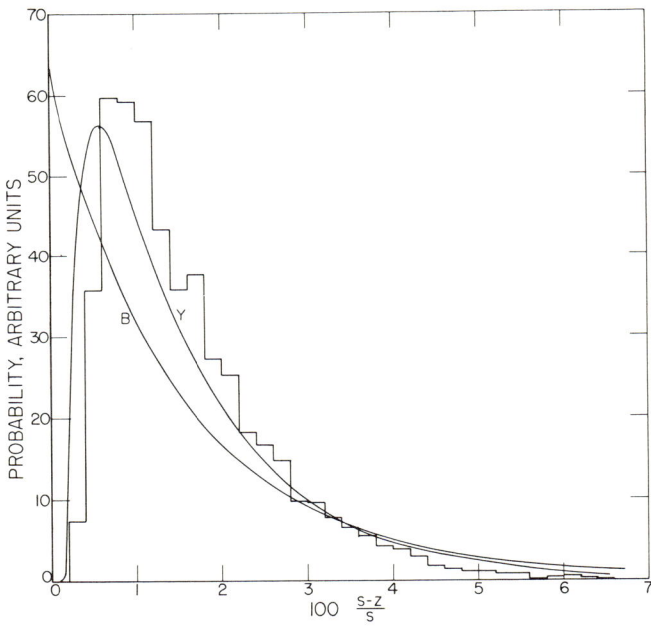

FIG. 20. Multiple-scattering detours. The pathlength distribution of protons slowed down in lead from 338.5 to 2.0 Mev is shown as a function of the percentage difference between the distance a traveled and the depth of penetration z. The histogram was obtained from 5000 histories generated according to Model I', taking into account multiple-scattering deflections, but not energy loss fluctuations. Curve B represents the theory of Bichsel and Uehling (1960), and curve Y a modified version of the theory of Yang (1951).

[3] The variable θ represents the obliquity of the track with respect to the z-axis, and is a stochastic function of the pathlength s. It is analogous to the variable ω used in Section III, A which represents the obliquity of the track with respect to the φ-axis in a single step of the condensed random walk. θ is used here rather than ω because we are concerned with the difference between pathlength and depth of penetration for long tracks consisting of many steps.

of "smoothed-out" trajectories such that the reduced multiple-scattering deflection $\vartheta = \theta/\chi_c \sqrt{B}$ of the Molière theory remains constant along the path. Thus they replace the stochastic dependence of θ on s' by a deterministic dependence, and set

$$s - z = \frac{1}{2} \int_0^s \theta^2(s') \, ds' = \frac{1}{2} \vartheta^2 \int_0^s \chi_c^2 B \, ds' = \frac{1}{2} \vartheta^2 \int_E^{E_0} \chi_c^2 B \left| \frac{dE}{ds} \right|^{-1} dE' \qquad (57)$$

The distribution of $s - z$ then follows from the Molière distribution of ϑ. As can be seen from Fig. 20, the result is a distribution that is reasonable for large $s - z$ differences, but which—incorrectly—peaks at $z = s$.

Yang (1951) has solved this problem in the small-angle diffusion approximation, without taking into account energy loss, and obtained a distribution which he approximated as

$$\left. \begin{aligned} L(q) \, dq &= \frac{2}{(\pi q^3)^{1/2}} (e^{-1/q} - 3e^{-9/q}) \, dq \, , \quad q \leqslant 2.0 \\ &= \frac{\pi}{4} e^{-\pi^2 q/16} \, dq \, , \quad q > 2.0 \end{aligned} \right\} \qquad (58)$$

where

$$q = \frac{2(s - z)}{\langle s - z \rangle_{\mathrm{av}}} . \qquad (59)$$

This distribution includes all particles regardless of their direction at the end of the path.

The Monte Carlo calculations pertain to 340-Mev protons slowed down to 2 Mev in a lead medium, and yield an average pathlength difference

$$100 \left\langle \frac{s - z}{s} \right\rangle_{\mathrm{av}} = 1.639 \pm 0.012$$

Inserting this result into Eq. (59) we have obtained a modified Yang distribution which, as shown in Fig. 20, is in good agreement with the Monte Carlo distribution, the peak being approximately at the right position. Thus it appears that the region of applicability of the Yang theory can be greatly extended when $\langle s - z \rangle_{\mathrm{av}}$ is obtained from an exact transport calculation, the form of his distribution remaining the same. Very similar findings have been made by Hebbard and Wilson (1955) for electron pathlength distributions.

Acknowledgments

The author is indebted to Dr. L. V. Spencer for many helpful discussions, to Dr. R. L. Mather for placing unpublished data at his disposal, and to Dr. J. Coyne for the loan of a FORTRAN program to compute the Mott scattering cross section.

VI. Appendix: Single and Multiple Scattering Theories

The following symbols will be used:

m = mass of electron
M = mass of proton
e = charge of electron
c = velocity of light
v = velocity
$\beta = v/c$
p = momentum

E = kinetic energy
τ = kinetic energy in units of rest mass
N = number of atoms per cm³
ρ = density of medium
Z = atomic number
A = atomic weight

A. Energy Loss

1. *Scattering of Electrons by Electrons*

In this and the next subsection we follow the presentation of Rohrlich and Carlson (1954). The Møller cross section for the scattering of electrons by electrons is

$$\frac{d\sigma}{d\epsilon} = \frac{C}{E}\left\{\frac{1}{\epsilon^2} + \frac{1}{(1-\epsilon)^2} + \left(\frac{\tau}{\tau+1}\right)^2 - \frac{2\tau+1}{(\tau+1)^2}\frac{1}{\epsilon(1-\epsilon)}\right\}, \quad \text{(A1)}$$

where ϵ is the energy transfer in units of E, and

$$C = \frac{2\pi e^4}{mv^2}. \quad \text{(A2)}$$

With this energy transfer there is associated an angular deflection ω' (in the laboratory system) such that

$$\sin^2 \omega' = \frac{4\epsilon}{\tau(1-2\epsilon) + \tau + 4}. \quad \text{(A3)}$$

The probability distribution of energy transfers $\epsilon > \epsilon_c$ is

$$g(\epsilon; \epsilon_c)\, d\epsilon = \frac{\dfrac{d\sigma}{d\epsilon}}{\displaystyle\int_{\epsilon_c}^{1/2} \dfrac{d\sigma}{d\epsilon}}\, d\epsilon = \frac{\dfrac{d\sigma}{d\epsilon}}{\dfrac{C}{E} H(\epsilon_c)}, \quad \text{(A4)}$$

where
$$H(\epsilon) = \frac{1}{\epsilon} - \frac{1}{1-\epsilon} + \left(\frac{\tau}{\tau+1}\right)^2 \left(\frac{1}{2} - \epsilon\right) - \frac{2\tau+1}{(\tau+1)^2} \log \frac{1-\epsilon}{\epsilon}. \quad (A5)$$

The normalization integral in Eq. (A4) extends only to $\epsilon = \frac{1}{2}$ because the outgoing electron of higher energy is, by definition, the primary electron. Note that $H(\frac{1}{2}) = 0$, and that the probability, per unit path length, of an inelastic scattering with fractional energy transfer $\epsilon > \epsilon_c$ is

$$\mu_c(E) = NZ \int_{\epsilon_c}^{1/2} \frac{d\sigma}{d\epsilon} d\epsilon = \frac{NZC}{E} H(\epsilon_c). \quad (A6)$$

2. *Stopping Power for Electrons and Positrons*

Next we consider the mean energy loss by ionization per unit path length (stopping power) resulting from collisions with energy transfers $\epsilon < \epsilon_c$,

$$-\left(\frac{dE}{ds}\right)_{\epsilon_c} = NZ \int_0^{\epsilon_c} \epsilon \frac{d\sigma}{d\epsilon} d\epsilon. \quad (A7)$$

The integral in Eq. (A7) must be evaluated separately for the intervals $0 \leq \epsilon \leq \epsilon'$ and $\epsilon' \leq \epsilon \leq \epsilon_c$ ($\epsilon' \ll \epsilon_c$). In the first interval the Møller cross section (for scattering by free electrons) does not apply, and the binding of the atomic electrons must be taken into account through the Bethe theory of stopping power, which predicts that for small ϵ'

$$\int_0^{\epsilon'} \epsilon \frac{d\sigma}{d\epsilon} d\epsilon = \frac{C}{E} \left\{ \log\left(\frac{2E^2\epsilon'(\tau+2)}{I^2}\right) - \beta^2 \right\}. \quad (A8)$$

The quantity I is the so-called mean ionization potential which can in principle be determined from atomic theory but in practice is obtained from experimental stopping power data. We have used the values $I = 163$ ev for aluminum, 305 ev for copper, 797 ev for gold, and 802 ev for lead.

In the second interval, the integral over the Møller cross section can be carried out. When the result is combined with Eq. (A8), the parameter ϵ' drops out, and one finds that

$$-\left(\frac{dE}{ds}\right)_{\epsilon_c} = NZC \left\{ \log \frac{E^2(\tau+2)}{2I^2} + f^-(\tau, \epsilon_c) - \delta \right\}, \quad (A9)$$

where
$$f^-(\tau, \epsilon_c) = -1 - \beta^2 + \left(\frac{\tau}{\tau+1}\right)^2 \frac{\epsilon_c^2}{2} + \frac{2\tau+1}{(\tau+1)^2} \log(1-\epsilon_c) \quad (A10)$$
$$+ \log[4\epsilon_c(1-\epsilon_c)] + \frac{1}{1-\epsilon_c}.$$

The parameter δ which has been inserted into Eq. (A9) represents the density effect, i.e., the reduction of the mean energy loss caused by the polarization of the medium. We have used values for δ presented by Nelms (1958), which were in turn derived from calculations of Sternheimer (1952, 1953).

For positrons, the Bhabha cross section for scattering by electrons must be used, and the upper limit of the fractional energy loss ϵ is one rather than one half. The stopping power for positrons (for $\epsilon_c = 1$) is obtained from Eq. (A9) with $f^-(\tau, \epsilon_c)$ replaced by

$$f^+(\tau, 1) = 2 \log 2 - \frac{\beta^2}{12} \left\{ 23 + \frac{14}{\tau+2} + \frac{10}{(\tau+2)^2} + \frac{4}{(\tau+2)^3} \right\}. \quad \text{(A11)}$$

3. Stopping Power for Protons

We have used the tabulations of Sternheimer (1959) who computed the stopping power for protons from the expression

$$-\frac{dE}{ds} = NZC \left\{ \log \left[\frac{2mv^2 W_{max}}{I^2(1-\beta^2)} \right] - 2\beta^2 - U - \delta \right\}, \quad \text{(A12)}$$

where

$$W_{max} = \frac{\tau(\tau+2) Mc^2}{\frac{M}{2m} + \frac{m}{2M} + \tau + 1} \quad \text{(A13)}$$

is the largest possible energy transfer in a collision between a proton and an atomic electron. U represents the so-called shell corrections taking into account the binding of the atomic electrons, and δ is again the density correction. Sternheimer's tabulated values were obtained with different values of the mean ionization potential I than those adopted in the present work, so that interpolation had to be used. Following a suggestion by Sternheimer (1960b) the interpolation was performed on the quantity $dE/ds\,A/Z$ which in very good approximation can be considered a function of $\log I$ only.

The mean pathlength traveled by the proton in the course of slowing down from energy E_0 to E is

$$\langle s \rangle_{av} = \int_E^{E_0} \frac{dE}{\left| \frac{dE}{ds} \right|}. \quad \text{(A14)}$$

The pathlength straggling has also been evaluated by Sternheimer (1960a) according to a theory of Bohr, from the expression

$$\langle s^2 \rangle_{av} - \langle s \rangle_{av}^2 = 4\pi e^4 N \int_E^{E_0} \frac{(1-\beta^2)^{-1}(1-\tfrac{1}{2}\beta^2) K\, dE}{1 + \frac{2m}{M}(1-\beta^2)^{-1/2} \left| \frac{dE}{ds} \right|^3}, \quad \text{(A15)}$$

with the parameter K taking into account the binding of atomic electrons. Sternheimer has tabulated the percentage pathlength straggling

$$p_s = 100 \sqrt{\langle s^2 \rangle_{av} - \langle s \rangle^2_{av}}/\langle s \rangle_{av} . \tag{A16}$$

Again interpolation was necessary. Following his suggestion (private communication) this was done assuming that p_s is a function of $\log I$ only.

4. *Distribution of Energy Losses*

If a particle with initial energy E_0 travels a pathlength s it will suffer an energy loss by ionization, $\varDelta E$, whose distribution has been derived by Landau (1944) for the case that $\varDelta E \ll E_0$. His result can be expressed as

$$W_I(\varDelta E) \, d(\varDelta E) = W_L(\lambda) \, d\lambda , \tag{A17}$$

where

$$\lambda = \frac{\varDelta E - \overline{\varDelta E}}{NZCs} + \log\left(\frac{E_0}{NZC_s}\right) - 1.116 , \tag{A18}$$

$\overline{\varDelta E}$ is the mean energy loss, and

$$W_L(\lambda) = \frac{1}{2\pi i} \int_{-i\infty+c}^{i\infty+c} \exp\left(u \log u + \lambda u\right) du \tag{A19}$$

is a universal function that has been evaluated by Landau and tabulated very completely by Börsch-Supan (1961).

Blunck and Leisegang (1950) have refined Landau's theory by considering in more detail the binding of the atomic electrons to which the particle transfers its energy. This results in a broadening of the energy loss distribution which is changed from $W_I(\varDelta E)$ to

$$W_l^*(\varDelta E) = \frac{1}{NZCsb \sqrt{\pi}} \int_{-\infty}^{\infty} W_I(\varDelta E - u) \exp\left(-\frac{u^2}{NZCsb}\right) du , \tag{A20}$$

where the broadening parameter b can be expressed (Blunck and Westphal 1951) as

$$b = \sqrt{q \overline{\varDelta E}} \, \frac{Z^{2/3}}{NZCs} , \tag{A21}$$

with $q \sim 20$ ev. A more detailed and complicated theory of the energy loss distribution has been developed by Symon and described by Rossi (1952).

B. Angular Deflections

1. Evaluation of the Parameters in Molière's Theory

Molière (1948) has given the following prescription for the evaluation of the parameters χ_c and B that enter into his distribution (Eq. 14 of the text) when energy loss is treated in the continuous-slowing-down approximation. It involves the computation of intermediate parameters χ_c', χ_a^2 and $\overline{\chi_a^2}$, according to the following prescription.

$$\chi_c'^2 = \frac{4\pi N Z^2}{p^2 v^2}, \tag{A22}$$

$$\chi_c^2 = \int_0^s \chi_c'^2[s'(E)]\, ds' = \int_E^{E_0} \chi_c'^2(E') \left|\frac{dE}{ds}\right|^{-1} dE', \tag{A23}$$

$$\chi_a^2 = a Z^{2/3} \left[1.13 + 3.76 \left(\frac{Z}{137\beta}\right)^2\right], \tag{A24}$$

[with $a = 6.8 \times 10^{-5}$ for electrons, $6.8 \times 10^{-5}\, (m/M)^2$ for protons] and

$$\log \overline{\chi_a^2} = \frac{1}{\chi_c^2} \int_E^{E_0} \chi_c'^2(E') \log \chi_1^2(E') \left|\frac{dE}{ds}\right|^{-1} dE'. \tag{A25}$$

The parameter B is obtained by solving the transcendental equation

$$B - \log B = \log \frac{\chi_c^2}{1.167 \overline{\chi_a^2}}. \tag{A26}$$

The Molière theory takes into account only elastic collisions against the Coulomb field of atomic nuclei, but disregards inelastic collisions with atomic electrons. The latter have been considered by Fano (1954) who obtained corrections to Molière's theory. He finds that if the multiply-scattered particles are electrons, one must replace Z^2 by $Z(Z + 1)$ in Eq. (A22), and that one must add a term

$$B' = (Z + 1)^{-1}\left\{\log\left[0.160 Z^{2/3}\left(1 + 3.33\left(\frac{Z}{137\beta}\right)^2\right)\right] - c_F\right\} \tag{A27}$$

on the right-hand side of Eq. (A26). For protons, one should leave Z^2 unchanged, but add to the right-hand side of Eq. (A26) a term

$$B'' = Z^{-1}\left\{\log\left[11.30 Z^{-4/3} \frac{\beta^2}{1 - \beta^2}\right] - c_F - \frac{1}{2}\beta^2\right\}. \tag{A28}$$

Fano has estimated the constant c_F to be -3.6 for hydrogen, -4.6 for lithium, -5.0 for oxygen, and -6.3 for lead. By interpolation we have obtained values for other elements, e.g., -5.2 for aluminum and -6.2 for gold. The Fano correction was derived without taking into account the energy loss of the multiply-scattered particle. Little error will be incurred by evaluating it with a value of β corresponding to an energy intermediate between E_0 and E.

2. Evaluation of the Goudsmit-Saunderson Theory

This section makes use of various procedures developed by Spencer (1955, 1959) that facilitate the numerical evaluation of the Goudsmit-Saunderson angular multiple-scattering distribution

$$A_{Gs}(\omega) = \sum_{l=0}^{\infty} \left(l + \frac{1}{2}\right) \exp\left\{-\int_0^s G_l(s')\, ds'\right\} P_l(\cos \omega). \quad \text{(A29)}$$

This Legendre series converges rather slowly, and the essential trick is to derive recursion relations with which a large number of the expansion coefficients can be computed accurately and easily.

a. Single-scattering cross section. The initial approximation is the Rutherford cross section for scattering of a charged particle by a nucleus of charge Ze,

$$\sigma_R = \frac{Z^2 e^4}{p^2 v^2 (1 - \cos \theta)^2}. \quad \text{(A30)}$$

The screening of the nuclear charge by orbital electrons can be taken into account approximately in various ways. We shall replace $(1 - \cos \theta)^2$ by $(1 - \cos \theta + 2\eta)^2$, where η is a screening parameter that can be obtained from the theory of Molière as

$$\eta = \frac{1}{4} \chi_a^2, \quad \text{(A31)}$$

with χ_a^2 given by Eq. (A24). An improved formula for η has been given by Nigam *et al.* (1959) but not used in the present calculation. As can be seen from Eq. (A24), η is a very small quantity.

An improvement over the Rutherford cross section is provided by the Mott cross section σ_M which pertains to the scattering by an unscreened nuclear charge and is obtained through the exact solution of the Dirac equation (Mott, 1929). The cross section is obtained as a rather slowly converging Legendre series in the deflection angle, and must be evaluated numerically. Following earlier calculations by Bartlett and Watson (1940), Doggett and Spencer (1956), and Sherman (1956) have made systematic tabulations. Instead of interpolating in these tables, we have used a FORTRAN computer program for the IBM 704 written by Dr. J. Coyne of the National Bureau of Standards. Typical ratios of the Mott to the Rutherford cross section for electrons and positrons, obtained with this program, are shown in Table XXI.

TABLE XXI

RATIO OF MOTT TO RUTHERFORD SCATTERING CROSS SECTION AT 0.5 MEV.

	Aluminum		Gold	
θ	Electrons	Positrons	Electrons	Positrons
0°	1.000	1.000	1.000	1.000
15°	1.019	0.961	1.083	0.919
30°	1.006	0.907	1.265	0.836
45°	0.962	0.839	1.508	0.757
60°	0.891	0.760	1.726	0.679
75°	0.862	0.732	1.851	0.605
90°	0.694	0.585	1.846	0.537
120°	0.481	0.419	1.472	0.421
150°	0.320	0.301	0.929	0.347
180°	0.260	0.259	0.678	0.320

Small deflections are particularly numerous and make an important contribution to the multiple-scattering angular distribution so that it is advantageous to take into account the analytical result that for small θ

$$\frac{\sigma_M}{\sigma_R} \sim 1 + \frac{\pi}{\sqrt{2}} \frac{Z\beta}{137} \cos \gamma \, (1 - \cos \theta)^{1/2}, \qquad (A32)$$

where

$$\cos \gamma = \mathrm{Re} \left\{ \frac{\Gamma\left(\frac{1}{2} - i\frac{Z}{137\beta}\right) \Gamma\left(1 + i\frac{Z}{137\beta}\right)}{\Gamma\left(\frac{1}{2} + i\frac{Z}{137\beta}\right) \Gamma\left(1 - i\frac{Z}{137\beta}\right)} \right\}. \qquad (A33)$$

On combining the Rutherford cross section, the screening correction, the analytical form of σ_M/σ_R at small angles, and tabulated values of the Mott cross section, we adopt the following form for the single scattering cross section.

$$\sigma(\theta) = \frac{Z^2 e^4}{p^2 v^2 (1 - \cos \theta + 2\eta)^2}$$
$$\times \left\{ 1 + \frac{\pi}{\sqrt{2}} \frac{Z\beta}{137} \cos \gamma (1 - \cos \theta + 2\eta)^{1/2} + h(\theta) \right\}, \qquad (A34)$$

where the function

$$h(\theta) = \frac{\sigma_M}{\sigma_R} - 1 - \frac{\pi}{\sqrt{2}} \frac{Z\beta}{137} \cos \gamma (1 - \cos \theta + 2\eta)^{1/2} \qquad (A35)$$

must be evaluated numerically.

Recently, Brown et al. (1961) have announced a numerical solution of the Dirac equation for scattering by a screened Coulomb potential. Their results will yield a cross section more accurate than Eq. (A34). Finally, we note that in Eq. (A34) Z^2 should be replaced by $Z(Z+1)$ to take into account inelastic scattering. A further small correction has been suggested by Spencer, involving multiplication of the cross section by a factor $1 + B'/\log 4\eta$, with B' given by Eq. (A27).

The preceding formulation applies only to electrons and positrons but not to protons. For fast protons one must consider not only spin and relativistic effects, but also the influence of nuclear interactions, and the modification of Coulomb scattering by the finite size of the nucleus.

b. *Expansion coefficients.* The cross section $\sigma(\theta)$ enters into the Goudsmit-Saunderson distribution in the form of coefficients

$$G_l = 2\pi N \int_0^\pi \sigma(\theta) \{1 - P_l(\cos \theta)\} \sin \theta \, d\theta. \qquad (A36)$$

For angular multiple-scattering distributions that are concentrated in the forward direction, a rather large number of G_l's may have to be evaluated. For example, in the example described in Table XXIV, the angular distribution multiplied by the solid-angle factor $\sin \omega$ peaks at 9.5° and as many as sixty terms in the Legendre expansion (A29) were used. To be accurate one should carry the integration in Eq. (A36) out analytically, particularly for large l. This can be done if the numerical part of the cross section, $h(\theta)$, is approximated by a polynomial, as follows:

$$h(\theta) = \sum_{j=1}^{J} h_j (1 - \cos \theta + 2\eta)^{j/2}. \qquad (A37)$$

This representation has been found to be accurate to within 1% or better, with $J = 5$. Combining Eqs. (A34, A36, and A37), one finds that G_l can be expressed as a linear combination of integrals of the form

$$p(m, l) = \int_{-1}^{1} (1 - x + 2\eta)^m \{1 - P_l(x)\} \, dx \qquad (A38)$$

for $m = -2, -3/2, -1, \ldots$, and for $l = 1, 2, 3, \ldots$. These integrals can be evaluated by simple recursion relations that can be derived from the properties of Legendre polynomials. We shall merely state them briefly, referring the reader to Spencer (1955) for an explanation of the methods by which they are derived.

The first basic recursion, which allows one to go from m to $m+1$, is

$$p(m+1, l) = (1 + 2\eta) p(m, l) + p(m, 1)$$
$$- \frac{l+1}{2l+1} p(m, l+1) - \frac{l}{2l+1} p(m, l-1).$$
(A39)

To start it, one needs to know $p(-2, l)$ and $p(-3/2, l)$ which can in turn be obtained from the following recursion relations:

$$\left. \begin{array}{l} p(-2, 1) = \log\left(1 + \dfrac{1}{\eta}\right) - (1+\eta)^{-1} \\[4pt] lp(-2, l+1) = (2l+1)(1+2\eta) p(-2, l) \\[4pt] \qquad\qquad - (l+1) p(-2, l-1) - (2l+1)(1+\eta)^{-1}, \quad l \geqslant 1 \end{array} \right\}$$
(A40)

and

$$\left. \begin{array}{l} p(-\tfrac{3}{2}, 1) = 2(2\tilde{\eta})^{3/2} (1+\tilde{\eta})^{-1} \\[4pt] p(-\tfrac{3}{2}, l+1) = \tilde{\eta} p(-\tfrac{3}{2}, l) + p(-\tfrac{3}{2}, 1), \quad l \geqslant 1 \end{array} \right\},$$
(A41)

where

$$\tilde{\eta} = 1 - 2\eta\{\sqrt{1 + 1/\eta} - 1\}.$$
(A42)

c. Energy dependence. The remaining task is to evaluate the integral

$$\int_0^s G_l(s') \, ds'$$
(A43)

in the continuous-slowing-down approximation. It is convenient to change variables from s to

$$t = \frac{r_0 - s}{r_0}$$
(A44)

where r_0 is the mean residual range, and make a corresponding change from $G_l(s)$ to $G_l^*(t)$. Spencer has shown that one can very accurately represent $G_l^*(t)$ by the expressions

$$G_1^*(t) = \frac{c_1}{t(t+c_2)}$$
(A45)

and
$$G_l^*(t) = G_1^*(t) \frac{G_l^*(1)}{G_1^*(1)}, \tag{A46}$$

where the two parameters c_1 and c_2 are obtained from a knowledge of $G_1^*(1)$ and $G_1^*(t)$. With this representation, one finds that

$$\int_0^s G_l(s')\,ds' = r_0 \int_i^1 G_l^*(t')\,dt'$$
$$= r_0 c_1 \frac{G_l^*(1)}{G_1^*(1)} \log\left\{\frac{t+c_2}{t(1+c_2)}\right\}. \tag{A47}$$

Tables XXII and XXIII contain examples of Goudsmit-Saunderson distributions for electrons and positrons, for progressively longer

TABLE XXII

ANGULAR DISTRIBUTION OF MULTIPLY-SCATTERED ELECTRONS.[a]

E/E_0	$2^{-1/32}$	$2^{-1/16}$	$2^{-1/8}$	$2^{-1/4}$	$2^{-1/2}$	2^{-1}	2^{-2}
E (Mev)	0.489	0.479	0.459	0.420	0.354	0.250	0.125
$r(E)/r(E_0)$	0.970	0.942	0.886	0.784	0.611	0.363	0.120

ω				Angular distribution.			
0°	2.86(1)	1.25(1)	5.60	2.62	1.29	7.16(−1)	5.20(−1)
15°	9.43	7.67	4.55	2.40	1.25	7.08(−1)	5.19(−1)
30°	9.65(−1)	2.16	2.52	1.86	1.13	6.85(−1)	5.17(−1)
45°	1.35(−1)	4.45(−1)	1.03	1.23	9.52(−1)	6.48(−1)	5.14(−1)
60°	3.42(−2)	1.06(−1)	3.63(−1)	7.16(−1)	7.56(−1)	6.02(−1)	5.10(−1)
75°	1.24(−2)	3.45(−2)	1.26(−1)	3.78(−1)	5.69(−1)	5.49(−1)	5.05(−1)
90°	5.58(−3)	1.44(−2)	4.91(−2)	1.90(−1)	4.10(−1)	4.95(−1)	5.00(−1)
120°	1.63(−3)	3.91(−3)	1.16(−2)	4.97(−2)	1.97(−1)	3.96(−1)	4.90(−1)
150°	6.88(−4)	1.63(−3)	4.56(−3)	1.82(−2)	1.01(−1)	3.28(−1)	4.83(−1)
180°	4.86(−4)	1.15(−3)	3.20(−3)	1.23(−2)	7.52(−2)	3.04(−1)	4.80(−1)

[a] For a pathlength in aluminum in which the energy is reduced from $E_0 = 0.5$ Mev to a value E, and the residual mean range from $r(E_0) = 0.2258$ gm/cm^2 to a value $r(E)$. Numbers in parentheses represent powers of ten.

pathlengths. Initially quite concentrated in the forward directions, the distributions rapidly become broader and practically isotropic when the particles have lost $\frac{3}{4}$ of their energy. Pertaining to the flux in an unbounded medium, they are more nearly isotropic, for long pathlengths, than the flux in the vicinity of the exit boundary of a foil, as can be verified through comparison with the results in Fig. 9. In Table XXIV,

Table XXIII

Angular Distribution of Multiply-Scattered Positrons.[a]

E/E_0	$2^{-1/32}$	$2^{-1/16}$	$2^{-1/8}$	$2^{-1/4}$	$2^{-1/2}$	2^{-1}	2^{-2}
E(Mev)	0.489	0.479	0.459	0.420	0.354	0.250	0.125
$r(E)/r(E_0)$	0.971	0.940	0.884	0.780	0.604	0.355	0.115
ω	\multicolumn{7}{c}{Angular distribution.}						
0°	3.08(1)	1.29(1)	5.85	2.74	1.36	7.48(−1)	5.29(−1)
15°	9.46	7.84	4.72	2.51	1.31	7.39(−1)	5.28(−1)
30°	8.64(−1)	2.13	2.55	1.92	1.17	7.11(−1)	5.25(−1)
45°	1.14(−1)	4.15(−1)	1.01	1.24	9.80(−1)	6.69(−1)	5.20(−1)
60°	2.82(−2)	9.44(−2)	3.38(−1)	7.02(−1)	7.66(−1)	6.15(−1)	5.14(−1)
75°	1.01(−2)	2.97(−2)	1.13(−1)	3.59(−1)	5.65(−1)	5.55(−1)	5.07(−1)
90°	4.52(−3)	1.22(−2)	4.23(−2)	1.73(−1)	3.97(−1)	4.93(−1)	5.00(−1)
120°	1.35(−3)	3.34(−3)	9.74(−3)	4.25(−2)	1.80(−1)	3.81(−1)	4.86(−1)
150°	6.11(−4)	1.47(−3)	3.95(−3)	1.51(−2)	8.70(−2)	3.06(−1)	4.75(−1)
180°	4.53(−4)	1.08(−3)	2.85(−3)	1.03(−2)	6.31(−2)	2.80(−1)	4.71(−1)

[a] For a pathlength in aluminum in which the energy is reduced from $E_0 = 0.5$ Mev to a value E, and the residual mean range from $r(E_0) = 0.2237$ gm/cm² to a value $r(E)$. Numbers in parentheses represent powers of ten.

Table XXIV

Convergence of Legendre-Sum for Multiple-Scattering Angular Distribution.[a]

l_{max}	ω = 0°	15°	30°	60°
10	2.33965(1)	1.10424(1)	1.48240	−2.79748(−1)
20	2.84052(1)	9.40448	9.60956(−1)	4.53502(−2)
30	2.85867(1)	9.43132	9.64998(−1)	3.41642(−2)
40	2.85896(1)	9.43133	9.64688(−1)	3.41821(−2)
50	2.85896(1)	9.43132	9.64685(−1)	3.41835(−2)

l_{max}	ω = 90°	120°	150°	180°
10	−1.38927(−1)	−3.84342(−2)	6.81425(−2)	7.03439(−1)
20	1.03224(−2)	−4.21503(−3)	−5.33409(−3)	3.54991(−2)
30	5.51196(−3)	1.73052(−3)	5.58971(−4)	1.11688(−3)
40	5.57968(−3)	1.62751(−3)	6.87469(−4)	4.93515(−4)
50	5.57903(−3)	1.62765(−3)	6.87881(−4)	4.86467(−4)
60	5.57904(−3)	1.62765(−3)	6.87874(−4)	4.86400(−4)

[a] Pertains to the electron angular distribution given in the second column of Table XXII ($E/E_0 = 2^{-1/32}$). Numbers in parentheses represent powers of ten.

the convergence of the Legendre series for the angular distribution is illustrated, as a function of the number of terms, l_{max}, retained in the expansion. Convergence is quite slow in the example, particularly at large angles. Thus at 150° the use of twenty terms would give a result which is ten times too large in absolute value and has the wrong sign, and almost sixty terms are required to provide convergence to six significant figures. The round-off error in the series summation has not yet been evaluated.

The cumulative form of the Goudsmit-Saunderson distribution, desirable for random sampling, can easily be obtained from the differential form, through replacing the Legendre polynomials $P_l(\cos \omega)$ in Eq. (A29) by

$$H_l(\cos \omega) = \int_{\cos \omega}^1 P_l(x) \, dx , \tag{A49}$$

and use of the recursion relation

$$\left. \begin{array}{l} H_0 = 1 - \cos \omega \\ H_1 = (1 - \cos^2 \omega)/2 \\ (l+1) H_l = (2l-1) \cos \omega H_{l-1} - (l-2) H_{l-2} , \quad l \geqslant 2 \end{array} \right\} , \tag{A50}$$

References

Agu, B. N. C., Burdett, T., and Matsukawa, E. (1958a). *Proc. Phys. Soc. London.* **71**, 201.
Agu, B. N. C., Burdett, T. A., and Matsukawa, E. (1958b). *Proc. Roy. Soc.* **72**, 727.
Archard, G. D. (1961). *J. Appl. Phys.* **32**, 1505.
Bartlett, J. H., and Watson, R. E. (1940). *Proc. Am. Acad. Arts Sci.* **74**, 54.
Berger, M. J. (1960). *Radiation Research* **12**, 422.
Bethe, H. A. (1935). *Proc. Roy. Soc.* **A150**, 129.
Bethe, H. A. (1953). *Phys. Rev.* **89**, 1256.
Bethe, H. A., Rose, M. E., and Smith, L. P. (1938). *Proc. Am. Phil. Soc.* **78**, 573.
Bichsel, H., and Uehling, E. A. (1960). *Phys. Rev.* **119**, 1670.
Birkhoff, R. D. (1958). *In* "Handbuch der Physik" (S. Flugge, ed.), Vol. 34, p. 53. Springer, Berlin.
Blanchard, C. H. (1951). *In* "Electron Physics" N.B.S. Circular 527, p. 9.
Blanchard, C. H., and Fano, U. (1951). *Phys. Rev.* **82**, 767.
Blunck, O., and Leisegang, S. (1950). *Z. Physik* **128**, 500.
Blunck, O., and Westphal, K. (1951). *Z. Physik* **130**, 641.
Börsch-Supan, W. (1961). *J. Research Natl. Bur. Standards* **65B**, 245.
Bothe, W. (1933). *In* "Handbuch der Physik" (H. Geiger ed.), Vol. 22, 2d ed., p. 1 Springer, Berlin.
Bothe, W. (1949). *Ann. Physik* **6**, 44.
Breitenberger, E. (1959). *Proc. Roy. Soc.* **250**, 514.
Brown, R. T., Lin, S. R., and Sherman, N. (1961). *Bull. Am. Phys. Soc.* **6**, 366.
Butcher, J. C., and Messel, H. (1960). *Nuclear Phys.* **20**, 15.

Buys, W. L. (1960). *Z. Physik* **157**, 478.
Carlson, B. G. (1955). Solution of the Transport Equation by *S* Approximations. Los Alamos Scientific Laboratory Report No. LA 1891.
Doggett, J. A., and Spencer, L. V. (1956). *Phys. Rev.* **103**, 1597.
Eyges, L. (1949). *Phys. Rev.* **76**, 264.
Fano, U. (1953). *Phys. Rev.* **92**, 328.
Fano, U. (1954). *Phys. Rev.* **93**, 117.
Fleischmann, W. (1960). *Z. Naturforsch.* **15a**, 1090.
Frank, H. (1959). *Z. Naturforsch.* **14a**, 247.
Goudsmit, S., and Saunderson, J. L. (1940). *Phys. Rev.* **57**, 24.
Hebbard, D. V., and Wilson, P. R. (1955). *Australian J. Phys.* **8**, 90.
Kahn, H. (1954). Application of Monte Carlo. U. S. AEC Report No. R-1237.
Kanter, H. (1957). *Ann. Physik* **20**, 144.
Koch, H. W., and Motz, J. W. (1959). *Rev. Modern Phys.* **31**, 920.
Landau, L. (1944). *J. Phys. USSR* **8**, 201.
Leiss, J. E., Penner, S., and Robinson, C. S. (1957). *Phys. Rev.* **107**, 1544.
Lewis, H. W. (1950). *Phys. Rev.* **78**, 526.
Lewis, H. W. (1952). *Phys. Rev.* **85**, 20.
MacCallum, C. (1960). *Bull. Am. Phys. Soc.* **5**, 379.
Mather, R., and Segrè, E. (1951). *Phys. Rev.* **84**, 191.
McGinnies, R. T. (1959). N.B.S. Circular 597.
Meister, H. (1958). *Z. Naturforsch.* **13a**, 809.
Miller, W. (1951). *Phys. Rev.* **82**, 452.
Molière, G. (1947). *Z. Naturforsch.* **2a**, 133.
Molière, G. (1948). *Z. Naturforsch.* **3a**, 78.
Mott, N. F. (1929). *Proc. Roy. Soc.* **A124**, 475.
Nelms, A. T. (1958). N.B.S. Circular 577 (Supplement).
Nigam, B. P., and Mathur, V. S. (1961). *Phys. Rev.* **121**, 1577.
Nigam, B. P., Sundaresan, M. K., and Wu, T. (1959). *Phys. Rev.* **115**, 491.
Roesch, W. C. (1954). Hanford Report No. HW 32121.
Rohrlich, F., and Carlson, B. C. (1954). *Phys. Rev.* **93**, 38.
Rossi, B. (1952). "High Energy Particles." Prentice-Hall Englewood Cliffs, New Jersey.
Schneider, D. O., and Cormack, D. V. (1959). *Radiation Research* **11**, 418.
Seliger, H. H. (1952). *Phys. Rev.* **88**, 408.
Seliger, H. H. (1955). *Phys. Rev.* **100**, 1029.
Sherman, N. (1956). *Phys. Rev.* **103**, 1601.
Sidei, T., Higasimura, T., and Kinosita, K. (1957). *Mem. Fac. Eng. Kyoto Univ.* **19**, 22.
Snyder, H. S., and Scott, (1949). *Phys. Rev.* **76**, 220.
Spencer, L. V. (1955). *Phys. Rev.* **98**, 1507.
Spencer, L. V. (1959). "Energy Dissipation by Fast Electrons." N. B. S. Monograph 1.
Spencer, L. V., and Fano, U. (1954). *Phys. Rev.* **93**, 1172.
Sternheimer, R. M. (1952). *Phys. Rev.* **88**, 851.
Sternheimer, R. M. (1953). *Phys. Rev.* **91**, 256.
Sternheimer, R. M. (1959). *Phys. Rev.* **115**, 137.
Sternheimer, R. M. (1960a). *Phys. Rev.* **117**, 485.
Sternheimer, R. M. (1960b). *Phys. Rev.* **118**, 1045.
Suzor, F., and Charpak, G. (1952). *J. phys. radium* **13**, 1.
Trump, J. G., and Van de Graaf, R. J. (1949). *Phys. Rev.* **75**, 44.
Wang, M. C., and Guth, E. (1951). *Phys. Rev.* **84**, 1092.
Wentzel, G. (1922). *Ann. Physik* **69**, 335.

WEYMOUTH, J. W. (1951). *Phys. Rev.* **84**, 766.
WILLIAMS, E. J. (1939). *Proc. Roy. Soc.* **169**, 531.
WILSON, R. (1950). *Phys. Rev.* **79**, 204.
WILSON, R. (1951). *Phys. Rev.* **84**, 100.
YANG, C. N. (1951). *Phys. Rev.* **84**, 599.

Monte Carlo Methods Applied to Configurations of Flexible Polymer Molecules

FREDERICK T. WALL, STANLEY WINDWER*, and PAUL J. GANS[†]

NOYES CHEMICAL LABORATORY, UNIVERSITY OF ILLINOIS
URBANA, ILLINOIS

I. Introduction 217
II. Monte Carlo Methods 220
   A. Direct Monte Carlo Procedure 220
   B. Chain Enrichment Method 226
   C. Biased Polymers 229
   D. Assembler Routines 232
III. Results and Conclusions 234
   A. General Discussion 234
   B. Empirical Analysis of Monte Carlo Calculations 237
   References 242

I. Introduction

FLEXIBLE HIGH-POLYMER MOLECULES are recognized as long chains of atoms that may be thousands of times larger than simple molecules. These long chains of atoms are held together by chemical bonds, about which a certain amount of rotation is generally possible. Such rotation gives rise to numerous possible chain configurations, some of which may be highly elongated and others of which may be tightly coiled. The fact that long polymer molecules are capable of exhibiting so many configurations suggests a statistical approach for their description. Many polymer properties, for example, the osmotic, diffusion, and light scattering coefficients, can be related to statistical averages such as the mean-square end-to-end separation, $\langle R_N{}^2 \rangle$, of the polymer chains (Flory, 1953). Although this is by no means the only statistical quantity of significance, our discussion will be principally directed toward deter-

---

* Present Address: Department of Chemistry, Adelphi College, Garden City, New York.
† Present Address: Department of Chemistry, New York University, New York 3, New York.

mining this particular quantity. Moreover, the mean-square end-to-end separation provides a good example for the use of Monte Carlo methods.

Kuhn (1934) proposed a model in which a long chain of noninterfering freely jointed rods was allowed to assume any random configuration. Since this is quite equivalent to the problem of a normal random walk (Chandrasaker, 1943), the mean-square end-to-end distance would be

$$\langle R_N^2 \rangle = NL^2 \tag{1}$$

where

$N \equiv$ the number of rods in the chain, or the number of steps in the random walk, and
$L \equiv$ the length of a single rod or step.

Taylor (1948) and Benoit (1948) corrected Kuhn's model so that successive elements of the polymer chain were not randomly oriented. Their modified solution of the original random-walk model did not depart from proportionality between $\langle R_N^2 \rangle$ and N, but it did involve a new proportionality constant a dependent on the specific restrictions imposed on the model; thus

$$\langle R_N^2 \rangle = aL^2 N \, . \tag{2}$$

When other restrictions of a similar nature were imposed on this model, for example, alternating fixed angles, hindered rotation, etc. (Flory, 1953; Hill, 1960), proportionality of $\langle R_N^2 \rangle$ to N remained unchanged; however, the proportionality constant a depended upon the limitations imposed.

Kuhn and Kuhn (1943) recognized that the normal random-walk model was severely deficient for it implied that the atoms could overlap without volume restriction. To avoid double occupancy in any region of space, each atom along a polymer chain must be placed in such a way that it does not overlap any other atom. To handle this problem effectively, a complete record of the entire chain is required. The restriction against multiple occupancy in space gives rise to what is known as the *excluded volume* problem.

When one considers the effect of excluded volume, it is conceivable that the functional dependence of $\langle R_N^2 \rangle$ on N as given above will be changed. One school of thought suggests that, for large N, the only effect of excluded volume is to increase the value of the constant a; therefore the ratio of $\langle R_N^2 \rangle / N$ will converge as N increases. Another school, however, suggests that the ratio does not approach a limit as N increases but that it diverges (Flory, 1949; James, 1953; Zimm *et al.* 1953).

The attempts of Flory (1949), Hermans (1950), Grimley (1952), and Bueche (1953) to treat the excluded volume effect as a small perturbation of the normal random walk were not entirely successful. Still other attempts, in which a distribution function was sought (Zimm et al., 1953), contained complicated functions, the evaluations of which are not presently feasible. In addition, estimates of upper and lower bounds of the mean-square end-to-end distance have been made (Frisch et al., 1951; Rubin, 1952) based on a method devised by Montroll (1950), who took into account short-range excluded volume effects.

Sykes (1961) recently obtained a count of all the possible configurations of nonintersecting walks on a plane square lattice up to 18 steps, essentially through use of a theorem that enabled him to calculate the number of walks recursively from a knowledge of the number of closed walks on a linear graph. (He also made calculations for other lattices.) Extension of his method to a greater number of steps becomes increasingly difficult because of sheer numerical complications. Except for this direct enumeration technique, which is good only for relatively small N, no analytical solution of a completely non-self-intersecting random-walk process has been accomplished. Domb and Sykes (1961) have, however, been able to extrapolate their data to large values of N in excellent agreement with those obtained by Monte Carlo methods.

King (1949) demonstrated that a solution of the excluded volume problem might be obtained by Monte Carlo methods. The Monte Carlo method is designed to obtain a numerical answer without knowledge of an analytical solution. For the polymer problem, it is recognized that there exists a set of non-self-intersecting N-step random-walk configurations that contains each and every configuration once and only once. If this set of configurations can be sampled in some unbiased fashion, and if the sample taken is large enough, then $\overline{R_N^2}$, the mean-square end-to-end distance averaged over the sample, can be made as close as desired to $\langle R_N^2 \rangle$, the average over the entire set. The sampling of this set is accomplished using automatic computing equipment that constructs N-step non-self-intersecting random walks, with each successive step chosen by a suitable Monte Carlo process.

While King utilized punched card calculating equipment, the low speed of which severely limited the size of his sample, Wall et al. (1954, 1955a, 1955b, 1957a, 1957b, 1959a), Rosenbluth and Rosenbluth (1955), and more recently Marcer (1960) have attacked the problem using high-speed digital computers with considerable success.

II. Monte Carlo Methods

A. DIRECT MONTE CARLO PROCEDURE

1. *Introduction*

The Monte Carlo method, applied in its simplest form to the study of polymer configurations, requires the generation of nonintersecting configurations in such a way that all valid samples occur with equal probability. This can be accomplished on a lattice by successively selecting random steps until a chain of the desired number of links has been constructed or until double occupancy occurs at some lattice point. The program for carrying out this generation process on a computer depends primarily on two operations, namely, the random selection of steps and the detection of double occupancy. The method of selecting steps depends in large measure upon the nature of the lattice employed, whereas the detection of double occupancy is primarily dependent upon the computer organization and the kinds of computer commands that are available.

2. *The Random Selection of Steps*

To illustrate the method of selecting random steps we shall describe the process as applied to the tetrahedral lattice. This is a particularly good lattice to use for the representation of polymer molecules since it provides a good approximation for chains of carbon atoms while possessing characteristics of sufficient generality to avoid trivial conclusions. Other types of lattices have also been studied and reference should be made to the relevant literature for details (Wall *et al.*, 1954, 1955a, 1955b, 1957a, 1957b, 1959a; Marcer, 1960).

An easy way to picture the tetrahedral lattice in a Cartesian coordinate system is to think first of the body-centered-cubic lattice. Upon placing any point of the body-centered lattice at the origin, the lattice can then be used to represent two interlocking tetrahedral lattices. Each of the eight corners of the cube surrounding the point at the origin can be regarded as the terminal of the first step in a tetrahedral chain. However, once the first link is chosen, the lattice is determined, and from then on only four of the eight points surrounding a given point represent permissable choices. Actually the eight vectors from a central point to the corners of a cube can be divided into two tetrahedral groups, one of which represents the negatives of the other. For convenience we shall let these vectors be $\sqrt{3}$ units in length so that their x, y, and z components are all ± 1. With this understanding, the two groups of vectors, represented by their x, y, and z components, can be written as indicated in Table I. The generation of a random walk now consists of

TABLE I

TETRAHEDRAL LATTICE VECTORS

Group 1	Group 2
111	$\bar{1}\bar{1}\bar{1}$
$\bar{1}\bar{1}1$	$11\bar{1}$
$\bar{1}1\bar{1}$	$1\bar{1}1$
$1\bar{1}\bar{1}$	$\bar{1}11$

the following process: assuming that a vector from Group 1 is chosen first, the possible steps are those illustrated in Box I of Figure 1. Possible

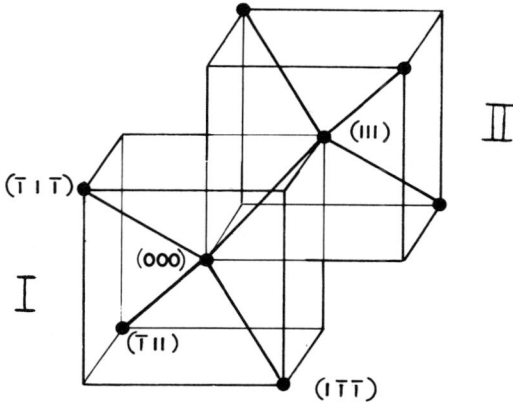

FIG. 1. A diagram showing the two sets of possible steps on a tetrahedral lattice.

second steps are illustrated in Box II. If the terminal point of any vector in Group 1 is translated to the origin (000), it can be seen from the figure that the available second steps now correspond to those of Group 2. For the third step the situation is reversed and a Group 1 step must be chosen, and so on, alternating groups at each step.

In making computations on an electronic computer that employs the binary number system, it is convenient to rewrite the vectors using 0 and 1 to denote the components $+1$ and -1, respectively. Hence the above mentioned groups would now be represented as shown in Table II.

To ensure that all non-self-intersecting configurations generated by

TABLE II

BINARY REPRESENTATION OF TETRAHEDRAL LATTICE VECTORS

Group 1	Group 2
000	111
110	001
101	010
011	100

this process are equally probable, every conceivable step must be chosen with equal probability. Moreover, if the chain of vectors is to be non-self-intersecting, a necessary but not sufficient condition that must be fulfilled is that a given vector never be immediately followed by its negative. If the backward step is always forbidden, and if we always choose equally between the three remaining steps, we ensure that all steps are equally probable, no matter when taken.

Upon excluding the backward step, it is clear from Table I that a given vector in one group may only be followed by any one of the three vectors in the other group that differs from the preceding vector in the sign of only one coordinate. This is equivalent to complementing only one bit of the corresponding representation in Table II. Thus the three possible trials for succeeding steps can be generated from the last step taken, by simply complementing, one at a time, the separate bit positions of the last step. If the bit positions are numbered zero, one and two, the problem of generating a random step can be reduced to the problem of generating random integers in the range zero to two, inclusive. Such a random number generator has been used effectively (Wall *et al.*, 1954, 1955a, 1957a) and is simple and fast in operation. Essentially the process first involves multiplying a random fraction by the integer three. The integer part of this result, which must lie in the range zero to two inclusive, is the desired random integer; the fractional part of the result is retained as the random fraction for the next selection (Lehmer, 1951). The bit position of the last vector corresponding to the generated random integer is then complemented to obtain the next step. The configuration resulting from the addition of this new step is now tested for double occupancy.

3. *Checking for Loops*

The method to be discussed below which is suitable for the detection of double occupancy requires representing the steps as vectors. On most binary computers it is feasible to code the vectors so that all three components are stored in a single machine word. The details, however, will depend almost entirely upon the computer being used. To illustrate the technique, we shall describe the method developed for use on the Illiac, an automatic digital computer at the University of Illinois.

The Illiac has a 40-bit word length and uses 2's complement notation for negative number representation. The eight possible vectors were coded as shown in Table III. The x part is the high order 14 bits;

TABLE III

THE EIGHT TETRAHEDRAL VECTORS

Step	Step vector		
	x	y	z
000	00000000000001	0000000000001	0000000000001
001	00000000000001	0000000000000	1111111111111
010	00000000000000	1111111111111	0000000000001
011	00000000000000	1111111111110	1111111111111
100	11111111111111	0000000000001	0000000000001
101	11111111111111	0000000000000	1111111111111
110	11111111111110	1111111111111	0000000000001
111	11111111111110	1111111111110	1111111111111

the y part the next 13 bits; and the z part, the low order 13 bits. Examination of the entries in Table III readily discloses that vector (000) plus vector (111) equals zero; vector (001) plus vector (110) equals zero, etc.; this is in accordance with the fact that the pairs of vectors are made up of opposites. Moreover, it is always true that if we start at the lattice point representing the last position reached in the generation procedure and sum the appropriate step vectors back toward the origin, the occurrence of a machine zero will indicate a double occupancy.

It should be pointed out that this is not the only possible vector method, nor is the vector representation of Table III unique. Finally the actual coding methods will depend upon the computing equipment available.

4. *A Monte Carlo Program*

A program illustrating the methods outlined above is shown in Flow Chart 1. (It is a simplification of programs developed earlier at the University of Illinois.) The purpose of this computer program is to construct and permanently record complete N step configurations, where N is a predetermined number.

Provision is made for three sets of storage locations in the memory.

The first will be called *vector storage*; it consists of N locations that will eventually contain vectors of the type listed in Table III. The second storage area will be called *step storage* and it consists of $N/10$ locations. These locations will hold the pertinent steps of Table II packed 10 to a location. Because of the characteristics of Illiac, time is saved if the generated configurations are punched out in the step form of Table II rather than in the vector form of Table III. The last storage area is composed of three *bit locations*, each of which will contain a single bit position of the step immediately under consideration.

During operation, the first portion of the program calls for reading in the desired value of N, which is stored for later use as an end constant. The next section performs two functions. It first clears all locations in step storage, vector storage, and the three bit locations to zero. This is equivalent to choosing the first step to be (000), but since the direction of the coordinate axes is arbitrary, such a choice of first step does not involve any real loss of generality. The second function of this section is to load the proper Table III vector (in this case, the first one) into the first location of vector storage.

In the third section of the program a random integer is generated and the corresponding bit location complemented. The three bits from the appropriate locations are now put together to form the next step, which is then properly packed into step storage. An eight-way fork is then executed on the numeric value of the generated step and the appropriate vector placed into vector storage. The vectors in vector storage are now summed in reverse order (beginning with the last added vector), with a check for zero made after every pair of additions. Since all loops must, for geometrical reasons, contain an even number of steps, there is no need for a zero-sum check after addition of vectors corresponding to an odd number of partial additions.

Further action of the program now depends upon the results of the zero checks. If no zero occurs, and hence no double occupancy is found, the number of steps so far generated is compared to N, the desired number, and if smaller, the program returns to the random-number generator. If N steps have been successfully assembled, the $N/10$ words of step storage are punched out and the program returns to the clearing routine. The entire generation procedure is now repeated.

If a loop is found, the entire configuration is discarded, after which the program returns to the clearing routine to start over again. This procedure is required because it is statistically incorrect just to ignore the step causing the loop by removing it. Simply discarding a loop-causing step, and trying again, erroneously increases the probability of that portion of the configuration attained up to the failure.

5. Difficulties in the Simple Monte Carlo Program

The major difficulty inherent in the simple Monte Carlo program just described is a result of the large number of failures caused by double occupancies. Wall and his co-workers (1954, 1955a, 1955b) have made an extensive study of this problem, and their results can be expressed, for large s, by the following equation[1]:

$$P_s = \frac{W_s}{W_0} = \exp(-\lambda s) \qquad (3)$$

where

$P_s \equiv$ the probability that a configuration will reach s steps without a double occupancy,

$W_s \equiv$ the number of walks completed after s steps,

$W_0 \equiv$ the total number of walks attempted, and

$\lambda =$ a parameter, called the attrition constant, depending only on the lattice and type of walk under consideration.

A quantity $s_{1/2}$ is defined as the number of steps required to make the probability of occurrence of double occupancy equal to one-half. Clearly

$$s_{1/2} = (\ln 2)/\lambda . \qquad (4)$$

Table IV lists values of λ and $s_{1/2}$ for various lattice types as obtained by Monte Carlo methods.[2]

It can be seen that, even for the tetrahedral lattice, for which $s_{1/2}$ is 16.9 steps, the chance of a given configuration attaining a length of the order of 100 steps is very small. Several techniques have been devised to minimize this difficulty, one of which is an enrichment technique which will be discussed in the next section.

Although attrition is the major problem to be overcome in attaining configurations of more than a few steps, a strictly computer oriented problem arises in connection with timing. The lengthiest part of the program, timewise, is the loop-checking routine. A calculation of the total time required to generate a single successful N-step configuration shows that, when failures are taken into account, the time increases

---

[1] This formula has been proved to be asymptotically correct by Hammersley and Morton (1954).

[2] Analytical calculations of λ for certain lattices have also been carried out by Domb and Sykes (1961) and by Hiley and Sykes (1961).

TABLE IV
ATTRITION PARAMETERS FOR VARIOUS LATTICES[a]

Lattice	λ	$s_{1/2}$
3 choice square	0.128	5.4
2 choice square	0.226	3.1
4 choice simple cubic	0.103	6.7
5 choice simple cubic	0.061	11.4
7 choice body-centered-cubic	0.070	9.9
11 choice face-centered-cubic	0.086	8.0
Tetrahedral	0.041	16.9
6 choice orthogonal 4 dimensional	0.051	14.0

[a] Wall *et al.*, (1955b, 1957a, 1959a).

exponentially with N. This emphasizes the need for a method that avoids discarding an entire configuration when a loop is discovered.

B. Chain Enrichment Method

1. *Description and Mathematical Derivation*

From the foregoing discussion it is clear that to achieve results applicable to molecules of practical size, it is necessary to overcome the high attrition of samples due to closure or the intersection of chains. As mentioned in Section II, A, 5 it has been shown that the attrition can be represented by

$$W_s = W_0 \exp(-\lambda s). \tag{5}$$

To avoid the effect of high attrition, Wall and Erpenbeck (1959a) developed an unbiased sample enrichment technique for generating walks of steps comparable in number to the degrees of polymerization of moderate sized macromolecules. This was accomplished in the following manner: first consider W_s chains of s steps, related to the number of walks started, W_0, by the attrition equation above. Let us further consider each of the W_s chains as p chains, that is to say, assign each chain a weight of p. Now treat each of these pW_s chains separately in generating their extensions. A certain number of these can be expected to reach $2s$ steps, which number will be

$$W_{2s} = pW_s \exp(-\lambda s) = pW_0 \exp(-2\lambda s). \tag{6}$$

Consider again each of the W_{2s} chains p times and once again add random vectors to each. The number reaching $3s$ steps will be

$$W_{3s} = pW_{2s} \exp(-\lambda s). \tag{7}$$

Continuing this process, we see in general that

$$W_{is} = p \exp(-\lambda s) W_{(i-1)s} = p^{i-1} W_0 \exp(-i\lambda s) \qquad (i = 1, 2, ...,). \tag{8}$$

To render this chain enrichment most effective, s and the integer p should be so chosen that

$$\exp(\lambda s) = p + \delta \tag{9}$$

where δ, a positive number, is made as small as possible. This ensures that on the average one or less of the pW_{is} chains will reach $(i+1)s$ steps. On the other hand, if δ is made negative, then

$$p \exp(-\lambda s) > 1, \tag{10}$$

and an explosion of samples will occur. This would obviously alleviate the problem of attrition but it would give rise to a poor sample of the set of possible configurations, since the explosion phenomenon would generate a large total number of configurations with a small variety of chain beginnings.

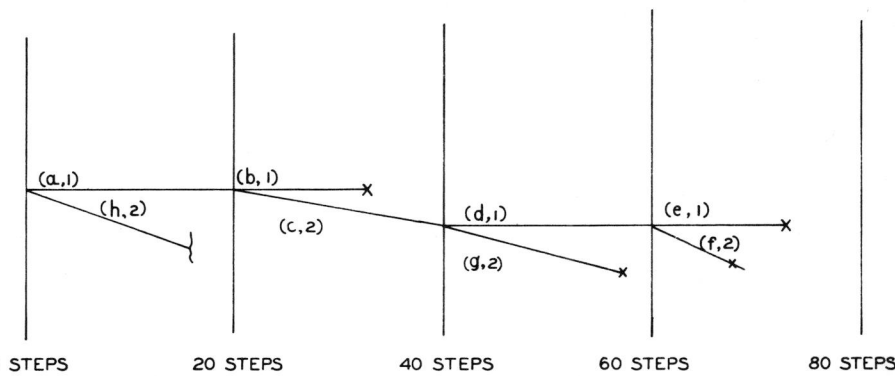

FIG. 2. Schematic representation of enrichment technique.

The enrichment technique is further illustrated in Fig. 2. Let us assign $s = 20$ and $p = 2$ for convenience of demonstration. Suppose that we have successfully constructed a segment of 20 steps $(a, 1)$ which contains no loops (see Fig. 2). We then proceed to grow another segment

(b, 1), but upon checking, a loop is discovered. At this stage, the complete sample is not discarded but the Monte Carlo computation reverts back to the end of the 20-step segment and an attempt is made to grow a new segment to 40 steps (c, 2). Let us suppose that this continues successfully through 60 steps (d, 1) and then fails twice when trying to go to 80 steps (e, 1) and (f, 2). The procedure then reverts back to 40 steps and fails (g, 2). At this point the generation process cannot start again at 20 steps since this has been tried twice, but must start from the origin (h, 2). A diagram of this type shows that for each 20-step segment completed ($s = 20$), two branches, successful or unsuccessful, are grown ($p = 2$).

2. Branching Routine

The program that will now be described utilizes the chain enrichment principle discussed in the preceding paragraphs. The procedure is illustrated in Flow Chart 2.

The first part of the general plan is identical to that described under the simple Monte Carlo program (Flow Chart 1). However, when a loop is found, instead of discarding the complete configuration, the program goes to what we shall call the S, P routine.

For purposes of illustration, let us suppose that $s = 40$ and $p = 5$. If N is the total length of polymer samples desired, then N/s additional locations known as the P counters (p_j's), are reserved in the memory of the computer. Let us further suppose that $N = 800$; in this instance 20 locations are reserved. Initially, the number p (that is to say, 5) is stored in the first P counter (p_0). Moreover, there is a segment counter which is increased by one each time a segment of s steps is completed. Upon completing a segment, the number p is systematically stored in the P counter corresponding to that segment (i.e., after the jth segment is completed the number five is stored in p_j). When a loop is found in segment j of the polymer, the program calls for subtracting one from the content of the p_{j-1} counter which is then tested for zero. If the resulting number is greater than zero, the segment has not been tried five times and the program goes to the random number generating routine. On the other hand, if the P counter check for zero shows that the segment has been tried five times, the program calls for subtracting one from the segment number, checking the segment number to see if it is negative, in which event the program goes to the clearing routine; otherwise the program continues as if a loop had just been found. This procedure continues until either a "negative" segment is reached or until a nonzero p_{j-1} counter is found.

When the polymer has grown to N steps (in our case 800) the program artificially creates a failure and goes to the S, P routine. This is done to fix an upper limit to size and to increase the number of samples with distinguishable final segments.

3. *Advantages and Difficulties*

With the choice of $p = 5$ and $s = 40$ for chains generated in a tetrahedral lattice, walks of 800 steps were easily obtained on the Illiac. The upper limit to the number of steps is determined by the memory capacity of the computer. Marcer (1960) has shown that by a simple extension of the above enrichment technique, walks of 2000 steps could be generated.

It should be noted that the choice of s and p, in relation to N, markedly influences the timing of the program. With N equal to 300, a typical run on the Illiac gave 256 segments in 1 hour for $s = 40$ and $p = 5$. On the other hand, with $s = 20$ and $p = 2$, 1700 segments were obtained in the same time period.

C. Biased Polymers

1. *Definition and Motivation*

Numerous investigators are interested in the effects of nearest neighbor interactions on polymer configurations. The physical chemical significance will not be discussed here, but it will be asserted that this is related to an important aspect of the statistical thermodynamics of polymers. For a statistical study of this problem to be effective, it is necessary to have polymer samples with a statistically significant number of intramolecular nearest neighbors. However, there are certain unusual configurations that may be energetically important, but which would occur very rarely in an unbiased Monte Carlo generation process. Examples of such configurations are those that are very tightly coiled and those that are almost completely uncoiled. Since such configurations may be most important under extreme conditions, a method of generation was sought that would yield an adequate sample of them.

With this goal in mind, we no longer select steps randomly with equal probability, but choose the steps in a biased way to achieve the desired result. This is done in such a fashion that the total bias for a given sample can be calculated and recorded. With this biasing factor known, the set of biased polymers can then be used in conjunction with a set of polymers generated without bias. This means, for example, that if a particular configuration occurs only once out of 100,000, we can find it once in a thousand tries if we introduce a 100 to 1 bias.

2. Derivation of Biasing Factors and Compatibility

The essential idea of the biased generation process in a tetrahedral lattice is to assign unequal probabilities for right (r), left (l), and trans (t) steps.

Suppose that we have so generated a biased polymer with a particular configuration in the tetrahedral lattice. The probability of obtaining such a configuration is

$$C^*(r, t, l, r, t, ..., t) = p^*(N_l, N_r, N_t \mid N) \cdot w^*(r, t, l, r, t ..., t) \quad (11)$$

$C^* \equiv$ probability of obtaining this configuration of N steps, with the first step r, the second t, etc., and

$p^* \equiv$ number of ways of obtaining N_r steps, N_l steps, N_t steps, out of a total of N steps, which is

$$\frac{N!}{N_r! \, N_l! \, N_t!} (p_r^*)^{N_r} (p_l^*)^{N_l} (p_t^*)^{N_t} \quad (12)$$

where

p_r^*, p_l^*, p_t^* are the probabilities of choosing a right, left, or trans step, respectively (for the unbiased case $p_r^* = p_l^* = p_t^* = \frac{1}{3}$),

$w^*(r, t, l, r, t, ..., t) =$ probability of obtaining the particular order indicated in a given configuration, and

r, l, t refer, as implied above, to the direction of the step on a tetrahedral lattice.

The frame of reference depends, of course, upon the last two steps (see Fig. 3). Thus a trans (t) step vector is exactly the same as the next to last step vector. It follows from the discussion above that

$$C^* = \frac{N!}{N_r! \, N_l! \, N_t!} (p_r^*)^{N_r} (p_l^*)^{N_l} (p_t^*)^{N_t} \cdot w^* . \quad (13)$$

Now suppose that this same configuration has been generated in an unbiased manner (say by the simple Monte Carlo method). The probability of so obtaining this configuration, C, is

$$C(r, t, l, r, t, ..., t) = p(N_r, N_l, N_t \mid N) \cdot w \quad (14)$$

where the terminology is the same as above (an asterisk denotes the biased case). Hence

$$C(r, t, l, r, t, ..., t) = \frac{N!}{N_r! \, N_l! \, N_t!} (p_r)^{N_r} (p_l)^{N_l} (p_t)^{N_t} \cdot w \quad (15)$$

where now $p_r = p_l = p_t = \frac{1}{3}$.

Dividing Eq. (13) into Eq. (15), we get

$$\omega = \frac{C}{C^*} = \left(\frac{p_r}{p_r^*}\right)^{N_r} \left(\frac{p_l}{p_l^*}\right)^{N_l} \left(\frac{p_t}{p_t^*}\right)^{N_t} \quad (16)$$

where ω equals the weighting factor. It should be noted that the combinatorial factor in the probability of obtaining a given order of steps is independent of bias and hence cancels out when one forms the quotient C/C^*; in other words, $w = w^*$.

Since all possible N step excluded volume configurations of the unbiased set have the same probability, we can arbitrarily set $C = 1$. Equation (16) then becomes

$$\omega = \left(\frac{p_r^*}{p_r}\right)^{-N_r} \left(\frac{p_l^*}{p_l}\right)^{-N_l} \left(\frac{p_t^*}{p_t}\right)^{-N_t} = (p_r^*)^{-N_r}(p_l^*)^{-N_l}(p_t^*)^{-N_t}(3)^{-N}. \quad (17)$$

This expression for ω is the weighting factor of the biased polymer in relation to the unbiased set. p_r^*, p_l^*, and p_t^* depend, of course, upon the bias incorporated into the program and N_r, N_l, and N_t are obtained in the computation by counting the number of right, left, and trans steps taken during the generation process.

3. Bias Routine

The program described in this section is based on recent work of the authors (1962). This differs from the previously described S, P routine primarily in the biasing technique and in the mode of addition

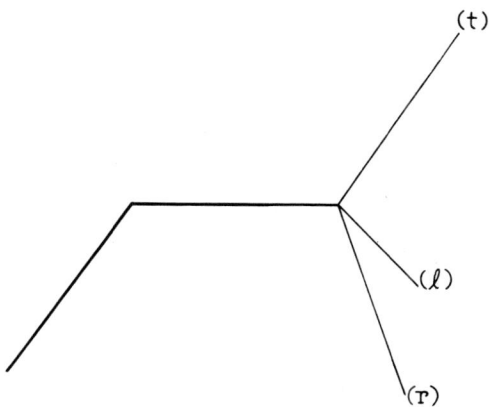

FIG. 3. Diagram showing the direction of steps (left, right, and trans) on a tetrahedral lattice.

of successive steps to the polymer. The other parts, that is to say, loop checking, storage of the polymer, counter, etc., are sufficiently similar to those of the previously described programs so that further discussion of them is unwarranted.

First of all, a random number is generated in a suitable range, let us say from 1 to 30 inclusive (see Flow Chart 3). This integer is employed in the bias routine which, depending on the bias in use, determines a left, right, or trans step. (Thus for a $\frac{2}{5}$ left, $\frac{2}{5}$ right, and $\frac{1}{5}$ trans bias, numbers in the range of 1 to 12 inclusive out of 1 to 30, could result in a left step, 13 to 24 in a right step, and 25 to 30 in a trans step.) These steps are coded in the machine as 1, 2, and 0. To identify a particular step as either left, right, or trans, it is necessary to know the previous two steps (see Fig. 3); therefore the previous two steps are continuously kept in a pair of set storage locations.

A table is stored in the memory of the computer to hold the eight possible steps of Table II. These eight steps are individually broken down into groups classified according to what the next to the last step might have been. One can then select a new step depending on whether it is denoted by 0, 1, or 2 as determined by the means described above. An eight-way switch in the program identifies the last step and a three-way switch then identifies the proper combination of the last two steps. From the integer selected (0, 1, or 2) the new step is determined.

The total numbers of left, right, and trans steps are stored in memory to be used later in calculating the bias factor by an assembly program. Checking for loops is accomplished in the same manner as in the previous program.

D. Assembly Routines

1. *Introduction*

Routines that produce non-self-intersecting random walks segmentally, such as the Wall-Erpenbeck program, suffer from two interrelated difficulties; they produce walks of all possible numbers of segments intermixed, and the segments of these walks are not produced consecutively. In operation these segments are stored in some secondary storage medium (a magnetic drum or tape) as they are produced. It is then the function of a separate program or subprogram to reread the secondary storage medium and produce correct configurations as output.

The general problem and its solution are readily illustrated. If the various segments are denoted by their segment numbers, and distinguished from each other by primes, then a typical output on the secondary storage medium might be

$$1\ 2\ 3\ 2'\ 2''\ 3''\ 1'''\ 2''' \ldots .$$

Thus segments 1 and 1''' represent possible first segments of s steps each, 2, 2', 2'', and 2''' second segments of s steps, etc. This output contains two one segment configurations (1 and 1'''); four configurations of two segments (1 2, 1 2', 1 2'', and 1''' 2'''), and two of three segments (1 2 3 and 1 2'' 3''). It should be noted that combinations like 1 2''', 1 2 3'', or 1 2' 3'' are not acceptable for there is no guarantee that 2''' is compatible with 1 or that 3'' is compatible with 1 2 or 1 2'. The correctness of this assertion becomes clear from an analysis of the S, P routine itself. Thus, to perform proper assembly, two pieces of information are needed. These are the order in which the segments were produced, which is presumed to be preserved in the secondary storage medium, and the segment number itself, which must be stored along with the actual segment configuration. The assembly program can also be used to sort the assembled polymer configurations to permit punching out in a group all configurations containing the same number of segments.

2. *An Assembly Method*

A procedure employed on the Illiac will be used to illustrate an assembly method. The secondary storage medium of the Illiac is a high-capacity magnetic drum. Sufficient space is set aside in the high-speed memory to hold the largest number of segments to be encountered in a single configuration. Assembly proceeds sequentially with the polymers of one segment being produced first, followed by configurations with successively larger numbers of segments.

Starting the reading of the magnetic drum with the first segment produced, segments are read into the high-speed memory and stored or not stored depending on their segment numbers. If configurations of n segments are being assembled, no configuration with segment number greater than n will be stored. If a segment number has a value less than or equal to n, that segment will replace any previously stored segment with the same segment number in the high speed memory. Output punching will occur whenever, and only when, the last segment number read is equal to n. Thus the entire configuration will be punched, from segment 1 up to and including segment n. This process continues until all stored segments have been scanned. After all segments have been scanned, the program is automatically reset to repeat the entire process, using, however, the number $n + 1$ instead of n. In this way all valid configurations are assembled and sorted into groups as required.

III. Results and Conclusions

A. GENERAL DISCUSSION

The earlier discussion in Sections I and II of this chapter was concerned with the methods that could be employed for generating possible polymer configurations. Such generation of conceivable configurations is, of course, a major step toward attaining a theoretical understanding of the structure of polymers. After the polymer configurations have been obtained, however, it is still necessary to compute certain data of practical interest. The programs to accomplish this are of a data processing character; the polymer configurations are utilized as data and are treated much as if they were determined experimentally in the laboratory. Indeed, Monte Carlo calculations can be regarded as machine experiments, and the data so obtained can be analyzed and subjected to empirical or such other treatments as may appear warranted. In this way, studies have been made of a number of polymer properties such as the mean-square end-to-end distance, the mean radius of gyration, the average number of nearest intramolecular polymer chain neighbors, etc. Fortunately, the representation used in the generation of configurations (that of Table II) is very convenient for the direct computation of quantities such as the square of the end-to-end separation.

Let us consecutively number the atoms in a polymer chain, starting with zero for the atom at the origin. Then let $\mathbf{d}_i$ be the vector from the origin to the atom numbered i in the configuration, and let $\mathbf{s}_i$ be the vector step from the $(i-1)$st atom to the ith atom. Then clearly

$$\mathbf{d}_{i+1} = \mathbf{d}_i + \mathbf{s}_{i+1} \tag{18}$$

and

$$d_{i+1}^2 = d_i^2 + s_{i+1}^2 + 2\mathbf{s}_{i+1} \cdot \mathbf{d}_i . \tag{19}$$

If x_i, y_i, and z_i are the components of $\mathbf{d}_i$, and a_{i+1}, b_{i+1}, and c_{i+1} are the corresponding components of $\mathbf{s}_{i+1}$, then

$$\mathbf{s}_{i+1} \cdot \mathbf{d}_i = a_{i+1}x_i + b_{i+1}y_i + c_{i+1}z_i . \tag{20}$$

Thus,

$$d_{i+1}^2 = d_i^2 + s_{i+1}^2 + 2(a_{i+1}x_i + b_{i+1}y_i + c_{i+1}z_i) . \tag{21}$$

The vector $\mathbf{s}_{i+1}$ has several special properties. First, its components, a_{i+1}, b_{i+1}, c_{i+1}, may only have either of two values, ± 1, as can be seen from Table I. Moreover, $s_i^2 = 3$, irrespective of i, a fact that follows from the first property. These facts can be used to simplify Eq. (21), which can then be written as

$$d_{i+1}^2 = d_i^2 + 3 + 2(\pm x_i \pm y_i \pm z_i). \tag{22}$$

From the above equation an algorithm can be established to govern the iterative formation of d_i^2 for any i and hence R_N^2, the square end-to-end distance of the completed chain, as follows: the components of

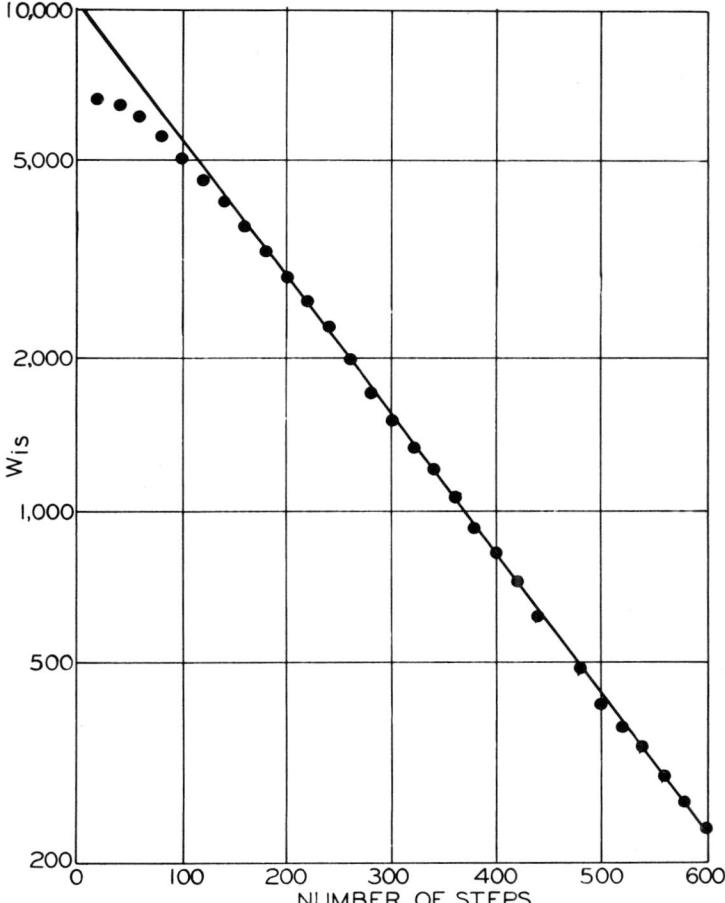

FIG. 4. Graph of ln W_{is} versus the number of steps for a tetrahedral lattice in which $s = 20$ and $p = 2$. The circles are the experimental points. (Figure reproduced by courtesy of the *Journal of Chemical Physics*.)

the total distance vector are each either added to or subtracted from the accumulator depending upon whether the corresponding component of the last added step in its binary representation (Table II) is 0 or 1, respectively. This result is doubled and added to the current value of d_i^2 plus three to obtain d_{i+1}^2. Finally, x_i, y_i, and z_i are augmented or diminished by 1, the choice again depending upon the presence of a 0 or 1 in the corresponding binary representation of the last step.

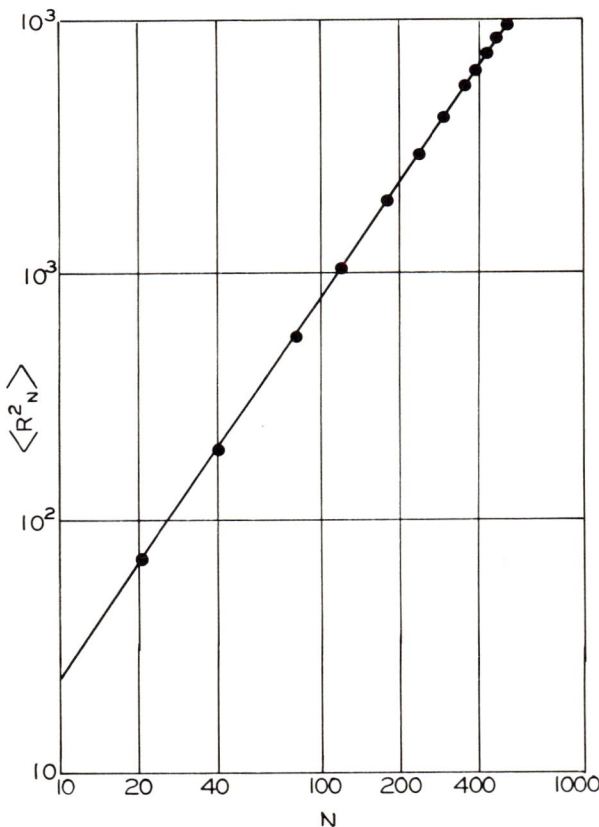

FIG. 5. Graph of ln $\langle R_N^2 \rangle$ versus ln N for a three-choice, two-dimensional square lattice. Each step is of unit length. (Figure reproduced by the courtesy of the *Journal Chemical Physics*.)

In actual practice the programming is best accomplished by means of an eight-way switch determined by the numeric value of the last added step. The time saved by this algorithm is considerable, since the three multiplications normally required to compute a given d_i^2 are replaced by five additions and a single digit shift.

Similarly, other properties can be computed by variations of this technique. If the x, y, z components of the distance $\mathbf{d}_i$ are accumulated, these sums divided by $N+1$, the number of atoms in the configuration, yield the vector position of the center of gravity. Another modification of the method enables one to calculate the radius of gyration, etc. (Wall and Erpenbeck, 1959b).

B. Empirical Analysis of Monte Carlo Calculations

From Eq. (8), it is clear that if $\ln W_{is}$ is plotted against the number of steps, a straight line should be obtained with slope equal to

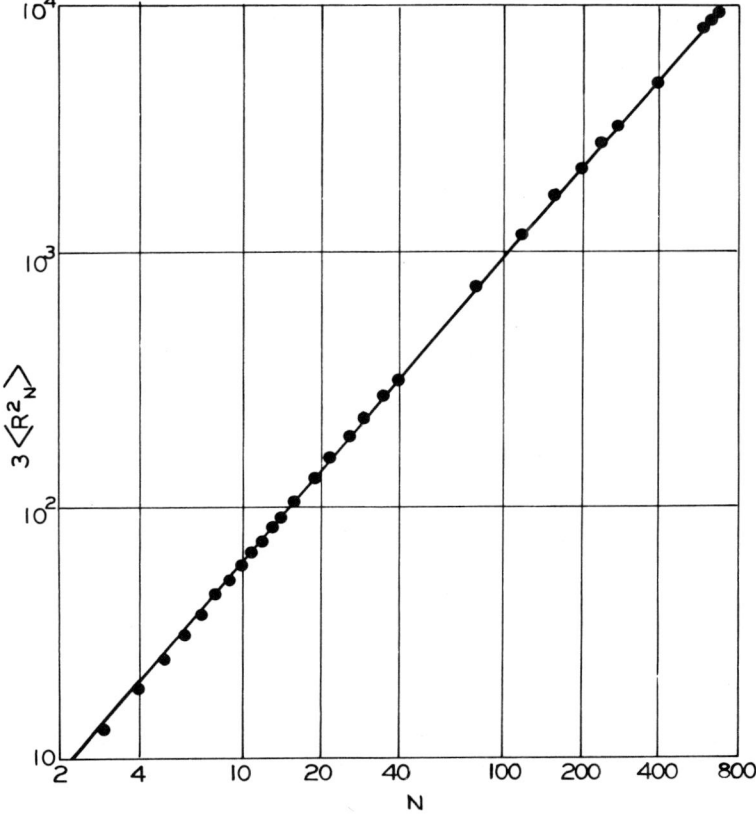

Fig. 6. Graph of $\ln 3 \langle R_N^2 \rangle$ versus $\ln N$ for a tetrahedral lattice. Each step is of unit length. (Figure reproduced by the courtesy of the *Journal of Chemical Physics*.)

$-\lambda + (1/s) \ln p$. Figure 4 is a plot of $\ln W_{is}$ versus is. The attrition constant λ is evaluated from the slope, and for the tetrahedral lattice it is found to be 0.041. Analysis of data for other types of lattices have been made and the results have been summarized in Table IV.

Figures 5 and 6 are plots of $\ln \langle R_N^2 \rangle$ versus $\ln N$ for the three-choice, two-dimensional square lattice and for the tetrahedral lattice.

For steps of unit length and for large N, the curves shown in these figures obey the following empirical equations:

$$\langle R_N^2 \rangle = 0.80\, N^{1.50} \quad \text{(2 dimensions)} \tag{23}$$

$$\langle R_N^2 \rangle = 1.40\, N^{1.18} \quad \text{(3 dimensions)}. \tag{24}$$

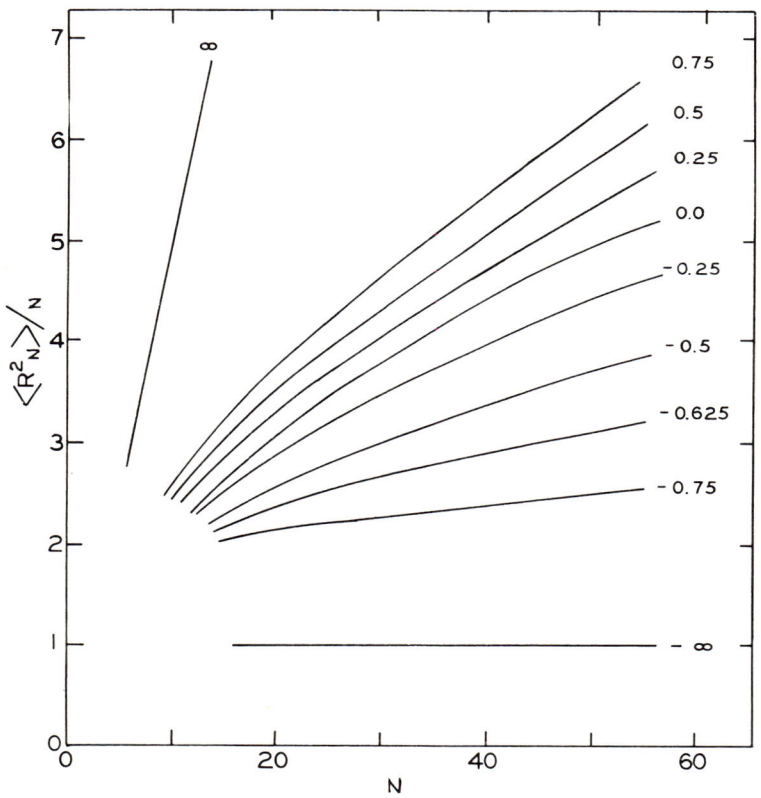

FIG. 7. Graphs of $\langle R_N^2 \rangle / N$ versus N for the two-choice square lattice for various values of ϵ/kT indicated next to the curves. (Figure reproduced by the courtesy of the *Annals of the New York Academy of Sciences*.)

From these results it appears that an equation of the form

$$\langle R_N{}^2 \rangle = aN^b \qquad (25)$$

should be expected to hold for real systems. From the apparent linearity of the curve in Fig. 6, one would expect that, if the exponent b is not constant, it would change only slowly with N.

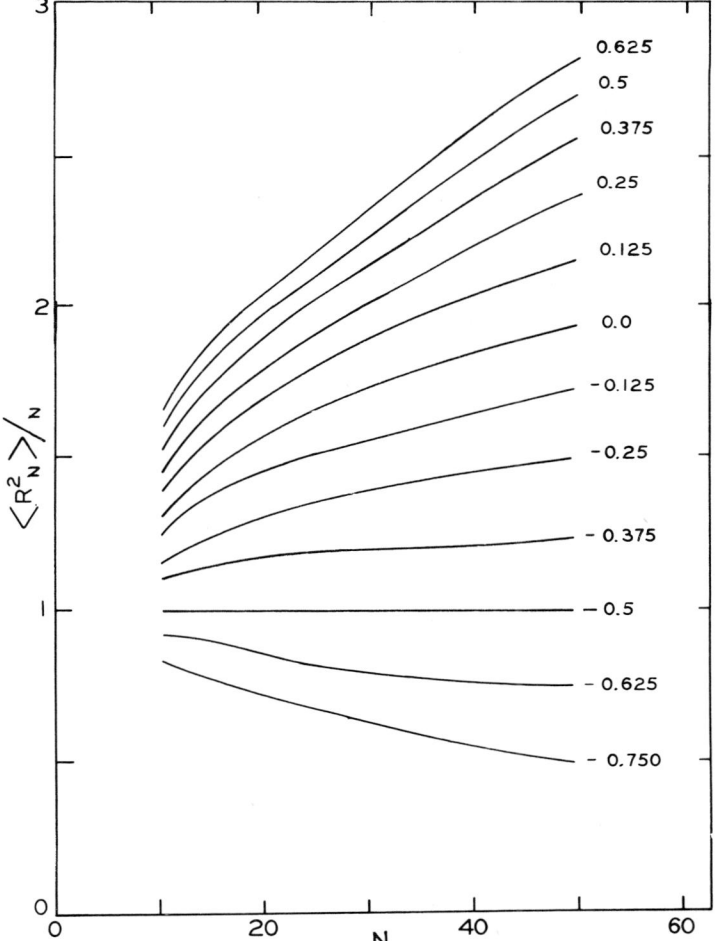

FIG. 8. Graphs of $\langle R_N{}^2 \rangle / N$ versus N for the four-choice cubic lattice for various values of ϵ/kT indicated next to the curves. (Figure reproduced by the courtesy of the *Annals of the New York Academy of Sciences*.)

Wall and Mazur (1961) studied the statistical mechanics of coiling-type polymers by taking into account intramolecular interactions. In approaching this study they assumed a more realistic potential function than had been implied in the previous Monte Carlo treatments. Not only was double occupancy forbidden, but they also stipulated a finite nonzero interaction energy for nonbonded atoms occupying adjacent lattice positions. Figures 7 and 8 show results obtained for mean-square end-to-end distances of polymer chains versus the number of steps in the chains for different values of ϵ/kT, where ϵ equals the potential energy of interaction of nonbonded nearest neighbors.

It will be seen that the chain lengths increase with increasing ϵ as would be expected. It is interesting to note that, for $\epsilon/kT = -0.50$ for the three-dimensional simple cubic lattice, the nearest neighbor attraction just compensates for the excluded volume effect. For a given ϵ, the temperature at which this occurs is known as the Flory Θ temperature.

The work of Wall and Mazur was limited to $\epsilon \leqslant 0.5kT$ because of scattering of data above 40 steps. The bias generation technique discussed in Section II, C was specifically developed to overcome this shortcoming. Presently, the authors are concerned with extending this work to larger N and for wider ranges of ϵ/kT.

KEY FOR FLOW CHART 1:
    i, j = counters
    T = temporary storage
    N = maximum number of steps
    v_k = vector corresponding to step whose numeric value is k

KEY FOR FLOW CHART 2:
    p_j = storage location of the p's
    j = segment number
    P = number of branches at each segment
    S = number of steps in a segment

POLYMER CONFIGURATIONS 241

FLOW CHART 1. Simple Monte Carlo.

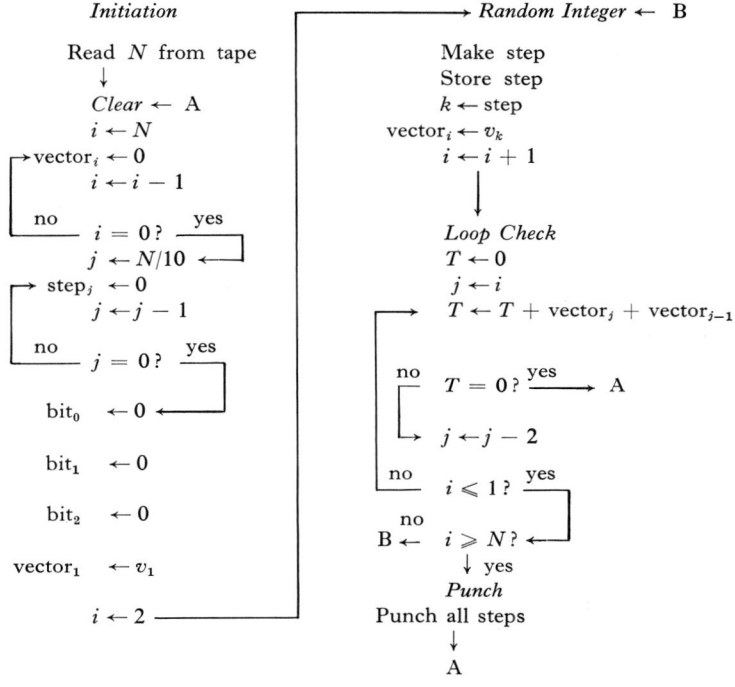

FLOW CHART 2. S, P routine.

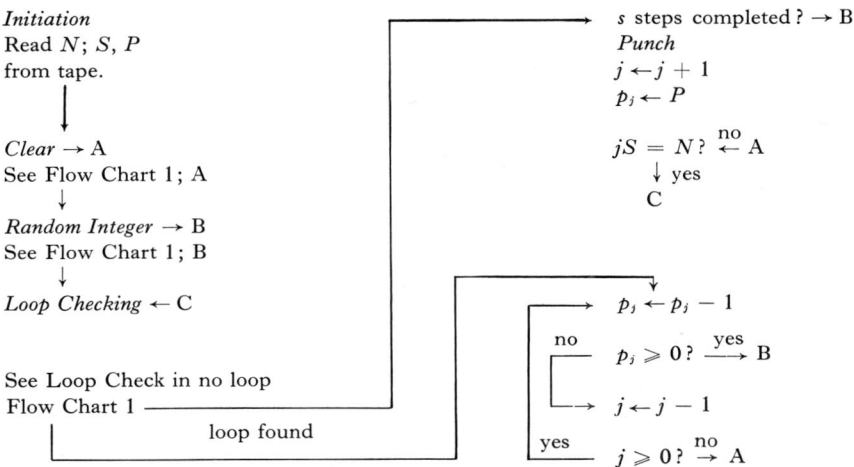

FLOW CHART 3. Biased polymers.

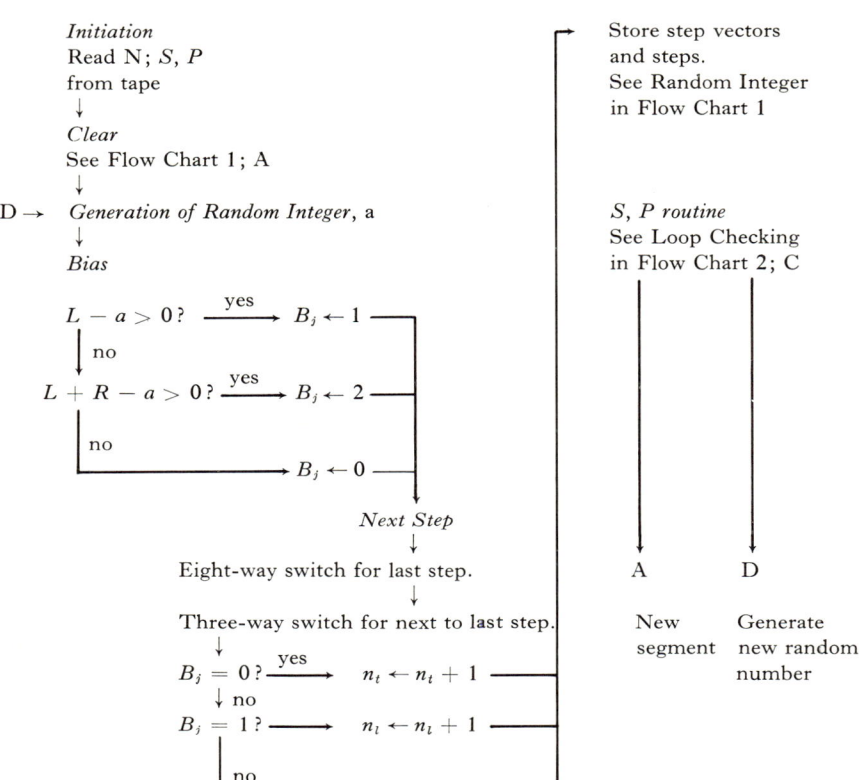

KEY: a = random integer generated
L = left constant
R = right constant
T = trans constant where
$a \leqslant R + L + T$
B_j = storage position of bias
(0 trans)
(1 left)
(2 right)

n_l = storage location of left bias counter
n_r = storage location of right bias counter
n_t = storage location of trans bias counter

plus definitions from Flow Chart 2

REFERENCES

BENOIT, H. (1948). *J. Polymer Sci.* **3**, 376.
BUECHE, F. (1953). *J. Chem. Phys.* **21**, 205.
CHANDRASAKER, S. (1943). *Revs. Modern Phys.* **15**, 3.
DOMB, C., and SYKES, M. F. (1961). *J. Math. Phys.* **2**, 63.
FLORY, P. J. (1949). *J. Chem. Phys.* **17**, 303.
FLORY, P. J. (1953). "Principles of Polymer Chemistry." Cornell Univ. Press, Ithaca, New York.
FRISCH, H. L., COLLINS, F. C., and FRIEDMAN, B. (1951). *J. Chem. Phys.* **19**, 1402.
GRIMLEY, T. B. (1952). *Proc. Roy. Soc.* **A212**, 339.
HAMMERSLEY, J. M., and MORTON, K. W. (1954). *J. Roy. Statistical Soc.* **B16**, 23.
HERMANS, J. J. (1950). *Rec. trav. chim.* **69**, 220.
HILEY, B. J., and SYKES, M. F. (1961). *J. Chem. Phys.* **34**, 1531.
HILL, T. (1960). "Introduction to Statistical Thermodynamics." Addison-Wesley, Massachusetts.
JAMES, H. M. (1953). *J. Chem. Phys.* **21**, 1628.
KING, G. W. (1949). Monte Carlo Method. *Nat. Bur. Standards Appl. Math. Ser.* **12**.
KUHN, W. (1934). *Kolloid-Z.* **68**, 2.
KUHN, W., and KUHN, H. (1943). *Helv. Chim. Acta.* **26**, 1394.
LEHMER, D. H. (1951). *Harvard Univ. Computation Lab. Ann.* **26**, 141.
MARCER, P. J. (1960). Ph. D. Thesis. Trinity College, Oxford.
MONTROLL, E. W. (1950). *J. Chem. Phys.* **18**, 734.
ROSENBLUTH, M. N., and ROSENBLUTH, A. W. (1955). *J. Chem. Phys.* **23**, 356.
RUBIN, R. J. (1952). *J. Chem. Phys.* **20**, 1940.
SYKES, M. R. (1961). *J. Math. Phys.* **2**, 52.
TAYLOR, W. J. (1948). *J. Chem. Phys.* **16**, 257.
WALL, F. T., and ERPENBECK, J. J. (1959a). *J. Chem. Phys.* **30**, 634.
WALL, F. T., and ERPENBECK, J. J. (1959b). *J. Chem. Phys.* **30**, 637.
WALL, F. T., and MAZUR, J. (1961). *Ann. Acad. Sci.* **89**, 608.
WALL, F. T., HILLER, L. A., Jr., and WHEELER, D. J. (1954). *J. Chem. Phys.* **22**, 1036.
WALL, F. T., HILLER, L. A., Jr., and ATCHISON, W. F. (1955a). *J. Chem. Phys.* **23**, 913.
WALL, F. T., HILLER, L. A., Jr., and ATCHISON, W. F. (1955b). *J. Chem. Phys.* **23**, 2314.
WALL, F. T., HILLER, L. A., Jr., and ATCHISON, W. F. (1957a). *J. Chem. Phys.* **26**, 1742.
WALL, F. T., RUBIN, R. J., and ISAACSON, L. M. (1957b). *J. Chem. Phys.* **27**, 186.
WALL, F. T., WINDWER, S., and GANS, P. J. (1962). *J. Chem. Phys.* **37**, 1461.
ZIMM, B. H., STOCKMAYER, W. H., and FIXMAN, M. (1953). *J. Chem. Phys.* **21**, 1716.

Monte Carlo Computations on the Ising Lattice

L. D. Fosdick

DIGITAL COMPUTER LABORATORY, UNIVERSITY OF ILLINOIS
URBANA, ILLINOIS

    I. Introduction 245
   II. The Ising Lattice 246
  III. Theory of the Monte Carlo Method for Estimating the Boltzmann Averages 249
  IV. Practical Considerations of the Computation 258
   V. The Square Ising Lattice 261
 VI. The Simple Cubic Lattice 265
VII. The Body-Centered Cubic Lattice 275
VIII. Estimation of the Critical Point 278
 IX. Conclusion 279
    References 280

I. Introduction

THE WORK DESCRIBED HERE is concerned with the use of a Monte Carlo sampling technique which was first introduced by Metropolis *et al.* (1953) in the computations of the equation of state of a hard-sphere gas. This technique was later used successfully by others (Rosenbluth and Rosenbluth, 1954; Wood and Parker, 1957; Wood and Jacobson, 1957; Salsburg *et al.*, 1959) in further computations on the equation of state of gases. The object of this Monte Carlo scheme is to provide estimates of the statistical mechanical Boltzmann averages of parameters of a system of interacting particles: thus, if $F(q)$ is some parameter of the system depending on its state q, and $E(q)$ is the energy of the state q, then this Monte Carlo technique provides an estimate of

$$\langle F \rangle = \frac{\int F(q)\, e^{-E(q)/kT}\, dq}{\int e^{-E(q)/kT}\, dq}. \tag{1}$$

The estimate is formed in the following way. One considers a system of N particles with the state q_t of the system at time t specified by the coordinates of the N particles. Using certain transition probabilities, $p_{qq'}$, giving the probability that state q at time t will be followed by state q' at time $t + 1$, a random sequence $q_1, q_2, q_3, ..., q_M$ of states is generated. The transition probabilities are chosen in such a way that the states in this sequence have a distribution which is approximately the Boltzmann distribution and consequently it is possible to obtain an estimate of $\langle F \rangle$ by forming

$$\langle F \rangle_M = \frac{1}{M} \sum_{t=1}^{M} F(q_t). \tag{2}$$

This method allows one to deal directly with the system at the microscopic level and it is therefore possible to obtain insights into the detailed behavior of the system which are sometimes hidden in more conventional analytical methods. It seems appropriate to describe this approach as a "mathematical experiment" because it is somewhat analogous to observing the parameters directly in the real physical system, as in a physical experiment. In the latter case nature provides the averaging, whereas in the mathematical experiment this is simulated on a model. This analogy holds in a statistical sense only, and the time development of the states does not necessarily correspond to the time development of the states in a real system.

In its present application this method is used to calculate estimates of the average values of short-range and long-range order parameters in an Ising lattice. The results of this work, and related work were reported (Fosdick, 1959; Ehrman *et al.* 1960) earlier by the author and the Ising lattice results are presented here again. A more extensive discussion of the theoretical basis for this method of computation than has appeared elsewhere is also presented here. Guttman (1961) has made similar computations on the body-centered cubic Ising lattice.

The computations described here were performed on the Illiac, a high-speed computing machine located at the Digital Computer Laboratory of the University of Illinois. The author was assisted in this work by J. R. Ehrman and D. C. Handscomb, and this work was supported in part by the Office of Naval Research.

II. The Ising Lattice

The Ising lattice was originally proposed as a model for ferromagnetism in 1925 and the reader is referred to Guttman (1956), Newell and Montroll (1953), and ter Haar (1960) for surveys of the work done on this model. The Ising lattice consists of a regular array of sites with each site occupied by a spin, or dipole, which is allowed only two orientations, "up" and "down." The orientation of the spin on a site is represented by the two-valued spin coordinate $\mu(k)$, where k denotes the kth lattice site, and the two values for $\mu(k)$, are $+1$ and -1. The configuration state of a lattice composed of N sites is represented by the N-tuple $\mu = \{\mu(1), \mu(2), ..., \mu(N)\}$. It is sometimes convenient to think of μ as a binary number of N places, each value of μ corresponding to a distinct configuration. The spins on the lattice sites are assumed to interact only with their nearest neighbors and with an external magnetic field, the energy for a particular configuration being given by

$$E_\mu = -J \sum_{k',k}^{(1)} \mu(k) \mu(k') + H \sum_k \mu(k) \qquad (3)$$

where the first summation extends over all nearest-neighbor pairs of sites (k', k) and the second sum extends over all sites in the lattice. When the coupling parameter J is positive the array of spins is called ferromagnetic since parallel nearest-neighbor spins are energetically favored, and when J is negative the array is called antiferromagnetic since antiparallel nearest-neighbor spins are energetically favored.

It is a well-known fact that the Ising lattice is a very poor model of a ferromagnetic substance. It is a somewhat better model for describing order-disorder phenomena in a binary alloy, and it has also been used to describe a gas, called the lattice gas. However, as a model for a physical system it suffers from a number of deficiencies and the primary reason for continued interest in the Ising lattice is that it is the simplest known model for a system of interacting particles exhibiting some of the important physical characteristics of such a system. In particular, this model exhibits a phase transition at a critical temperature T_c. Below T_c there are long-range correlations between the spins described as long-range order, above T_c these long-range correlations vanish but short-range correlations persist. At the critical temperature the specific heat becomes infinite in a two-dimensional Ising lattice and probably also in a three-dimensional Ising lattice; strictly speaking, this infinity appears only in the limit as the number of spins in the lattice becomes infinite.

The correlations between the spins in the lattice, that is the order in the lattice, is described by certain parameters. The short-range order in the lattice is described here by the parameter $f_\mu(j)$, equal to the fraction of jth neighbor sites which are occupied by an antiparallel pair of spins in the μth configuration, thus

$$f_\mu(j) = \frac{1}{\alpha(j) N} \sum_{k',k}^{(j)} [1 - \mu(k) \mu(k')] \qquad (4)$$

where $\alpha(j)$ is the number of jth neighbors of a lattice site, and the summation extends over all pairs of sites k and k', where k and k' are jth neighbors. The long-range order in the lattice is described by the parameter S_μ, where

$$S_\mu = \frac{1}{N} \left| \sum_k \mu(k) \right|, \tag{5}$$

the summation extending over all lattice sites.

The simplicity of the Ising model is deceptive. An exact treatment of the statistics of the two-dimensional Ising lattice finally appeared in 1944, (Onsager, 1944; see also Kaufman, 1949; Kaufman and Onsager, 1949) a long while after the model was originally proposed, and the mathematical machinery required for this is formidable. An exact treatment of the statistics of the three-dimensional lattice does not exist. Some time ago H. M. James suggested to the author the possibility of applying the Monte Carlo method to the Ising model following the ideas of Metropolis, Rosenbluth, and Teller in their application of the Monte Carlo method to the hard-sphere gas. This was the starting point of the work described here.

The computations described here are concerned with obtaining estimates of the average values for the order parameters where the averages are computed with respect to the Boltzmann distribution; thus, the average value of the short-range order parameter is given by

$$f(j) = \sum_\mu f_\mu(j) \exp\left(-\frac{E_\mu}{kT}\right) \bigg/ \sum_\mu \exp\left(-\frac{E_\mu}{kT}\right), \tag{6}$$

and the average value of the long-range order parameter is given by

$$S = \sum_\mu S_\mu \exp\left(-\frac{E_\mu}{kT}\right) \bigg/ \sum_\mu \exp\left(-\frac{E_\mu}{kT}\right) \tag{7}$$

where the summations extend over all μ. Since the dimensionless quantities J/kT and H/kT appear in the exponent E_μ/kT it will be convenient to define the parameters

$$K = \frac{J}{kT}, \qquad L = \frac{H}{kT}. \tag{8}$$

When $L = 0$ we note that $f(1)$ is, excepting an additive constant, proportional to the average configurational energy of the lattice. Also when $L = 0$, S can be identified as the spontaneous magnetization per spin

(Yang, 1952). This can be seen by noting that for a lattice composed of an infinite number of spins a very small magnetic field removes all states from the ensemble average for which the total spin is antiparallel to the field. For a finite lattice this of course is not strictly true and the error in this approximation will be worst in the neighborhood of T_c.

Three lattice geometries have been considered in these computations: the square lattice, the simple cubic lattice, and the body-centered cubic lattice. Since exact results were known for the square lattice it provided a means for testing the accuracy of the computations. For the square lattice and body-centered cubic lattice estimates of $f(1)$ were obtained. For the simple cubic lattice estimates of $f(1)$ and $f(2)$ were obtained. Estimates of the long-range order S were obtained for the simple cubic lattice and the body-centered cubic lattice. The computations were performed for an extensive range of values for K and L.

It is assumed throughout these computations that periodic boundary conditions are imposed on the system. Thus the opposite edges, or surfaces, of the lattice are joined by nearest-neighbor bonds; in the square lattice the sites on the right edge are nearest neighbors of the corresponding sites on the left edge and the sites on the top edge are nearest neighbors of the corresponding sites on the bottom edge, as if the lattice were on the surface of a toroid.

III. Theory of the Monte Carlo Method for Estimating the Boltzmann Averages

In this discussion we will let $F(\mu)$ represent a parameter of the lattice depending on the configurational state and the Boltzmann average will be represented by $\langle F \rangle$, thus

$$\langle F \rangle = \sum_\mu F(\mu) \exp\left(-\frac{E_\mu}{kT}\right) \Big/ \sum_\mu \exp\left(-\frac{E_\mu}{kT}\right). \tag{9}$$

The evaluation of any term in the above sums is a simple matter once μ is given, and although it is certainly possible, in principle, to evaluate each of the 2^N terms in numerator and denominator and form the sums, it is clearly not a practical technique even for the fastest computers. For example, with a square lattice having ten sites on an edge and 1 μsec (a gross underestimate) to evaluate each term in the sum, the time to evaluate the 2^{100} terms would be about 3×10^{16} years.

The complete evaluation of all of the terms in the summations might be replaced by a randomly selected subset of terms to form an estimate of $\langle F \rangle$. Suppose that there is a random process for selecting values of μ where the probability $p(\mu)$ for selecting any μ is 2^{-N}; that is, every configuration has equal probability of being selected. Suppose that the sequence of μ's selected is $\mu_1, \mu_2, ..., \mu_M$ where the subscript denotes the sequential order of selection, and not the value, of μ; thus, in such a sequence it is possible to have $\mu_i = \mu_j$. An estimate of $\langle F \rangle$ is then given by

$$\langle F \rangle_M = \sum_{t=1}^{M} F(\mu_t) \exp\left(-\frac{E_{\mu_t}}{kT}\right) \bigg/ \sum_{t=1}^{M} \exp\left(-\frac{E_{\mu_t}}{kT}\right). \tag{10}$$

This technique is called simple Monte Carlo sampling. It is not likely that this technique would be very fruitful because the principal contribution to the average will usually come from a very small fraction of the 2^N configurations and a reasonable estimate would require such an enormous value for M that the evaluation of the estimate would still be beyond the capability of any computer.

The poor results that one can expect from simple Monte Carlo sampling can be indicated more explicitly in the following way. When the configurations are equally probable there will be a very large probability for selecting configurations with the long-range order parameter S, Eq. (5), equal to, or almost equal to zero. Regarding the product SN as the sum of N independent random variables [see Eq. (5)], each with average value 0 and variance 1, the central limit theorem can be used to obtain the probability of selecting a configuration with $S > \epsilon$:

$$\Pr\{S > \epsilon\} = 1 - \frac{1}{\sqrt{2\pi}} \int_{-\sqrt{N}\epsilon}^{\sqrt{N}\epsilon} \exp\left(-\frac{y^2}{2}\right) dy. \tag{11}$$

For temperatures below the critical temperature it is known that the Boltzmann average of the long-range order rapidly rises to values in the neighborhood of one, but it is clear from the above formula that the probability of selecting a configuration with S in the neighborhood of 1 with N of the order of 100 is vanishingly small. Hence, one would never expect good estimates below the critical temperature in a reasonable amount of computing time.

A second and more promising technique for the evaluation of the sums in Eq. (9) is importance sampling (Kahn, 1956). The essential characteristic of importance sampling is that the unbiased selecting of values for μ in simple Monte Carlo sampling is replaced by a biased selection scheme, where the bias is designed to favor those values of μ which make the important contributions to the right side of Eq. (9).

To be more explicit, consider a probability function $p(\mu)$ giving the probability of selecting configuration μ. Suppose that a sequence of μ's, $\mu_1, \mu_2, ..., \mu_M$ is selected by a random process where the probability for selecting the tth configuration in the sequence, μ_t, is given by $p(\mu)$, independent of t. An estimate of $\langle F \rangle$ is now given by

$$\langle F \rangle_M{}^* = \sum_{t=1}^{M} F(\mu_t) \exp\left(-\frac{E_{\mu_t}}{kT}\right) [p(\mu)]^{-1} \bigg/ \sum_{t=1}^{M} \exp\left(-\frac{E_{\mu_t}}{kT}\right) [p(\mu)]^{-1} \quad (12)$$

where the asterisk is used to denote the fact that the μ's have been selected with probability $p(\mu)$. Let us now consider the choice of $p(\mu)$.

We are interested in finding $p(\mu)$ which will give an estimate $\langle F \rangle_M{}^*$ which is good according to some reasonable criterion. A natural criterion is based on the magnitude of the average value of the quantity $[(\langle F \rangle_M{}^*/\langle F \rangle) - 1]^2$, where the average is understood to be taken over the ensemble of possible values for $\langle F \rangle_M{}^*$; this quantity is simply the variance of the reduced random variable $\langle F \rangle_M{}^*/\langle F \rangle$. It is obvious that it is desirable, insofar as statistical error is concerned, to make the variance as small as possible.

We shall now determine the probability function $p(\mu)$, which is optimum according to the above criterion, that is, the $p(\mu)$ which yields a minimum variance. Let the parameter γ denote a particular sequence $\mu_1, \mu_2, ..., \mu_M$ used in the evaluation of $\langle F \rangle_M{}^*$ and let Γ denote the set of all γ. Let $m(\mu; \gamma)$ be an integer in the interval $(0, M)$ which, for a particular γ, is equal to the number of times the value μ appears in the sequence γ. It follows from the law of large numbers that $m(\mu; \gamma)/M \to p(\mu)$ as $M \to \infty$. We define

$$\epsilon_M(\mu; \gamma) = m(\mu; \gamma) - p(\mu) M, \quad (13)$$

and it follows that we have $\epsilon_M(\mu; \gamma)/M \to 0$ as $M \to \infty$. It is clear from these definitions that $m(\mu; \gamma)$ is a random variable in the space Γ with a binomial distribution; it has average value $p(\mu) M$ and variance $p(\mu)(1 - p(\mu)) M$. Also, the random variable $\epsilon(\mu; \gamma)$ has average value zero and variance $p(\mu)(1 - p(\mu)) M$. With these parameters we may express $\langle F \rangle_M{}^*$ as follows:

$$\langle F \rangle_M{}^* = \frac{\sum F(\mu) \exp\left(-\frac{E_\mu}{kT}\right) \left[\frac{p(\mu) M + \epsilon_M(\mu; \gamma)}{p(\mu)}\right]}{\sum_\mu \exp\left(-\frac{E_\mu}{kT}\right) \left[\frac{p(\mu) M + \epsilon_M(\mu; \gamma)}{p(\mu)}\right]} \quad (14)$$

where the summations extend over all values of μ. Using Eq. (9), one easily obtains from the above expression

$$\frac{\langle F\rangle_M^*}{\langle F\rangle} - 1 = \frac{\left(\sum_\mu F(\mu)\exp\left(-\frac{E_\mu}{kT}\right)\cdot\frac{\epsilon_M(\mu;\gamma)}{p(\mu)M}\Big/\sum_\mu F(\mu)\exp\left(-\frac{E_\mu}{kT}\right)\right.}{1 + \left[\sum_\mu \exp\left(-\frac{E_\mu}{kT}\right)\frac{\epsilon_M(\mu;\gamma)}{p(\mu)M}\Big/\sum_\mu \exp\left(-\frac{E_\mu}{kT}\right)\right]}$$
(15)

The calculation of the average of $[(\langle F\rangle_M^*/\langle F\rangle) - 1]^2$ over the space Γ from the above expression is complicated by the presence of the random variable $\epsilon_M(\mu;\gamma)$ in the denominator. Because of the asymptotic behavior of $\epsilon_M(\mu;\gamma)/M$, it is clear that the denominator approaches the value one as M becomes infinite. For the computation of the average of $[(\langle F\rangle_M^*/\langle F\rangle) - 1]^2$ we shall make the approximation that the denominator is one. The result which we then obtain must, strictly speaking, be understood in a probabilistic sense as follows. For given δ one can determine a probability

$$\omega = \omega\left\{\left|\frac{\epsilon_M(\mu;\gamma)}{p(\mu)M}\right| > \delta\right\} \quad \text{(all } \mu\text{)};$$
(16)

i.e., the probability of the event represented by the expression within the brackets. Now pick δ small enough so that we may make the approximation that terms of order δ are negligible. When the average of $[(\langle F\rangle_M^*/\langle F\rangle) - 1]^2$ is computed with this approximation, then the result is true with probability $1 - \omega$. This argument can be made more specific by noting that, to a good approximation, we have

$$1 - \omega = \frac{1}{\sqrt{2\pi}}\int_{-\delta\cdot\sqrt{p(\mu)M/(1-p(\mu))}}^{+\delta\cdot\sqrt{p(\mu)M/(1-p(\mu))}} \exp\left(-\frac{y^2}{2}\right) dy.$$
(17)

Furthermore, we note that as $M \to \infty$ our result will be true with probability 1.

Using the foregoing approximation with the understanding that the result is to be interpreted in the probabilistic sense described above, we have for the average of $[(\langle F\rangle_M^*/\langle F\rangle) - 1]^2$, denoted by V_M,

$$V_M = \frac{\frac{1}{M}\sum_\mu \left(\frac{F(\mu)}{\langle F\rangle} - 1\right)^2 \left(\frac{1-p(\mu)}{p(\mu)}\right)\cdot\exp\left(-\frac{2E_\mu}{kT}\right)}{\left(\sum_\mu \exp\left(-\frac{E_\mu}{kT}\right)\right)^2},$$
(18)

where the following relations have been used:

$$\text{average of } \epsilon_M(\mu; \gamma) \epsilon_M(\mu'; \gamma) = \begin{cases} 0; & (\mu \neq \mu') \\ Mp(\mu)(1 - p(\mu)); & (\mu = \mu'). \end{cases} \quad (19)$$

The choice of $p(\mu)$, call it $p_0(\mu)$, which minimizes V_M^* is obtained by differentiation with respect to p subject to the constraint

$$\sum_\mu p(\mu) = 1. \quad (20)$$

Carrying out this calculation in the usual way one obtains

$$p_0(\mu) = \frac{\left|\dfrac{F(\mu)}{\langle F \rangle} - 1\right| \exp\left(-\dfrac{E_\mu}{kT}\right)}{\sum_\mu \left|\dfrac{F(\mu)}{\langle F \rangle} - 1\right| \exp\left(-\dfrac{E_\mu}{kT}\right)}. \quad (21)$$

This result, as was to be expected, depends explicitly on $\langle F \rangle$. Since $\langle F \rangle$ is the quantity being estimated we clearly cannot use this result for $p(\mu)$. On the other hand, if a good guess of $\langle F \rangle$ is available one could use it in place of $\langle F \rangle$ in the formula for $p_0(\mu)$ and one might construct an iterative procedure in which successive estimates of $\langle F \rangle$ are used to form new $p_0(\mu)$'s. In such a calculation it would be necessary to construct a separate iterative process for each parameter $F(\mu)$ and this is a potential disadvantage. However, this procedure would lead to a minimum variance estimate and it might be useful in problems where fluctuations in the estimates are large and cannot easily be reduced by increasing the number of samples. Notice also that only the relative probability is necessary for the evaluation of $\langle F \rangle_M^*$, Eq. (12), hence it would not be necessary to compute explicitly the denominator in $p_0(\mu)$. A sampling process for estimating Boltzmann averages based on these ideas has not ever been used so far as the author is aware, but it seemed appropriate to indicate this alternative to the usual scheme, presented below, which has become so popular that one may forget that there are other and perhaps better alternatives.

The choice for $p(\mu)$ which has been used in the equation of state computations by Metropolis, Rosenbluth, Wood, and others, and which has also been used in the present computations is

$$p(\mu) = \frac{\exp\left(-\dfrac{E_\mu}{kT}\right)}{\sum_\mu \exp\left(-\dfrac{E_\mu}{kT}\right)}. \quad (22)$$

This is a natural choice since it represents precisely the distribution of the ensemble that is used in computing $\langle F \rangle$. With this value for $p(\mu)$ it follows from Eq. (12) that the estimate of $\langle F \rangle$ is given by

$$\langle F \rangle_M^* = \frac{1}{M} \sum_{t=1}^{M} F(\mu_t), \qquad (23)$$

and it follows from Eq. (18) that the variance of this estimate is given approximately by

$$V_M = \frac{\frac{1}{M} \sum_\mu \left[\frac{F(\mu)}{\langle F \rangle} - 1\right]^2 \exp\left(-\frac{E_\mu}{kT}\right)}{\sum_\mu \exp\left(-\frac{E_\mu}{kT}\right)}, \qquad (24)$$

where we have made the approximation $1 - p(\mu) = 1$. It is interesting to compare this variance with the variance V_M^0 for the optimum $p(\mu) = p_0(\mu)$. In the same approximation as above we have from Eqs. (18) and (21)

$$V_M^0 = \frac{\frac{1}{M} \left[\sum_\mu \left|\frac{F(\mu)}{\langle F \rangle} - 1\right| \exp\left(-\frac{E_\mu}{kT}\right)\right]^2}{\left[\sum_\mu \exp\left(-\frac{E_\mu}{kT}\right)\right]^2}. \qquad (25)$$

Hence the difference $V_M - V_M^0$ represents the variance of

$$\frac{1}{\sqrt{M}} \left|\frac{F(\mu)}{\langle F \rangle} - 1\right|$$

with respect to the Boltzmann distribution. This suggests that the optimum sampling scheme might be particularly advantageous near the critical temperature where the specific heat, and hence the variance in the energy, becomes infinite.

Let us now consider the realization of the importance sampling scheme using $p(\mu)$ given by Eq. (22). This realization is not simple because there does not appear to exist the equivalent of the "urn" containing configurations with the desired distribution from which samples can be drawn. In place of this a Markov process is used in which a sequence of configurations is generated having the property that in the limit as the length of the sequence becomes infinite, the probability of occurrence of configuration μ is given by $p(\mu)$ of Eq. (22). The presentation below requires some familiarity with the theory of

Markov processes and the reader is referred to Feller (1950) for a comparatively simple discussion of this theory and to Doob (1953) for a more thorough discussion.

In the usual fashion for describing the evolution of a Markov process we shall use a parameter t, called the time which takes on the sequential values 0, 1, 2, ... The Markov process generates a sequence, or chain, of states μ_1, μ_2, ... starting from an initial state μ_0 where, as before, the subscript identifies the value of t and not the value of μ. The Markov process is characterized by the transition probabilities $p_{\mu\mu'}$, where $p_{\mu\mu'}$ is equal to the conditional probability for finding the system in the state μ' at time $t+1$ given that the system is in the state μ at time t. In these considerations it will be true that $p_{\mu\mu'}$ is independent of t. The probabilities satisfy the normalization condition

$$\sum_{\mu'} p_{\mu\mu'} = 1 \tag{26}$$

and if $\psi_t(\mu)$ is the probability for finding the system in state μ at time t, then the probability for finding the system in the state μ' at time $t+1$ is given by

$$\psi_{t+1}(\mu') = \sum_{\mu} p_{\mu\mu'} \psi_t(\mu) . \tag{27}$$

It is convenient to use matrix notation to describe the system of equations represented by Eq. (27). The set of transition probabilities is represented by the matrix P with elements $p_{\mu\mu'}$, called a stochastic matrix by virtue of the conditions $p_{\mu\mu'} \geq 0$ and $\sum_{\mu'} p_{\mu\mu'} = 1$. This matrix clearly has 2^N rows and 2^N columns. Let Ψ_t be the row vector having the 2^N components $\psi_t(\mu)$, then the system of equations represented by Eq. (27) is written

$$\Psi_{t+1} = \Psi_t P , \tag{28}$$

and it follows that the probability distribution at time t in terms of the distribution at time $t=0$ is given by

$$\Psi_t = \Psi_0 P^t . \tag{29}$$

The realization of the importance sampling scheme used here is based on the fact that for an appropriate choice of P one has

$$\lim_{t \to \infty} \Psi_t = \Psi \tag{30}$$

where the components of Ψ are

$$\psi(\mu) = \frac{\exp(-E_\mu/kT)}{\sum_\mu \exp(-E_\mu/kT)}, \qquad (31)$$

and this limit does not depend on Ψ_0. Moreover, it is possible to apply the law of large numbers and we have

$$\lim_{M\to\infty} \frac{1}{M} \sum_{t=1}^{M} F(\mu_t) = \sum_\mu F(\mu)\,\psi(\mu). \qquad (32)$$

It is important to recognize that this expression implies that one can form an estimate of $\langle F \rangle$ by taking a simple average of F over the configurations of a single Markov chain. Although the forms of the expressions are identical, this is not the same kind of calculation implied by Eq. (23) because there it was assumed that each μ_t was chosen with the probability $p(\mu)$ given by Eq. (22) while in the present computation a correlation exists between successive configurations and the desired probability distribution appears only in the limit $t \to \infty$.

We now consider the choice of P which leads to the above behavior. This choice is most easily described by outlining the process used for generating the Markov chain, then it will be verified that the requisite conditions are satisfied. In these considerations we will assume that the sites of the lattice are numbered $1, 2, \ldots, N$ in some sequence which is arbitrary with respect to the geometry of the lattice. Now assume that at the time t the lattice is in configuration μ_t, then configuration μ_{t+1} is generated as follows. Starting with site number 1 and proceeding sequentially through all of the sites a random process, described now in terms of the kth site, takes place. Let ΔE represent the change in the energy of this system that would result from the reversal of the orientation of the spin on the kth site; then if $\Delta E \leqslant 0$ the spin orientation is reversed [i.e., $\mu(k) \to -\mu(k)$], but if $\Delta E > 0$, then the spin orientation is reversed with probability $e^{-\Delta E/kT}$. This completes consideration of site k, next a similar treatment is given to site $k+1$, and so forth. Finally, after treatment of site N, the resulting configuration is called μ_{t+1}.

Let $p_{\mu\mu'}^{(t)}$ denote an element of P^t. In order to prove that the limits indicated by Eqs. (30) and (32) exist and are unique it is necessary and sufficient that for some $t \geqslant 1$, $p_{\mu\mu'}^{(t)} > 0$ for all μ, μ'. Moreover, under this condition,

$$\lim_{t\to\infty} p_{\mu\mu'}^{(t)} = \psi(\mu) \qquad (33)$$

where this limit is uniquely determined by the equation

$$\Psi P = \Psi ; \tag{34}$$

i.e., Ψ is an eigenvector of P with eigenvalue one.

First it will be shown that the existence condition is satisfied by the Markov process that has been described. The matrix P can be written as the product $P(1) P(2) \ldots P(N)$, where $P(k)$ is the stochastic matrix of transition probabilities corresponding to the reversal of the spin on site k; hence, the μ, μ' element of $P(k)$ is the probability for the transition $\mu \to \mu'$ when the random process is applied to the spin on site k. It is easily seen that this matrix will link pairs of configurations differing from each other only in the reversal of spin k. It is important to recognize that these linked pairs are connected by nonzero transition probabilities, although it is possible for $p_{\mu\mu}(k) = 0$. A set of states which has the property that every state in the set is a consequent of all of the other states in the set is called an ergodic class. It follows from these definitions that the matrix $P(k)$ represents a division of the 2^N states into 2^{N-1} ergodic classes. Now consider the matrix product $P(k) P(k+1)$. The matrix $P(k+1)$ also defines a set of 2^{N-1} ergodic classes and it is obvious that all of these classes must be distinct from those defined by $P(k)$. It follows that the matrix product $P(k) P(k+1)$ defines a set of 2^{N-2} classes; if (μ, μ') is a class in $P(k)$ and (μ, μ'') and (μ', μ''') are two classes in $P(k+1)$ then $(\mu, \mu', \mu'', \mu''')$ is an ergodic class in $P(k) P(k+1)$. By the obvious extension of this argument to the product $P = P(1) P(2) \ldots P(N)$ it follows that all of the states are in a single ergodic class. It is evident from this argument that for any pair of states μ, μ' it is possible to find a t such that $p_{\mu\mu'}^{(t)} > 0$. However, it remains to be shown that for a particular t, $p_{\mu\mu'}^{(t)} > 0$ for all μ, μ'. To do this it is now only necessary to show that there are no cyclically moving classes in P; that is, it is necessary to show that P does not lead to a periodic behavior wherein the system starting in some configuration μ at time t_0 can only return to μ when $t = at' + t_0$ where $t' > 1$ and ($a = 1, 2, 3, \ldots$). Since P represents a single ergodic class this behavior is clearly impossible if P contains a single nonzero diagonal element. The diagonal element $p_{\mu\mu}$ of P where μ is the configuration of minimum energy is nonzero, therefore P does not have any cyclically moving classes. This completes the proof that there exists a $t \geqslant 1$ such that $P_{\mu\mu'}^{(t)} > 0$ for all μ, μ', and therefore the limits indicated in Eqs. (30), (32), and (33) exist.

It remains to be shown that the limit Ψ in Eq. (30) has the components given by Eq. (31). Because of the uniqueness of this limit it suffices to show that a vector with components given by Eq. (31) is an eigenvector of P with eigenvalue one. This can be done by demonstrating that Ψ is an eigenvector with eigenvalue one of $P(k)$ for all k, it will follow that this is also true for the product $P(1) P(2) \ldots P(N)$. Consider the matrix $P(k)$ and assume that (μ, μ') is an ergodic class in $P(k)$. If $E_\mu \geq E_{\mu'}$, then $p_{\mu\mu'}(k) = 1$, all the other elements in the same row are zero; if $E_\mu < E_{\mu'}$, then

$$p_{\mu\mu'}(k) = \exp\left[-(E_{\mu'} - E_\mu)/kT\right] \text{ and } p_{\mu\mu}(k) = 1 - p_{\mu\mu'}(k),$$

the remaining elements in the row being equal to zero. It is now easy to verify that for Ψ with components given by Eq. (31) we have $\Psi P(k) = \Psi$, independent of k, hence Ψ is also an eigenvector of P with eigenvalue one.

IV. Practical Considerations of the Computation

There are three important points on which the practical success of the method described above depends: the accuracy of an estimate based on a small value of M compared with that obtained in the limit $M \to \infty$; the accuracy of an estimate based on a small value of N compared with that obtained in the limit $N \to \infty$; the effect of the phase transition. The initial computations which were performed on the square lattice were designed to examine these points. Before going on to a description of the square lattice computations there are several general comments about these three points which should be made.

The accuracy of an estimate based on a small value of M depends on the rate at which the initial configuration is "forgotten" and the statistical fluctuations that are to be expected in a sample of finite size. The rate at which the initial configuration is forgotten depends on the eigenvalues of P. All of the eigenvalues of P excepting the one associated with the eigenvector Ψ have modulus less than one. It is clear that this must be so if $\lim_{t \to \infty} \Psi_0 P^t$ exists independent of Ψ_0. If Λ is the set of these $n-1$ eigenvalues and ρ is the modulus of the maximum eigenvalue Λ, then it is obvious from an expansion of Ψ_0 along the eigenvectors of P that the rate at which the effect of the initial configuration dies out increases as ρ decreases. From the discussion in the last section it is evident that P is not unique; that is, there are other Markov processes that one could construct which would lead to the desired limits. In view of this freedom one could consider the possibility of selecting P in such a way as to minimize ρ but we have made no attempt to do this. It is not at all obvious that a choice of P based on such a criterion will lead to a shorter total computation time, since the resultant computation time for each step in the Markov chain might be very large. The present choice of P has the practical advantage that its realization as a computer program is very simple and the computation time for each step in the Markov chain is correspondingly short.

A quantitative estimate of the statistical fluctuations that are to be expected in the finite Markov chain has not been made. One can get a qualitative picture of what to expect from a form of the central limit theorem which can be applied to the Markov process represented by P: Given that

$$\lim_{M \to \infty} \left\langle \left[\frac{1}{\sqrt{M}} \sum_{t=1}^{M} (F(\mu_t) - \langle F \rangle) \right]^2 \right\rangle = \sigma_1^2 \qquad (35)$$

exists and $\sigma_1^2 > 0$, then for any Ψ_0

$$\lim_{M \to \infty} \Pr \left\{ \frac{1}{\sqrt{M}} \sum_{t=1}^{M} (F(\mu_t) - \langle F \rangle) \leq \lambda \right\} = \frac{1}{\sigma_1 \sqrt{2\pi}} \int_{-\infty}^{\lambda} \exp\left(-\frac{y^2}{2\sigma_1^2}\right) dy \qquad (36)$$

uniformly in λ, where $\Pr\{\}$ is the probability of the event indicated within the brackets. The hypotheses under which this theorem applies are satisfied by P: it is sufficient that P be a finite dimensional stochastic matrix, with all states in the same ergodic class and no cyclically moving classes. It follows from this theorem that the estimates

$$\langle F \rangle_M = \frac{1}{M} \sum_{t=1}^{M} F(\mu_t) \qquad (37)$$

will approach a normal distribution, and the error in $\langle F \rangle_M$ is, loosely speaking, proportional to $M^{-1/2}$. The variance of this normal distribution depends on $F(\mu)$ and P and this dependence can be indicated as follows. We note that

$$\left\langle \left[\frac{1}{\sqrt{M}} \sum_{t=1}^{M} (F(\mu_t) - \langle F \rangle) \right]^2 \right\rangle$$

$$= \langle F^2 \rangle - \langle F \rangle^2 + \frac{2}{M} \sum_{t=1}^{M} \sum_{t'=t+1}^{M} (\langle F(\mu_t) F(\mu_{t'}) \rangle - \langle F \rangle^2). \qquad (38)$$

The first pair of terms on the right are independent of P, they depend only on the statistics of $F(\mu)$ with respect to the Boltzmann distribution. The remaining part of the right side of this equation, the double summation, represents the dependence of the variance on correlations between states of the chain. It is easy to verify that each term in the sum is positive, less than $\langle F^2 \rangle - \langle F \rangle^2$, and tends to zero as $t' - t$ increases. It follows that the variance σ_1^2 will always be larger than $\langle F^2 \rangle - \langle F \rangle^2$ and it approaches this minimum value as the range of the correlations is reduced. Notice that this is consistent with the result in Eq. (24) which was obtained under the hypothesis that no correlation existed between samples: Eq. (38) is an approximation to σ_1^2 and it must be multiplied by $1/M\langle F \rangle^2$ to represent an approximation to the variance of the reduced estimate $\langle F \rangle_M/\langle F \rangle$.

We define a correlation length τ_0, where τ_0 is the largest value of τ satisfying the inequality

$$\langle F(\mu_t) F(\mu_{t+\tau}) \rangle - \langle F \rangle^2 > \epsilon, \qquad (39)$$

where ϵ is some small number representing the notion that for smaller ϵ we may neglect the correlations. In this approximation we have the inequality

$$\sigma_1^2 < (2\tau_0 + 1)(\langle F^2 \rangle - \langle F \rangle^2), \qquad (40)$$

and it follows that an approximation to the standard deviation of the estimate $\langle F \rangle_M$ is given by

$$\sum_M = (2\tau_0 + 1)^{1/2} \cdot \left(\frac{\langle F^2 \rangle - \langle F \rangle^2}{M} \right)^{1/2}. \qquad (41)$$

This formula provides a means for estimating the error in $\langle F \rangle_M$.

Following the same argument as before, concerning the decay of the effect of the initial configuration, a reduction in ρ will lead to a smaller variance since it will decrease τ_0. This provides a second reason why it might be desirable to try to select P in such a way as to minimize ρ, or at least to make it smaller.

Let us now turn to the second point indicated above. For obvious reasons the number of spins in the lattice cannot be made arbitrarily large. The practical limitation is one of time, not memory space: as N increases the computing time goes beyond reasonable bounds long before the memory space is exhausted. Although this remark is based on our experience with the ILLIAC which has an electrostatic storage unit of 1024 words and a magnetic drum storage unit of 12,800 words, with each word containing 40 binary places, it seems evident that time would still be the limiting factor on any of the large computers. Generally speaking, it was found to be impractical to perform extensive computations on lattices that had more than about one thousand sites. It is well known from the various approximations that have been used to treat the Ising model that rather accurate computations can be made at very low temperatures and very high temperatures when only a small number of sites in the lattice are given detailed consideration. On the other hand, these approximations become increasingly bad as the critical temperature is approached because the cooperative effects of large numbers of spins become significant. The computations on the square lattice provided a means for determining empirically the effect of lattice size on the accuracy of the estimates.

Regardless of the size of the lattice it is to be expected that the fluctuations in the system will become large as the critical temperature is approached, and these will appear as larger variances in our estimates. In particular, since the specific heat becomes infinite as the critical temperature is approached (and $N \to \infty$) it is to be expected that large fluctuations will be observed in the configuration energy, or, correspondingly, in the short-range order parameter. It is important to recognize that these fluctuations are in a sense natural, they are in the physics of the problem and they have nothing to do with the method of computation. The difficulties imposed on the computation by these fluctuations are analogous to the difficulties imposed on a real physical experiment in which measurements are being taken close to a phase transition.

Before closing this section one other point of a slightly different nature deserves some comment. When these computations on the Ising model were first considered there was some concern about the effect of systematically going through the sites in the manner that has been indicated. The discussion in the last section has shown that the limits exist regardless of the order in which the sites are numbered. Nevertheless, on a short-term basis one might suspect the possibility of some kind of quasi-periodic behavior and a selection of sites on a random basis might be preferred because it would tend to eliminate such a possibility. However, from the point of view of program simplicity and computation time the systematic scheme is preferable and it was used in these computations. The results of the square lattice computations were examined carefully for evidences of periodicity and no such evidence was detected.

V. The Square Ising Lattice

The state of the lattice was represented by binary numbers in the computer, with each binary digit corresponding to a lattice site and the value of that digit representing the orientation of the spin on the site. It was convenient to have the geometrical organization of the binary digits correspond directly to the geometry of the lattice, hence rows in the lattice were represented by words in the computer and successive rows by successive words. The numbering of the sites for determining the order in which the sites were sequentially treated in one time step of the Markov process was as follows: beginning at the left end of the top row the sites were numbered sequentially 1, 2, ..., $\sqrt{N}$ and in the next adjacent row, starting at the left end, $\sqrt{N} + 1$, $\sqrt{N} + 2$, ..., $2\sqrt{N}$, etc.

Computations were performed for three different lattice sizes: 10×10, 20×20, and 37×37, i.e., $N = 100$, 400, and 1369, respectively. Since the configuration of one row of the lattice was represented by the bits in a single word, 40 sites in a row represented an upper limit; however, for various reasons related to coding details it was decided to restrict the maximum length of a row to 37 sites instead. As mentioned earlier, periodic boundary conditions were imposed so that the opposite edges of the lattice were joined by nearest-neighbor bonds.

Two quite different initial configurations were used. In one set of computations the configuration at $t = 0$ was completely ordered, with $\mu(k) = 1$ for all k. In another set of computations the initial configuration was highly disordered; this configuration was generated by assigning the values of $\mu(k)$ with a random process in which the probability for $\mu(k) = 1$ was equal to $\frac{1}{2}$.

Random numbers, which we shall denote by ξ, for these computations were generated by the mid-square process (Metropolis, 1956). The probability distribution for these numbers is uniform on the interval $(0, 1)$. Of course, strictly speaking, the numbers are not truly random and the probability distribution is not truly uniform but for our purposes we shall treat them as such. The random initial configuration was simply a set of $\sqrt{N}$ ξ's with each ξ truncated to $\sqrt{N}$ places. Random numbers were also used in determining whether or not to reverse the orientation of a spin. During the generation of the Markov sequence of configurations the spin on an individual site is always reversed if the resultant change in energy is negative or zero. However, if this energy change ΔE is positive, then the spin is reversed with probability $e^{-\Delta E/kT}$. This is realized in the following way. A random number ξ is generated, and if $\xi < e^{-\Delta E/kT}$ the spin is reversed, otherwise it is not reversed. The exponential $e^{-\Delta E/kT}$ was not computed each time this test was made since this is a relatively time consuming operation. Instead a table of possible values of $e^{-\Delta E/kT}$ was prepared at the very beginning of the entire calculation. All of these computations were made with zero magnetic field, so this table only needs two entries since there are only two cases in which the energy will increase when the spin is reversed:

all nearest neighbors are parallel to the spin under consideration in which case e^{-8K} is required; and three nearest neighbors are parallel to the spin under consideration in which case e^{-4K} is required. It was convenient in these computations to introduce a parameter

$$x = e^{-2K}. \tag{42}$$

The computations were performed for different values of x above and below the critical point which is located at $x_c = 0.4142$.

In these computations estimates of the parameter $f(1)$ were made, and the results were then compared with the exact results for an infinite lattice. The computations were divided into two parts. In the first part a Markov sequence of configurations was generated and during this computation the value of $f_\mu(1)$ at the completion of each time step was recorded on punched tape; hence, the output tape contained a sequence $f_{\mu_1}(1), f_{\mu_2}(1), \dots$. The length of these sequences was not constant, they varied somewhat with the amount of computing time that was available. Later, this output tape was read by another program which computed averages and standard deviations. It is intuitively evident that if some small number of the initial configurations corresponding to $t = 1, 2, \dots, k$ are omitted from the averaging one is likely to get a better estimate in averaging over the resultant chain of length $M - k$ than if the entire chain is included; of course when M is very large this would have little effect but it is obviously desirable to use an M which is as small as possible for results of a given accuracy. By definition when k is about equal to τ_0 most of the bias due to the initial configuration would be eliminated by this truncation. For this reason averages of $f_\mu(1)$ over fixed numbers of configurations beginning at various points along the chain were computed. In each of these computations the "standard deviation" σ of the average for the ensemble was computed. Thus, we computed

$$\langle f_\mu(1) \rangle_M = \frac{1}{M} \sum_{t=k}^{k+M-1} f_{\mu_t}(1), \tag{43}$$

and

$$\sigma = \left[\frac{1}{M}\left[\left(\frac{1}{M}\sum_{t=k}^{k+M-1} f_{\mu_t}^2(1)\right) - \langle f_\mu(1)\rangle^2{}_M\right]\right]^{1/2}. \tag{44}$$

It is to be noted that this number is not an estimate of Σ_M given by Eq. (41), but we do have the following approximation to Σ_M,

$$\sum_M \approx (2\tau_0 + 1)^{1/2} \sigma, \tag{45}$$

to serve as an estimate of the error in the averages.

The computing time required to generate a new configuration, that is, to make one time step in the Markov sequence was approximately two seconds for the 10 × 10 lattice. This time varies linearly with the number of sites in the lattice.

Some of the results of these computations are displayed in Table I. In this table averages over ensembles of length $M = 128$, with $k = 1 + (j - 1)\,128$ are shown: the parameter $f(1, j)$ is used to designate

TABLE I

VALUES OF $f(1, j)$ OBTAINED FOR THE SQUARE LATTICE. THE EXACT VALUE FOR $f(1)$ APPEARS AT THE END OF THE RIGHT-HAND COLUMN FOR EVERY x

x	j	10×10 lattice		20×20 lattice		37×37 lattice	
		Initially ordered	Initially disordered	Initially ordered	Initially disordered	Initially ordered	Initially disordered
0.30	1	0.0209±0.0020	0.0428±0.0054	0.0222±0.0010	0.0760±0.0045	0.0229±0.0006	0.0444±0.0037
	2	0.0238±0.0023	0.0240±0.0023	0.0220±0.0009	0.0220±0.0012	0.0230±0.0005	0.0228±0.0006
	3	0.0246±0.0021	0.0246±0.0021	0.0234±0.0011	0.0228±0.0013	0.0223±0.0006	0.0227±0.0006
	4	0.0199±0.0019	0.0194±0.0019				
	5	0.0204±0.0019	0.0200±0.0018				
	6	0.0208±0.0020	0.0212±0.0021			$f(1)=0.0223$ (exact)	
0.40	1	0.0956±0.0044	0.1338±0.0074	0.1043±0.0026	0.1198±0.0041	0.1026±0.0015	0.1497±0.0037
	2	0.1083±0.0043	0.1194±0.0061	0.0992±0.0021	0.1007±0.0023	0.1155±0.0013	0.1104±0.0017
	3	0.0896±0.0045	0.1052±0.0052	0.1027±0.0026	0.1076±0.0027	0.1112±0.0016	0.1192±0.0013
	4	0.0884±0.0044	0.0947±0.0046				
	5	0.0912±0.0044	0.0940±0.0050				
	6	0.0867±0.0041	0.0886±0.0043			$f(1)=0.1074$ (exact)	
0.43	1	0.1650±0.0050	0.1702±0.0058	0.1656±0.0036	0.1879±0.0041	0.1819±0.0017	0.1859±0.0028
	2	0.1568±0.0053	0.1802±0.0066	0.1748±0.0032	0.1841±0.0030	0.1833±0.0015	0.1968±0.0014
	3	0.1768±0.0070	0.1690±0.0061	0.1708±0.0030	0.1630±0.0032	0.1759±0.0018	0.1958±0.0014
	4	0.1436±0.0056	0.1621±0.0055				
	5	0.1479±0.0055	0.1501±0.0058				
	6	0.1640±0.0061	0.1403±0.0052			$f(1)=0.1900$ (exact)	
0.45	1	0.2085±0.0057	0.2024±0.0058	0.2145±0.0028	0.2091±0.0029	0.2127±0.0018	0.2186±0.0022
	2	0.1858±0.0055	0.2170±0.0059	0.2162±0.0032	0.2236±0.0024	0.2236±0.0011	0.2293±0.0012
	3	0.2061±0.0062	0.2131±0.0056	0.2718±0.0027	0.2305±0.0024	0.2271±0.0011	0.2279±0.0014
	4	0.1892±0.0051	0.1800±0.0057	0.2316±0.0024	0.2256±0.0030		
	5	0.1968±0.0048	0.1986±0.0052				
	6	0.2018±0.0058	0.2110±0.0055			$f(1)=0.2245$ (exact)	
0.50	1	0.2820±0.0053	0.2904±0.0045	0.2833±0.0026	0.2898±0.0021	0.2784±0.0016	0.2838±0.0018
	2	0.2808±0.0052	0.2812±0.0043	0.2770±0.0021	0.2778±0.0021	0.2846±0.0013	0.2863±0.0011
	3	0.2788±0.0043	0.2756±0.0049	0.2827±0.0022	0.2839±0.0022	0.2830±0.0012	0.2838±0.0018
	4	0.2682±0.0054	0.2841±0.0049				
	5	0.2721±0.0046	0.2756±0.0045				
	6	0.2804±0.0039	0.2906±0.0042			$f(1)=0.2834$ (exact)	
0.60	1	0.3616±0.0038	0.3652±0.0036	0.3568±0.0022	0.3583±0.0018	0.3554±0.0014	0.3561±0.0012
	2	0.3606±0.0036	0.3560±0.0038	0.3533±0.0017	0.3582±0.0020	0.3562±0.0009	0.3574±0.0010
	3	0.3620±0.0033	0.3607±0.0037	0.3564±0.0015	0.3543±0.0020	0.3562±0.0009	0.3556±0.0009
	4	0.3500±0.0040	0.3571±0.0036				
	5	0.3570±0.0037	0.3597±0.0037				
	6	0.3506±0.0038	0.3504±0.0041			$f(1)=0.3570$ (exact)	
0.70	1	0.3976±0.0035	0.4176±0.0038	0.3964±0.0035	0.4022±0.0016	0.4001±0.0035	0.4076±0.0011
	2	0.4122±0.0035	0.4118±0.0033	0.4055±0.0020	0.4062±0.0017	0.4068±0.0010	0.4078±0.0009
	3	0.4135±0.0028	0.4020±0.0034	0.4037±0.0015	0.4049±0.0016	0.4027±0.0008	0.4048±0.0010
	4	0.3918±0.0034	0.3924±0.0033				
	5	0.4099±0.0036	0.4172±0.0036				
	6	0.4088±0.0035	0.3998±0.0042			$f(1)=0.4056$ (exact)	
0.80	1	0.4330±0.0041	0.4356±0.0033	0.4301±0.0048	0.4512±0.0016	0.4116±0.0070	0.4413±0.0008
	2	0.4376±0.0031	0.4312±0.0030	0.4424±0.0015	0.4475±0.0016	0.4425±0.0010	0.4421±0.0008
	3	0.4472±0.0030	0.4331±0.0036	0.4426±0.0015	0.4367±0.0018	0.4441±0.0007	0.4371±0.0009
	4	0.4500±0.0037	0.4424±0.0033				
	5	0.4485±0.0028	0.4493±0.0029				
	6	0.4345±0.0029	0.4498±0.0030			$f(1)=0.4432$ (exact)	

these averages. The initial configuration corresponds to $k = 0$, and hence is not included in $f(1, 1)$. These results are presented in the form $f(1, j) \pm \sigma$ where σ is given by Eq. (44). In this table the exact result (Kaufman and Onsager, 1949) for an infinite lattice is shown for comparison.

These results indicate that it is possible and practical to obtain estimates of the parameter $f(1)$ which are accurate to a few per cent. As anticipated earlier the largest fluctuations appear in the neighborhood of the critical point. It appears that in most cases the estimate for $j = 2$ is comparatively good and the bias introduced by the initial configuration is not noticeable in a comparison of results for the different initial configurations. It is interesting to note that this is true at low temperatures where one might expect the chain starting from a configuration of complete disorder would take a long time to order itself: the results indicate that a comparatively long time is taken for this but already at $j = 2$ the value of $f(1, j)$ is in good agreement with the exact result. If one assumes $\tau_0 = 50$ it is seen that the estimates of $f(1, j)$ are within Σ_M [given by approximation (45)] of the exact value excepting the two results at $x = 0.3$ for the initially disordered lattices when $j = 1$.

VI. The Simple Cubic Lattice

The Illiac program which was prepared to do the computations on the simple cubic lattice is more elaborate than the one which was prepared for the square lattice computations. An important part of the simple cubic lattice program is a test to determine a point in the sequence of configurations at which the computation of the ensemble averages is to commence. This test, called the convergence test, is explained below.

The problem of obtaining a reasonably accurate result without using enormous amounts of computing time depends partly, as indicated earlier, on the rate of convergence of the generated ensemble to a Boltzmann distribution. The work on the square lattice has shown that in many cases the convergence is quite rapid. With the square lattice the exact solution provides a guide for checking the ensemble averages, but when the exact solution is not known a rule must be made for selecting the point at which the averaging may commence. The problem of constructing such a rule is tricky. One might consider the successive values of a parameter, such as the short-range order parameter, and commence averaging where this sequence appears to be steady in some sense. This is dangerous for there may be two, or more, sets of states each of which has a relatively high probability of occurring, but which are linked by small transition probabilities. Such a situation would result in fairly steady sequential values for certain parameters except that here and there the apparent equilibrium value of the parameter might change abruptly. It may happen that these abrupt changes occur so infrequently that they would not even occur in a very long (on the computational time scale) computation. In this instance an average taken over the apparently steady sequence might give a very incorrect result. Since the region in phase space which contributes significantly to the ensemble average becomes broader near a phase transition it is to be expected that this phenomenon is likely to occur in such a region. It is true that no such difficulties were ever apparent in the two-dimensional lattice computations but similar difficulties have been encountered in the hard-sphere equation of state computations.

The rule which has been adopted for the present computations is characterized by the fact that two statistically independent sets of results are developed and the point of convergence is established when these results agree within a given margin of error. Two distinct lattices are used to develop two statistically independent sequences of configurations. One of the sequences starts from a configuration of complete order, while the other starts from a disordered configuration, just like the two initial configurations discussed in the square lattice computations. The difference between the present case and the former is that now the two sequences are developed simultaneously and hence may be compared, one with the other, as the two sequences are developed. We denote the sequence starting from a configuration of complete order as the low-temperature (LT) sequence. Correspondingly, the sequence starting from a disordered configuration is called the high-temperature (HT) sequence. The generation of the ensemble is divided into three stages. In the first stage an LT sequence of M_0 configurations and an HT sequence of M_0 configurations are developed. Three parameters associated with each configuration are held in the computer store: $f_\mu(1), f_\mu(2)$, and S_μ. In the second stage the average of $f_\mu(1)$ taken over the last M_0 configurations is computed for each sequence and similarly for $f_\mu(2)$: these are designated $[f(1)]_{LT}$, $[f(1)]_{HT}$, $[f(2)]_{LT}$, and $[f(2)]_{HT}$. The convergence test is passed when the following two inequalities are satisfied for the first time:

$$\frac{|[f(1)]_{LT} - [f(1)]_{HT}|}{[f(1)]_{LT} + [f(1)]_{HT}} < \epsilon_1, \qquad (46)$$

$$\frac{|[f(2)]_{LT} - [f(2)]_{HT}|}{[f(2)]_{LT} + [f(2)]_{HT}} < \epsilon_2, \qquad (47)$$

where ϵ_1 and ϵ_2 are small positive numbers. If both conditions are not satisfied a new configuration is generated for the LT sequence, and for the HT sequence. The oldest information in the sequence, namely, that pertaining to the configuration which occurred at time $t - M_0$ is thrown out to make space for the information on the new configuration. Unlike the square lattice computations the parameters for each configuration are not punched on tape in order to save time. Thus, only the order parameters for the last M_0 configurations in each chain are retained. As each new configuration is added to the LT sequence and to the HT sequence the convergence test is repeated. When the convergence test is finally passed the third stage of the computations begins. All of the values for $f_\mu(1)$ in the two sequences are collected into one sum, $\Sigma f_\mu(1)$, and similarly for $f_\mu(2)$ and S_μ. As each new configuration is generated in each sequence the values for $f_\mu(1)$, $f_\mu(2)$, and S_μ are added to the corresponding sum. When ΔM new configurations in each sequence have been generated then the sums are divided by $2(M_0 + \Delta M) = R$, the number of configurations in the ensemble, to obtain the final estimates of the quantities $f(1)$, $f(2)$, and S. In the third stage of the computations the sum of squares of each of the parameters is also generated in order to calculate the standard deviations.

This process does not ensure against the possibility of obtaining an erroneous answer because of the chains being trapped in a set of metastable states. It can be expected, however, that the chance of detecting such a situation is better than it would be if only one sequence was considered. Of course, if memory space and computing time permit, then one can extend this method to include a still larger number of independent chains.

The model for almost all of the computations has 8 sites on an edge, and thus contained a total of 512 sites. Some computations were done on a model with 16 sites on an edge but because of the large amounts of computing time required this work was rather limited. The amount of time required to complete one iteration for the $8 \times 8 \times 8$ array was about six seconds and it was approximately eight times this for the $16 \times 16 \times 16$ array. As with the square lattice, periodic boundary conditions were always imposed. The sites were selected systematically for detailed consideration in an analogous fashion to the method used with the square lattice: successive sites in a row were treated, then successive rows and finally successive planes.

In these computations and in the body-centered cubic computations a different scheme for generating random numbers was used. The random number sequence used in these computations was generated from the following recurrence formula:

$$\xi_{n+5} = 7\xi_{n+4} + \xi_{n+3} - 4\xi_{n+2} + 3\xi_{n+1} + \pi(\xi_n) \tag{48}$$

268   L. D. FOSDICK

Table II

SUMMARY OF RESULTS FROM THE SIMPLE CUBIC LATTICE. THE SUPERSCRIPT "a" ON THE ID NUMBER INDICATES THE LARGE (16 × 16 × 16) LATTICE

ID	K	L	S	f(1)	f(2)	M_0	ΔM	Total
1	0.5000	0	0.9919±0.0004	0.0079±0.0003	0.0080±0.0003	50	50	156
2	0.4500	0	0.9823±0.0009	0.0169±0.0007	0.0174±0.0008	25	25	55
3	0.4000	0	0.9699±0.0014	0.0280±0.0010	0.0293±0.0011	25	25	62
4	0.3380	0	0.9424±0.0011	0.0516±0.0006	0.0549±0.0007	100	300	403
5	0.2857	0	0.8653±0.0060	0.1073±0.0021	0.1181±0.0026	50	50	102
6	0.2500	0	0.7440±0.0025	0.1853±0.0011	0.2096±0.0013	100	300	400
7	0.2381	0	0.6551±0.0060	0.2286±0.0024	0.2623±0.0029	50	50	161
8[a]	0.2381	0	0.6486±0.0024	0.2313±0.0010	0.2652±0.0013	50	50	100
9	0.2273	0	0.4976±0.0090	0.2841±0.0024	0.3315±0.0031	50	50	116
10	0.2174	0	0.3054±0.0123	0.3301±0.0026	0.3891±0.0033	50	50	100
11[a]	0.2174	0	0.1883±0.0091	0.3405±0.0016	0.4013±0.0020	50	50	100
12	0.2000	0	0.1603±0.0040	0.3713±0.0007	0.4344±0.0008	100	300	400
13[a]	0.2000	0	0.0781±0.0062	0.3708±0.0013	0.4341±0.0017	50	50	100
14	0.1667	0	0.0819±0.0022	0.4047±0.0005	0.4639±0.0005	100	300	400
15	0.1430	0	0.0686±0.0018	0.4211±0.0004	0.4755±0.0003	100	300	400
16	0.1250	0	0.0563±0.0030	0.4329±0.0007	0.4818±0.0006	50	50	100
17	0.2500	0.0625	0.8184±0.0032	0.1449±0.0017	0.1599±0.0020	50	50	102
18	0.2273	0.0625	0.7354±0.0044	0.1958±0.0021	0.2192±0.0025	50	50	101
19	0.2000	0.0625	0.5529±0.0062	0.2847±0.0024	0.3237±0.0028	50	50	100
20	0.1667	0.0625	0.3284±0.0071	0.3678±0.0018	0.4179±0.0021	50	50	100
21	0.1250	0.0625	0.1834±0.0052	0.4203±0.0009	0.4620±0.0008	50	50	100
22	0.0500	0.0625	0.1025±0.0086	0.4709±0.0009	0.4946±0.0007	25	25	50
23	−0.1000	0.0625	0.0680±0.0112	0.5525±0.0008	0.4893±0.0010	25	25	50
24	0.338	0.1250	0.9601±0.0010	0.0372±0.0007	0.0388±0.0008	50	50	128
25	0.2857	0.1250	0.9158±0.0018	0.0746±0.0012	0.0797±0.0014	50	50	121
26	0.2500	0.1250	0.8683±0.0023	0.1116±0.0014	0.1210±0.0016	50	50	105
27	0.2381	0.1250	0.8383±0.0022	0.1340±0.0013	0.1459±0.0015	50	50	105
28	0.2273	0.1250	0.8094±0.0025	0.1533±0.0014	0.1681±0.0017	50	50	105
29	0.2174	0.1250	0.7678±0.0054	0.1788±0.0024	0.1972±0.0028	50	50	100
30	0.2000	0.1250	0.6917±0.0053	0.2249±0.0023	0.2502±0.0028	50	50	100
31	0.1667	0.1250	0.5303±0.0050	0.3084±0.0019	0.3445±0.0022	50	50	100
32	0.1250	0.1250	0.3415±0.0045	0.3880±0.0013	0.4286±0.0014	50	50	100
33[a]	0.1250	0.1250	0.3369±0.0019	0.3892±0.0006	0.4297±0.0007	50	50	100
34	0.0750	0.1250	0.2145±0.0060	0.4424±0.0010	0.4722±0.0010	25	25	50
35	−0.1000	0.1250	0.0920±0.0096	0.5505±0.0009	0.4871±0.0011	25	25	50
36	−0.2000	0.1250	0.0613±0.0023	0.6266±0.0023	0.4342±0.0031	25	25	50
37	0.2857	0.2500	0.9390±0.0013	0.0562±0.0009	0.0587±0.0010	50	50	109
38	0.2381	0.2500	0.8966±0.0018	0.0909±0.0012	0.0969±0.0013	50	50	103
39	0.2000	0.2500	0.8127±0.0024	0.1548±0.0014	0.1670±0.0016	50	50	102
40	0.1250	0.2500	0.5655±0.0034	0.3082±0.0015	0.3341±0.0016	50	50	100
41	0.0750	0.2500	0.3965±0.0053	0.3938±0.0016	0.4175±0.0016	25	25	50
42	0.0625	0.2500	0.3678±0.0050	0.4091±0.0013	0.4303±0.0013	25	25	50
43	0.0250	0.2500	0.2896±0.0054	0.4486±0.0012	0.4586±0.0011	25	25	50
44	−0.0500	0.2500	0.1964±0.0038	0.5056±0.0007	0.4798±0.0006	50	50	100
45	−0.1000	0.2500	0.1566±0.0038	0.5394±0.0007	0.4785±0.0007	50	50	100
46	−0.1500	0.2500	0.1264±0.0040	0.5729±0.0008	0.4674±0.0008	50	50	100
47	−0.2000	0.2500	0.1030±0.0037	0.6171±0.0013	0.4366±0.0015	50	50	100
48	−0.2250	0.2500	0.0829±0.0025	0.6673±0.0034	0.3840±0.0041	25	25	61
49	−0.2500	0.2500	0.0564±0.0014	0.7960±0.0024	0.2274±0.0029	50	50	122
50	−0.2750	0.2500	0.0423±0.0019	0.8655±0.0023	0.1460±0.0026	25	25	67
51	−0.3000	0.2500	0.0297±0.0017	0.9033±0.0020	0.1040±0.0023	25	25	81
52	−0.3500	0.2500	0.0173±0.0007	0.9504±0.0010	0.0519±0.0011	50	50	106
53	−0.5000	0.2500	0.0064±0.0006	0.9876±0.0006	0.0126±0.0006	25	25	79
54	0.1875	0.7500	0.9473±0.0014	0.0498±0.0016	0.0513±0.0011	25	25	62
55	0.1000	0.7500	0.8424±0.0026	0.1405±0.0016	0.1452±0.0017	25	25	52
56	0.0500	0.7500	0.7450±0.0030	0.2165±0.0019	0.2216±0.0019	25	25	51
57	−0.0250	0.7500	0.5793±0.0041	0.3381±0.0015	0.3324±0.0016	25	25	50
58	−0.1000	0.7500	0.4350±0.0058	0.4377±0.0017	0.4001±0.0013	25	25	50
59	−0.1500	0.7500	0.3624±0.0064	0.4915±0.0018	0.4206±0.0015	25	25	50
60	−0.1875	0.7500	0.3143±0.0059	0.5331±0.0021	0.4223±0.0016	25	25	50
61	−0.2500	0.7500	0.2199±0.0029	0.6561±0.0037	0.3416±0.0038	25	25	55
62	−0.3000	0.7500	0.1157±0.0021	0.8374±0.0028	0.1613±0.0030	25	25	89
63	−0.3750	0.7500	0.0469±0.0014	0.9408±0.0012	0.0589±0.0012	25	25	67
64	0.338	1.250	0.9921±0.0004	0.0078±0.0003	0.0079±0.0003	50	50	101
65	0.2000	1.250	0.9834±0.0007	0.0163±0.0005	0.0165±0.0005	50	50	101
66	0.0500	1.250	0.9053±0.0014	0.0902±0.0019	0.0911±0.0014	50	50	100
67	−0.0500	1.250	0.7670±0.0020	0.2103±0.0012	0.2065±0.0012	50	50	100
68	−0.1000	1.250	0.6778±0.0023	0.2840±0.0011	0.2697±0.0011	50	50	100
69	−0.1500	1.250	0.5852±0.0027	0.3578±0.0011	0.3239±0.0009	50	50	100
70	−0.2000	1.250	0.5016±0.0033	0.4261±0.0012	0.3598±0.0008	50	50	100
71	−0.2500	1.250	0.4221±0.0035	0.4960±0.0013	0.3754±0.0008	50	50	100

TABLE II (Continued)

ID	K	L	S	f(1)	f(2)	M_0	ΔM	Total
72	−0.2750	1.250	0.3786±0.0061	0.5384±0.0027	0.3679±0.0021	25	25	50
73	−0.3000	1.250	0.3131±0.0022	0.6216±0.0026	0.3169±0.0023	50	50	129
74	−0.3250	1.250	0.2282±0.0033	0.7358±0.0032	0.2280±0.0028	25	25	52
75	−0.3500	1.250	0.1690±0.0020	0.8088±0.0020	0.1711±0.0018	50	50	116
76	−0.4000	1.250	0.0961±0.0020	0.8940±0.0017	0.0993±0.0015	25	25	58
77	−0.5000	1.250	0.0298±0.0007	0.9686±0.0006	0.0307±0.0006	50	50	134
78	0.2000	0.4000	0.8834±0.0022	0.1035±0.0015	0.1091±0.0017	25	25	64
79	0.1750	0.3500	0.8171±0.0033	0.1543±0.0018	0.1645±0.0021	25	25	79
80	0.1500	0.3000	0.7121±0.0047	0.2251±0.0025	0.2430±0.0028	25	25	95
81	−0.0500	0.1000	0.1080±0.0123	0.5233±0.0008	0.4964±0.0009	25	25	50
82	−0.1000	0.2000	0.1364±0.0086	0.5443±0.0011	0.4821±0.0012	25	25	50
83	−0.1500	0.3000	0.1544±0.0076	0.5687±0.0013	0.4675±0.0014	25	25	50
84	−0.2000	0.4000	0.1604±0.0052	0.6035±0.0018	0.4341±0.0015	25	25	51
85	−0.2500	0.5000	0.1260±0.0025	0.7397±0.0040	0.2806±0.0049	25	25	60
86	−0.3000	0.6000	0.0863±0.0017	0.8622±0.0021	0.1404±0.0023	25	25	87
87	−0.0500	0.2000	0.1679±0.0084	0.5133±0.0012	0.4871±0.0011	25	25	50
88	−0.1000	0.4000	0.2468±0.0067	0.5182±0.0016	0.4629±0.0013	25	25	50
89	−0.1500	0.6000	0.2915±0.0059	0.5245±0.0018	0.4410±0.0013	25	25	50
90	−0.2000	0.8000	0.3204±0.0055	0.5362±0.0020	0.4168±0.0015	25	25	50
91	−0.2500	1.0000	0.3260±0.0055	0.5665±0.0024	0.3823±0.0017	25	25	50
92	−0.3000	1.2000	0.2889±0.0030	0.6466±0.0039	0.3046±0.0036	25	25	111
93	−0.0500	0.4000	0.3023±0.0052	0.4759±0.0014	0.4536±0.0013	25	25	50
94	−0.1000	0.8000	0.4638±0.0051	0.4226±0.0015	0.3886±0.0012	25	25	50
95	−0.1500	1.2000	0.5635±0.0051	0.3736±0.0018	0.3357±0.0014	25	25	50
96	−0.2000	1.6000	0.6321±0.0042	0.3313±0.0016	0.2958±0.0012	25	25	50
97	−0.2500	2.0000	0.6898±0.0038	0.2906±0.0015	0.2574±0.0014	25	25	50
98	−0.3000	2.4000	0.7282±0.0033	0.2595±0.0014	0.2314±0.0012	25	25	50

where $\pi(\xi_n)$ denotes a permutation of the 40 bit binary number ξ_n:

$$\xi_n = X_0 X_1 \cdots X_{39}, \tag{49}$$

$$\pi(\xi_n) = X_0 X_4 X_5 \cdots X_{39} X_1 X_2 X_3. \tag{50}$$

The starting values for the sequence are the decimal integers:

$$\xi_0 = -135, 194, 964, 053,$$
$$\xi_1 = -196, 183, 218, 426,$$
$$\xi_2 = +429, 116, 103, 143, \tag{51}$$
$$\xi_3 = +134, 943, 841, 955,$$
$$\xi_4 = +438, 699, 540, 310.$$

This method for generating random numbers had the advantage that it was faster than the mid-square process, requiring only 0.6 msec per random number compared with 1 msec for the mid-square process.

The results of this computation are compiled in Table II. Order parameters S, $f(1)$, and $f(2)$ are shown as functions of K in Figs. 1, 2, and 3, respectively. In Table II the numbers in the first column are simply identification numbers (ID). The next two columns contain the energy parameters K and L. In the last three columns the numbers M_0 and ΔM appearing there have already been defined and the numbers

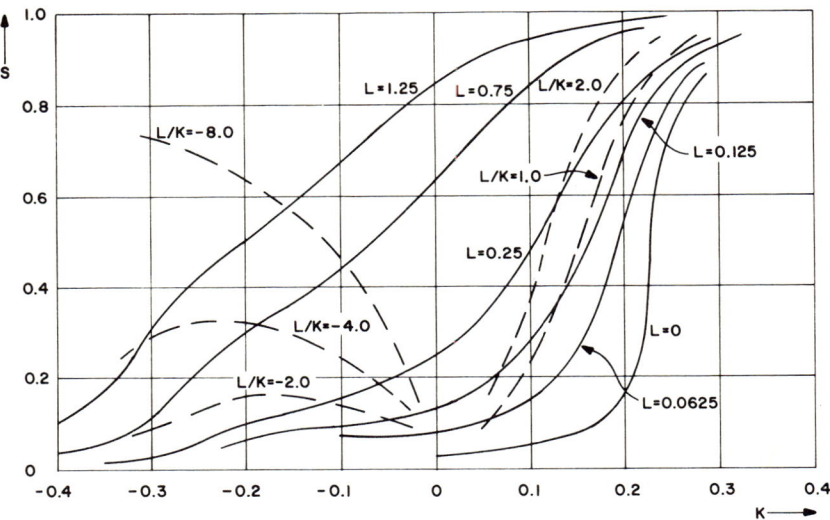

Fig. 1. Long-range order S in the simple cubic lattice shown as a function of K for different L (solid curves) and L/K (dashed curves).

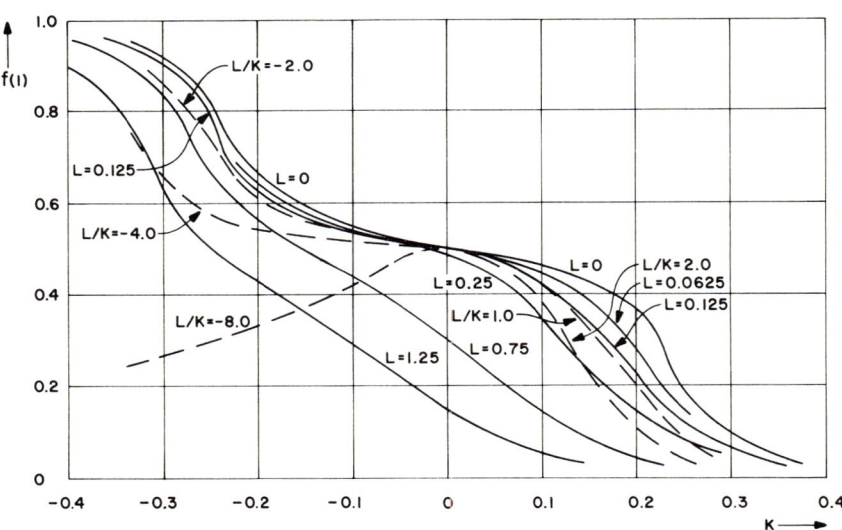

Fig. 2. First-neighbor short-range order parameter $f(1)$ in the simple cubic lattice shown as a function of K for different L (solid curves) and L/K (dashed curves).

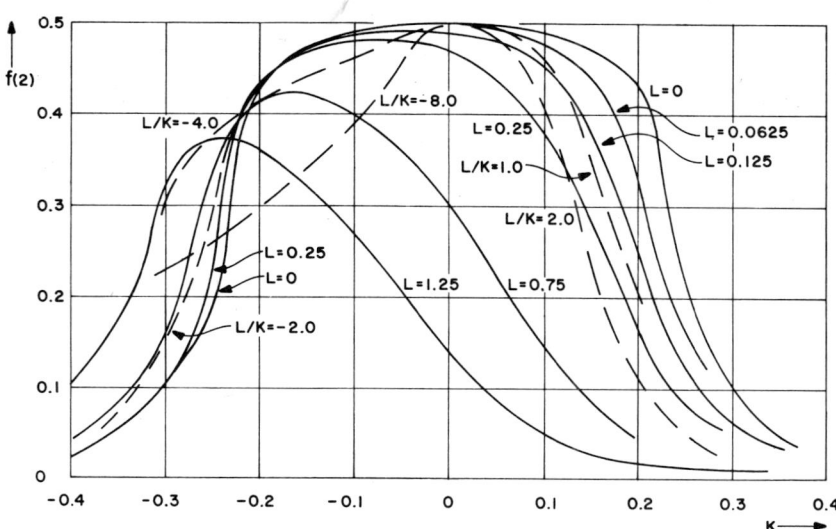

FIG. 3. Second-neighbor short-range order parameter $f(2)$ in the simple cubic lattice shown as a function of K for different L (solid curves) and L/K (dashed curves).

in the column headed "total" are the values for t, the Markov chain time parameter, at the completion of the computation. Hence, in row 1, the values $M_0 = 50$, $\Delta M = 50$, total $= 156$ mean that 50 configurations were generated in the LT sequence and 50 configurations were generated in the HT sequence before the convergence tests began (since $M_0 = 50$); next, after convergence 50 more configurations were generated in each sequence (since $\Delta M = 50$) to make a total of 200 configurations in the ensemble; since the total, the final value of t, was 156 it follows that the convergence test was passed upon completion of the one hundred and sixth iteration and therefore the first 56 configurations in each sequence were discarded. In the columns following K and L the order parameters S, $f(1)$, and $f(2)$ are presented. The spread indicated for each order parameters is the standard deviation of the mean; it is equivalent to σ, Eq. (44), used in the square lattice computations. Using the approximation (45), an estimate of the error in the results is given by $(2\tau_0 + 1)^{1/2}\sigma$ where $\tau_0 = t - (M_0 + \Delta M)$. The results are listed in order of decreasing K (i.e., increasing temperature for fixed J) in groups in which L is fixed. Near the end of the table, starting at $ID = 78$, some results are grouped together in which L/K is fixed: note that $L/K = H/J$, the ratio of the external field coupling energy to the nearest-neighbor coupling energy, which is independent of T. In the figures the results are plotted for fixed L (solid lines) and for fixed L/K (dashed lines). Although the order parameters can be computed analytically at $K = 0$, the curves for constant L/K have not been extended through $K = 0$ because computations were not performed in the region of $K = 0$ and an extrapolation of these curves from the available data did not seem appropriate.

It will be noted that three different values for M_0 appear in the table: 100, 50, and 25. In the very first (chronologically) computations the large value of M_0 was used together with $\Delta M = 300$, but because of the large amounts of computing time being absorbed it was decided to set M_0 and ΔM both at 50. Still later, for reasons of economizing on computing time the still smaller values $M_0 = 25$ and $\Delta M = 25$ were introduced. In the first two cases the convergence test parameters were given the value 0.02: $\epsilon_1 = \epsilon_2 = 0.02$. In the last case, in an attempt to compensate for the relatively small value of $M_0 = 25$ the parameters were given the value 0.01: $\epsilon_1 = \epsilon_2 = 0.01$.

Since the size effect can be expected to be most significant in the zero field case in the neighborhood of the apparent critical temperature, some computations for a $16 \times 16 \times 16$ array were made in this region. These have identification numbers 8, 11, and 13, and when compared against the corresponding results for the $8 \times 8 \times 8$ system, it will be noted that there is a marked difference in the value of S for the two cases. The differences in the results obtained for the short-range order parameters, on the other hand, are relatively slight. Hence, it appears that although the estimate of S is a crude approximation of its value for the infinite system in this region, the estimates of $f(1)$ and $f(2)$ are rather good. In the case of $L = 0.125$ a computation was made with the large lattice ($ID = 33$) in the region where S can be expected to be most sensitive to size effects. Comparison of these results with those for the $8 \times 8 \times 8$ lattice shows that the difference in the results obtained for S, as well as for $f(1)$ and $f(2)$, is slight. Hence, for this value and higher values of L the estimates of S can be expected to be fairly a good approximation to the value for an infinite system.

In the antiferromagnetic region the value $L/K = -6$ is a critical one. For L/K greater in magnitude than this value the external field coupling dominates the nearest-neighbor coupling and at low temperatures the system tends to the state in which all spins are parallel to the external field. For L/K smaller in magnitude than this value, the nearest-neighbor coupling dominates, and at low temperatures the system tends to the state in which all nearest-neighbor spins are antiparallel. The series of computations at $L/K = -4$ and $L/K = -8$ illustrate the alternate behavior in the order parameters as the parameter K tends to are plotted for fixed L (solid lines) and for fixed L/K (dashed lines). Although the order parameters can be computed analytically at $K = 0$, the curves for constant L/K have not been extended through $K = 0$ because computations were not performed in the region of $K = 0$ and an extrapolation of these curves from the available data did not seem appropriate.

It will be noted that three different values for M_0 appear in the table: 100, 50, and 25. In the very first (chronologically) computations the large value of M_0 was used together with $\Delta M = 300$, but because of the large amounts of computing time being absorbed it was decided to set M_0 and ΔM both at 50. Still later, for reasons of economizing on computing time the still smaller values $M_0 = 25$ and $\Delta M = 25$ were introduced. In the first two cases the convergence test parameters were given the value 0.02: $\epsilon_1 = \epsilon_2 = 0.02$. In the last case, in an attempt to compensate for the relatively small value of $M_0 = 25$ the parameters were given the value 0.01: $\epsilon_1 = \epsilon_2 = 0.01$.

Since the size effect can be expected to be most significant in the zero field case in the neighborhood of the apparent critical temperature, some computations for a $16 \times 16 \times 16$ array were made in this region. These have identification numbers 8, 11, and 13, and when compared against the corresponding results for the $8 \times 8 \times 8$ system, it will be noted that there is a marked difference in the value of S for the two cases. The differences in the results obtained for the short-range order parameters, on the other hand, are relatively slight. Hence, it appears that although the estimate of S is a crude approximation of its value for the infinite system in this region, the estimates of $f(1)$ and $f(2)$ are rather good. In the case of $L = 0.125$ a computation was made with the large lattice ($ID = 33$) in the region where S can be expected to be most sensitive to size effects. Comparison of these results with those for the $8 \times 8 \times 8$ lattice shows that the difference in the results obtained for S, as well as for $f(1)$ and $f(2)$, is slight. Hence, for this value and higher values of L the estimates of S can be expected to be fairly a good approximation to the value for an infinite system.

In the antiferromagnetic region the value $L/K = -6$ is a critical one. For L/K greater in magnitude than this value the external field coupling dominates the nearest-neighbor coupling and at low temperatures the system tends to the state in which all spins are parallel to the external field. For L/K smaller in magnitude than this value, the nearest-neighbor coupling dominates, and at low temperatures the system tends to the state in which all nearest-neighbor spins are antiparallel. The series of computations at $L/K = -4$ and $L/K = -8$ illustrate the alternate behavior in the order parameters as the parameter K tends to

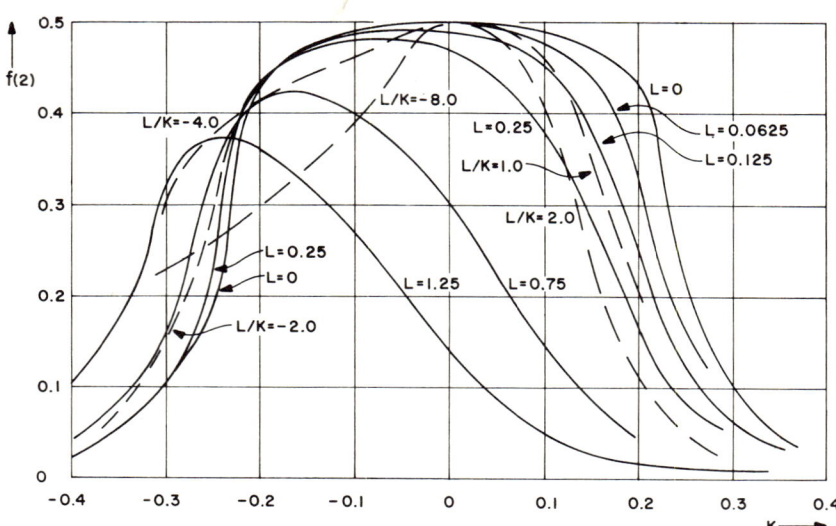

FIG. 3. Second-neighbor short-range order parameter $f(2)$ in the simple cubic lattice shown as a function of K for different L (solid curves) and L/K (dashed curves).

in the column headed "total" are the values for t, the Markov chain time parameter, at the completion of the computation. Hence, in row 1, the values $M_0 = 50$, $\Delta M = 50$, total $= 156$ mean that 50 configurations were generated in the LT sequence and 50 configurations were generated in the HT sequence before the convergence tests began (since $M_0 = 50$); next, after convergence 50 more configurations were generated in each sequence (since $\Delta M = 50$) to make a total of 200 configurations in the ensemble; since the total, the final value of t, was 156 it follows that the convergence test was passed upon completion of the one hundred and sixth iteration and therefore the first 56 configurations in each sequence were discarded. In the columns following K and L the order parameters S, $f(1)$, and $f(2)$ are presented. The spread indicated for each order parameters is the standard deviation of the mean; it is equivalent to σ, Eq. (44), used in the square lattice computations. Using the approximation (45), an estimate of the error in the results is given by $(2\tau_0 + 1)^{1/2}\sigma$ where $\tau_0 = t - (M_0 + \Delta M)$. The results are listed in order of decreasing K (i.e., increasing temperature for fixed J) in groups in which L is fixed. Near the end of the table, starting at $ID = 78$, some results are grouped together in which L/K is fixed: note that $L/K = H/J$, the ratio of the external field coupling energy to the nearest-neighbor coupling energy, which is independent of T. In the figures the results large negative values (i.e., as $T \to 0$ for J equal to a negative constant). It is interesting to notice that a relatively large number of iterations had to be performed in the computation with $ID = 92$ before the convergence condition was satisfied; in computations 85 and 86 a similar situation is noticed. The reason for this is that we are near the critical magnitude of K while below the critical magnitude of L/K. In the region $K < 0$, $-6K > L \geqslant 0$ there are two states of minimum energy (the two states in which all neighbors are antiparallel), just as there are on the line $K > 0, L = 0$. Thus, there will again be a critical value of K, around which configurations consisting of mixtures of these two states will tend to persist. As the line $L/K = -6$ is approached, a third configuration also assumes importance—that in which all spins are parallel to the external field. This explains why computation 92 ($L/K = -4$) is even slower in convergence than computations 85 and 86 ($L/K = -2$). It is significant that the existence of this situation is strongly brought to one's attention because of the rules which have been set up for convergence testing. In a test based on the examination of one sequence of configurations there is a greater chance that one would fail to observe this near-critical situation since the sequence might remain entirely in one set of configurations during the computations. Furthermore, it should be noticed that the standard deviations in these cases do not indicate anything unusual. One can infer from the small standard deviations that once the convergence conditions were satisfied both the sequences remained in the one class of states which was most probable.

For regions in which series expansions of the long-range order and short-range order can be used, a comparison with results obtained from the present Monte Carlo method is possible. The series given in Newell and Montroll (1953) have been evaluated for a few cases and a comparison with the Monte Carlo results is shown in Table III.

TABLE III

COMPARISON OF RESULTS OBTAINED FROM THE MONTE CARLO CALCULATION WITH RESULTS OBTAINED FROM EVALUATION OF SERIES EXPRESSIONS FOR THE SIMPLE CUBIC LATTICE

	S		$f(1)$	
$K(L=0)$	Monte Carlo	Series	Monte Carlo	Series
0.5000	0.9919	0.9945	0.0079	0.0054
0.4000	0.9699	0.9795	0.0280	0.0195
0.3380	0.9424	0.9504	0.0516	0.0453
0.2000	...	...	0.3713	0.3756
0.1665	...	...	0.4048	0.4050
0.1430	...	...	0.4210	0.4217

274　　　　　　　　　　　　　L. D. FOSDICK

Table IV

Summary of Results from the Body-Centered Cubic Lattice. The Superscript "a" on the ID Number Indicates the Small (Four Unit Cells on an Edge) Lattice

ID	K	L	S	f(1)	M_0	ΔM	Total	ID	K	L	S	f(1)	M_0	ΔM	Total
1	0.3	0	0.9804 ±0.0007 / 0.9800 ±0.0018	0.0189 ±0.0007 / 0.0188 ±0.0012	50	50	101	25	0.125	0.25	0.7361 ±0.0047 / 0.7379 ±0.0042	0.2130 ±0.0032 / 0.2119 ±0.0027	20	30	50
2[a]	0.225	0	0.8966 ±0.0060 / 0.8931 ±0.0129	0.0908 ±0.0045 / 0.0866 ±0.0064	50	50	100	26	0.1	0.25	0.6084 ±0.0052 / 0.6056 ±0.0053	0.2905 ±0.0029 / 0.2946 ±0.0030	20	30	50
3	0.225	0	0.9067 ±0.0019 / 0.9032 ±0.0032	0.0831 ±0.0015 / 0.0847 ±0.0016	50	50	103	27	0.05	0.25	0.3789 ±0.0055 / 0.3712 ±0.0069	0.4089 ±0.0023 / 0.4108 ±0.0024	20	30	50
4	0.1875	0	0.8023 ±0.0028 / 0.7850 ±0.0095	0.1607 ±0.0018 / 0.1662 ±0.0034	50	50	101	28	0.175	0.35	0.9222 ±0.0037 / 0.9198 ±0.0020	0.0720 ±0.0030 / 0.0741 ±0.0018	20	30	50
5[a]	0.175	0	0.7127 ±0.0109 / 0.6717 ±0.0202	0.2173 ±0.0063 / 0.2251 ±0.0088	50	50	100	29	0.15	0.35	0.8718 ±0.0033 / 0.8795 ±0.0027	0.1143 ±0.0026 / 0.1081 ±0.0023	20	30	50
6	0.175	0	0.6884 ±0.0050 / 0.6697 ±0.0102	0.2251 ±0.0027 / 0.2321 ±0.0035	50	50	109	30	0.125	0.35	0.8135 ±0.0034 / 0.8076 ±0.0040	0.1596 ±0.0026 / 0.1649 ±0.0029	20	30	50
7	0.1625	0	0.5216 ±0.0091 / 0.5246 ±0.0103	0.2981 ±0.0033 / 0.2982 ±0.0032	50	50	100	31	0.1	0.35	0.7105 ±0.0042 / 0.7164 ±0.0040	0.2329 ±0.0027 / 0.2293 ±0.0026	20	30	50
8	0.15	0	0.2159 ±0.0132 / 0.1602 ±0.0101	0.3734 ±0.0027 / 0.3850 ±0.0017	50	50	101	32	0.05	0.35	0.4970 ±0.0051 / 0.4943 ±0.0049	0.3621 ±0.0025 / 0.3613 ±0.0024	20	30	50
9	0.1375	0	0.0938 ±0.0095 / 0.1091 ±0.0071	0.4030 ±0.0023 / 0.4099 ±0.0013	50	50	100	33	−0.05	0.35	0.2451 ±0.0060 / 0.2463 ±0.0085	0.4922 ±0.0023 / 0.4916 ±0.0018	20	30	50
10	0.125	0	0.0784 ±0.0071 / 0.0684 ±0.0051	0.4179 ±0.0017 / 0.4223 ±0.0013	50	50	100	34	−0.1	0.35	0.1849 ±0.0050 / 0.1870 ±0.0059	0.5304 ±0.0020 / 0.5351 ±0.0017	20	30	50
11[a]	0.1	0	0.1366 ±0.0103 / 0.1433 ±0.0107	0.4430 ±0.0030 / 0.4376 ±0.0029	50	50	100	35	−0.15	0.35	0.1387 ±0.0033 / 0.1360 ±0.0028	0.5954 ±0.0021 / 0.6010 ±0.0034	20	30	51
12	0.1	0	0.0441 ±0.0038 / 0.0484 ±0.0036	0.4404 ±0.0012 / 0.4430 ±0.0010	50	50	100	36	−0.2	0.35	0.0681 ±0.0024 / 0.0646 ±0.0021	0.8300 ±0.0050 / 0.8501 ±0.0028	20	30	63
13[a]	0.05	0	0.1031 ±0.0098 / 0.1169 ±0.0074	0.4635 ±0.0036 / 0.4714 ±0.0022	50	50	100	37	−0.25	0.35	0.0296 ±0.0011 / 0.0268 ±0.0015	0.9386 ±0.0047 / 0.9461 ±0.0017	20	30	51
14	0.05	0	0.0453 ±0.0062 / 0.0318 ±0.0024	0.4684 ±0.0026 / 0.4767 ±0.0018	50	50	101	38	−0.05	0.5	0.3431 ±0.0057 / 0.3467 ±0.0065	0.4591 ±0.0025 / 0.4601 ±0.0157	20	30	50
15	0.2	0.05	0.8809 ±0.0029 / 0.8823 ±0.0031	0.1044 ±0.0023 / 0.1030 ±0.0024	20	30	58	39	−0.15	0.75	0.3060 ±0.0054 / 0.3044 ±0.0036	0.5278 ±0.0032 / 0.5321 ±0.0167	20	30	50
16	0.1625	0.05	0.7136 ±0.0056 / 0.7137 ±0.0052	0.2190 ±0.0032 / 0.2185 ±0.0030	20	30	56	40	0.05	1.0	0.8729 ±0.0025 / 0.8725 ±0.0025	0.1174 ±0.0021 / 0.1180 ±0.0022	20	30	55
17	0.15	0.05	0.6190 ±0.0093 / 0.6018 ±0.0073	0.2715 ±0.0042 / 0.2769 ±0.0036	20	30	50	41	−0.1	1.0	0.5127 ±0.0067 / 0.5133 ±0.0088	0.3947 ±0.0044 / 0.3945 ±0.0063	20	30	50
18	0.125	0.05	0.3135 ±0.0102 / 0.3165 ±0.0130	0.3869 ±0.0027 / 0.3853 ±0.0039	20	30	50	42	−0.15	1.0	0.4107 ±0.0056 / 0.4119 ±0.0074	0.4715 ±0.0036 / 0.4738 ±0.0044	20	30	50
19	0.1	0.05	0.1530 ±0.0076 / 0.1516 ±0.0094	0.4351 ±0.0014 / 0.4330 ±0.0023	20	30	50	43	−0.2	1.0	0.2704 ±0.0035 / 0.2734 ±0.0028	0.6472 ±0.0055 / 0.6472 ±0.0037	20	30	56
20	0.15	0.15	0.7657 ±0.0047 / 0.7663 ±0.0047	0.1906 ±0.0031 / 0.1902 ±0.0028	20	30	51	44	−0.25	1.0	0.1299 ±0.0038 / 0.1222 ±0.0025	0.8472 ±0.0054 / 0.8604 ±0.0028	20	30	52
21	0.125	0.15	0.6051 ±0.0062 / 0.6113 ±0.0063	0.2857 ±0.0034 / 0.2836 ±0.0030	20	30	53	45	−0.3	1.0	0.0655 ±0.0066 / 0.0548 ±0.0016	0.9247 ±0.0091 / 0.9409 ±0.0016	20	30	50
22	0.1	0.15	0.4388 ±0.0077 / 0.4441 ±0.0086	0.3652 ±0.0031 / 0.3632 ±0.0033	20	30	50	46	−0.25	1.25	0.2003 ±0.0078 / 0.1921 ±0.0025	0.7744 ±0.0091 / 0.7877 ±0.0028	20	30	50
23	0.05	0.15	0.2338 ±0.0064 / 0.2349 ±0.0066	0.4517 ±0.0018 / 0.4472 ±0.0016	20	30	50	47	−0.2	2.0	0.6734 ±0.0057 / 0.6728 ±0.0064	0.2991 ±0.0051 / 0.3002 ±0.0055	20	30	50
24	0.15	0.25	0.8371 ±0.0034 / 0.8338 ±0.0034	0.1416 ±0.0025 / 0.1430 ±0.0024	20	30	51								

VII. The Body-Centered Cubic Lattice

The computations for this lattice are not as extensive as those for the simple cubic lattice. Only S and $f(1)$ have been estimated. The convergence test is the same as the one used with the simple cubic lattice except that since only S_μ and $f_\mu(1)$ are calculated during each iteration, just the first inequality (46) is required to be satisfied. The value of ϵ_1 was set at 0.02. In the third stage of the generation of the configuration sequence (i.e., after the convergence condition was satisfied) the results for the two independent sequences were not combined but instead they were left separate. Thus, two separate estimates of the average were computed, one based on the HT sequence and the other based on the LT sequence. The LT sequence for positive K was started with an initial configuration in which all of the spins were parallel to one another and parallel to the external field. When K was negative the initial configuration for the LT sequence was one in which all nearest-neighbors were antiparallel. In the simple cubic lattice computations the LT sequence always started with a configuration in which all spins were parallel to each other and to the external field, even when K was negative.

In most of the calculations a model consisting of eight unit cells on an edge, and therefore 1024 sites, was used. In a few cases a smaller lattice having four unit cells on an edge was used. Twenty seconds was required to complete one iteration on the larger lattice. This program was not quite as efficient in its use of computer time as the simple cubic lattice program.

The sites were selected for detailed consideration in a systematic fashion. The two sublattices of the system were processed separately: that is, all the "center" sites were first processed sequentially, then all "corner" sites were processed sequentially to complete one time step in the Markov chain.

The results are compiled in Table IV and displayed graphically in Figs. 4 and 5. In Table IV the upper value for S and for $f(1)$ is obtained from the LT sequence and the lower value for S and for $f(1)$ is obtained from the HT sequence. In the initial calculations the larger samples with $M_0 = 50$ and $\Delta M = 50$ were computed, but to conserve on computer time this was later reduced to $M_0 = 20$ and $\Delta M = 30$. In all cases the covergence parameter had the value $\epsilon_1 = 0.02$.

Comparison of the results for the smaller lattice, having four unit cells on an edge, with those for the larger lattice shows again the relatively strong dependence of the estimate of long-range on lattice size. On the other hand, the short-range order estimate again is not so sensitive to

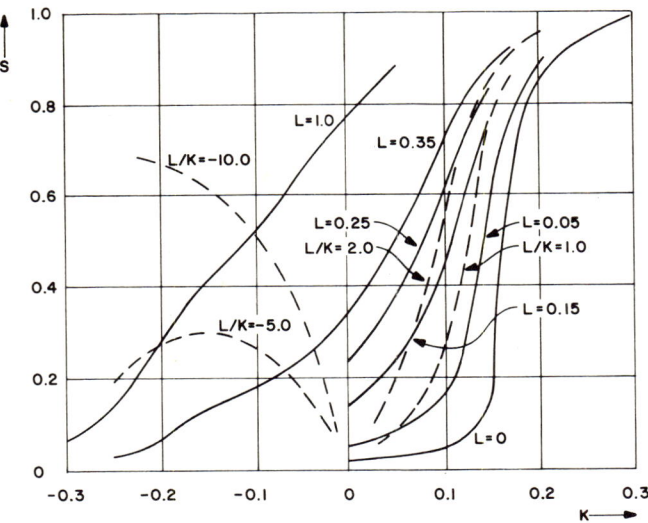

Fig. 4. Long-range order S in the body-centered cubic lattice shown as a function of K for different L (solid curves) and L/K (dashed curves).

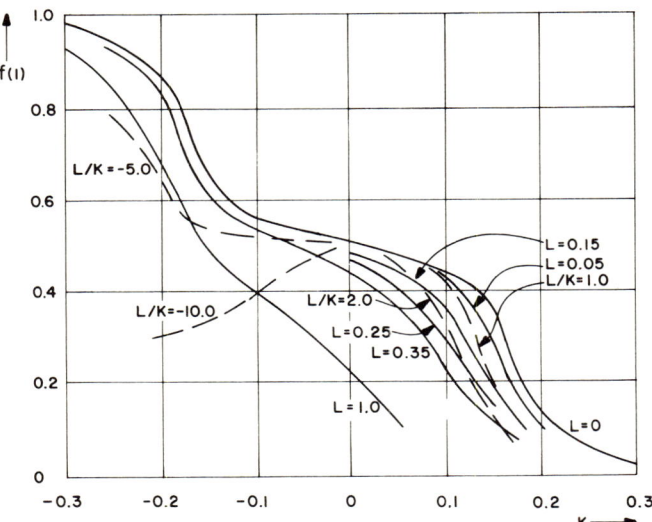

Fig. 5. First-neighbor short-range order parameter $f(1)$ in the body-centered cubic lattice shown as a function of K for different L (solid curves) and L/K (dashed curves).

change in lattice size. Comparison of results for the LT sequences with those for the HT sequences indicates the differences between the pairs of results are usually not larger than 5σ. This difference is frequently a little larger than the error estimate given by $(2\tau_0 + 1)^{1/2}\sigma$ where $\tau_0 = t - (M + \Delta M)$. In Table V there is a comparison of the Monte Carlo results with those obtained from series expressions (Newell and Montroll, 1953).

TABLE V

COMPARISON OF RESULTS OBTAINED FROM THE MONTE CARLO CALCULATION WITH RESULTS OBTAINED FROM EVALUATION OF SERIES EXPRESSIONS FOR THE BODY-CENTERED CUBIC LATTICE

	S		$f(1)$	
$K(L=0)$	Monte Carlo	Series	Monte Carlo	Series
0.300	0.9804 0.9800	0.9806	0.0189 0.0188	0.0188
0.150	...	...	0.3734 0.3850	0.3955
0.050	...	...	0.4684 0.4767	0.4742

It is well known (Newell and Montroll, 1953) that the Ising model is equivalent to a model of a binary substitutional alloy and from this viewpoint one can compare the present results with the results of the measurement of long-range order in β-brass made by Chipman and Warren (1950). This comparison appears in Fig. 6. The results of the

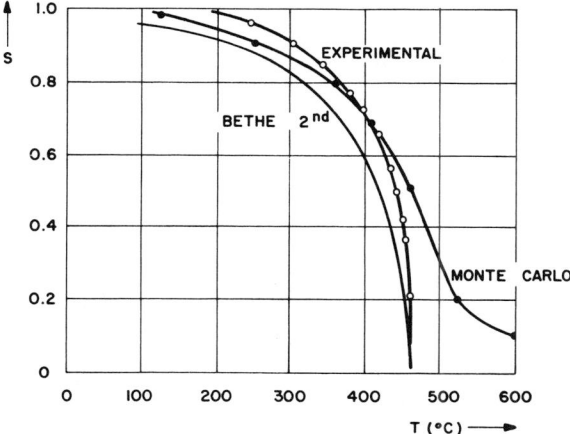

FIG. 6. Comparison of long-range order with experimental results on β-brass by Chipman and Warren, and with the results of the Bethe second approximation.

Bethe (1935) second approximation, according to Chipman and Warren, are also shown. In this figure the results have been normalized so as to make T_c coincide with the experimentally observed value of 465°C. For this normalization we set $K_c = 0.1616$; this choice is discussed in the next section. The Monte Carlo results seem to fall closer to the experimental results than do the results of the Bethe second approximation at low temperatures. However, the relative position of the Monte Carlo curve depends on the normalization and the apparent deviation of the Monte Carlo curve from the Bethe curve and the experimental curve must be viewed with this in mind. Experimental measurements of the short-range order which may be compared with the Monte Carlo results do not seem to be available. Suoninen and Warren (1958) measured the short-range order in AgZn but at only one temperature and the estimated error is 25%.

VIII. Estimation of the Critical Point

The finite size of the model precludes the existence of a true critical point. Nevertheless the model does exhibit a behavior in a fairly narrow range of K which is similar to that associated with a critical point or "Curie point" in these systems: the long-range order in the ferromagnetic case falls off sharply to very small values and the specific heat, as indicated by the standard deviation of the short-range order, appears to go through a maximum. It would be desirable to obtain from this information on a finite lattice an estimate of the critical point K_c. One procedure for obtaining such an estimate would be to record the values of K at which the specific heat becomes maximal as the size N of the system is increases. When we treat these results as a function of $1/N$ and extrapolate to $1/N = 0$, the desired estimate is provided. Unfortunately, this procedure demands such enormous amounts of computing time that it does not seem to be practical at the present time. Since it is convenient to have some definition of the apparent critical point simply so that it may be discussed without ambiguity we use here the value of K at which $S = \frac{1}{2}$; the value of K defined in this fashion is denoted by $K_{1/2}$. This parameter has the virtue that it can be computed relatively accurately from a small number of calculations of S in the neighborhood of the apparent critical point and there is no need to make the disagreable subjective decision which attends extrapolation of the long-range order above K_c to zero, or extrapolation of the inverse susceptibility below K_c to zero. In addition to this it seems reasonable to expect that $K_{1/2}$ will be close to K_c for the following two reasons.

The abrupt vanishing of the long-range order at K_c, which one expects to find in three-dimensional lattices on the basis of the exact results for the infinite two-dimensional square lattice, implies that $K_{1/2}$ for an infinite three-dimensional lattice will lie very close to K_c. Yang's (1952) exact calculation of the long-range order in an infinite two-dimensional lattice yields a value for the fractional error $(K_{1/2} - K_c)/K_c$ of about 1.6×2^{-9}. The tighter coupling of the three-dimensional lattice should make this error still smaller. The second reason is that the portion of the long-range order curve in the neighborhood of $S = \frac{1}{2}$ is, relative to the "tail" of this curve, fairly insensitive to changes in lattice size and it is therefore expected that $K_{1/2}$ for the finite lattice of the model used here lies close to $K_{1/2}$ and hence close to K_c for an infinite lattice. Unfortunately, a quantitative estimate of this is lacking, but the qualitative behavior of our results indicates that this conjecture is reasonable. For the simple cubic lattice, performing a linear interpolation between $ID = 7$ and $ID = 9$ of Table II, we find

$$K_{1/2} = 0.2275 \quad \text{(simple cubic).} \tag{52}$$

For the body-centered cubic lattice, performing a linear interpolation between $ID = 7$ and $ID = 8$ of Table IV (averaging the upper and lower values of S first), we find

$$K_{1/2} = 0.1616 \quad \text{(body-centered cubic).} \tag{53}$$

This figure has been identified as K_c for the normalization of the long-range order curve of Fig. 6 which was introduced in the preceding section.

IX. Conclusion

It is apparent from these results that this Monte Carlo method can be used effectively to compute order parameters in an Ising lattice. These results are believed to be among the most accurate that are now available. Where a direct comparison is possible our results are in good agreement with those obtained by Guttman (1961). However, fluctuations in the immediate neighborhood of the critical point and the small value of N still make it difficult to get really accurate estimates of the order parameters, especially the long-range order parameter, in this temperature region. This in turn makes it impossible to reliably determine the order of the phase transition at the critical point. It is possible that a modified version of the present method along the lines indicated in Sections III and IV for minimizing the variance might be more successful near the critical point. However, it is evident that what is really needed is some scheme for extending the computations to very large values of N. With the fastest computing machines now under construction it is possible that systems one hundred times as large as those treated here could be handled. But instead of obtaining estimates for larger systems by the brute force approach of simply increasing N and repeating the above computations, it would be more satisfying to devise a subtler scheme. Such a scheme, for example, might be one whereby through a sequence of successive approximations estimates valid for $N \to \infty$ could be obtained, with each approximation requiring explicit consideration of only a small number of lattice sites. Though it is not clear how to do this, or if it can be done, it is certainly clear that a successful effort in this direction would be very valuable for the Ising lattice problem, and cooperative phenomena in general.

References

Bethe, H. A. (1935). *Proc. Roy. Soc.* **A150**, 552.
Chipman, D., and Warren B. (1950). *J. Appl. Phys.* **21**, 696.
Doob, J. L. (1953). "Stochastic Processes." Chap. V. Wiley, New York.
Ehrman, J. R., Fosdick, L. D., and Handscomb, D. C. (1960). *J. Math. Phys.* **1**, 547.
Feller, W. (1950). "Probability Theory and Its Applications." Chap. 15. Wiley, New York.
Fosdick, L. D. (1959). *Phys. Rev.* **116**, 565.
Guttman, L. (1956). *In* "Solid State Physics" (F. Seitz and D. Turnbull, eds.), Vol. 3, p. 145. Academic Press, New York.
Guttman, L. (1961). *J. Chem. Phys.* **34**, 1024.
Kahn H. (1956). *In* "Symposium on Monte Carlo Methods" (H. A. Meyer ed.) p. 146. Wiley, New York.
Kaufman B. (1949). *Phys. Rev.* **76** 1232.
Kaufman B. and Onsager L. (1949). *Phys. Rev.* **76**, 1244.
Metropolis, N., Rosenbluth, A. W., Rosenbluth, M. N., Teller, A. H., and Teller E. (1953). *J. Chem. Phys.* **21**, 1087.
Metropolis N. (1956). *In* "Symposium on Monte Carlo Methods" (H. A. Meyer, ed.), p. 3. Wiley, New York.
Newell, G. F., and Montroll, E. W. (1953). *Revs. Modern Phys.* **25**, 353.
Onsager, L. (1944). *Phys. Rev.* **65**, 117.
Rosenbluth, M. N., and Rosenbluth, A. W. (1954). *J. Chem. Phys.* **22**, 881.
Salsburg, Z. W., Jacobson, J. D., Fickett, W., and Wood, W. W. (1959). *J. Chem. Phys.* **30**, 65.
Suoninen, E., and Warren, B. E. (1958). *Acta Met.* **6**, 172.
ter Haar, D. (1960). "Elements of Statistical Mechanics," Chapter XII. Holt, Rinehart, and Winston, New York.
Wood, W. W., and Jacobson, J. D. (1957). *J. Chem. Phys.* **27**, 1207.
Wood, W. W., and Parker, F. R. (1957). *J. Chem. Phys.* **27**, 720.
Yang, C. N. (1952). *Phys. Rev.* **85**, 808.

A Monte Carlo Solution of Percolation in the Cubic Crystal

J. M. Hammersley
OXFORD UNIVERSITY
OXFORD, ENGLAND

I. Introduction 281
II. Schemes for Estimating $P(p)$ by Monte Carlo Methods 283
   A. Scheme A 283
   B. Scheme B 284
   C. Scheme C 285
III. Choice of Machine 288
IV. Details of the Mercury Calculation 290
V. Monte Carlo Refinements and Numerical Results 295
   References 298

I. Introduction

PERCOLATION PROCESSES ARISE in a wide variety of practical applications, and something is known of their general nature and qualitative properties (Broadbent and Hammersley, 1957; Hammersley, 1957a, b, 1959, 1961) but, at present, their quantitative properties are quite beyond the resources of mathematical theory. Monte Carlo methods can, however, provide quantitative results, and to date a few simple two-dimensional percolation problems have yielded to this treatment. The present paper (brought up to date at the proof stage by footnotes and additional references) is an account of the first successful Monte Carlo attack on a full-scale three-dimensional problem. At first sight this three-dimensional problem seems to demand a prohibitive amount of Monte Carlo computing, say about 50 years' work on an IBM 704. Nevertheless, by suitable computational tricks one can reduce the computing time from years to hours. Although the present paper is self-contained and relates to just one particular percolation process, the methods described may be modified to deal with other percolation problems. The paper is based in part on an unpublished lecture delivered to the British Association at Dublin on September 5, 1957. The computations themselves were performed in July 1958 on the Ferranti Mercury computer at the United Kingdom Atomic Energy Research Establishment, Harwell.

If a lump of porous material is put in a bucket of water, will the interior of the material get wet and, if so, to what extent? The material may be visualized as a network of interconnecting pores, some of which are large enough to convey water and some so small that water cannot penetrate along them. The numerical results will naturally depend upon the structure of this network of pores. Here we shall suppose that the network has a simple regular structure called the cubic crystal. This crystal consists of (i) *atoms*, situated at the points with integer coordinates in three-dimensional Euclidean space, and (ii) *bonds*, which join nearest-neighbor atoms (i.e., atoms one unit of distance apart). The bonds play the role of the pores and the atoms provide the interconnections between pores. Thus, for example, since each atom has six nearest neighbors, a given end of a given pore will be connected to the ends of five other pores. We shall suppose that each pore has, independently of all other pores, a probability p of being large enough to transmit water. A bond will be called undammed or dammed according to whether it represents a pore which will or will not transmit water. Thus each bond has a probability p of being undammed and a probability $1-p$ of being dammed. Of course, an undammed bond can transmit fluid in either direction. We can think of the lump of porous material as a large block of atoms and bonds hewn from the cubic crystal. When this block is immersed, water is supplied to all atoms on the boundary of the block and it spreads from the boundary into the interior along any bonds which happen to be undammed. Let $P(p)$ denote the proportion of the interior atoms which eventually become wet as a result of this spreading. Our problem is to determine $P(p)$ as a function of p for $0 \leqslant p \leqslant 1$. In particular, it is known from the general theory of percolation processes that there exists a critical probability p_0 such that $P(p) = 0$ for $p < p_0$ and $P(p) > 0$ for $p > p_0$; and we shall want to know the numerical value of p_0, inasmuch as the interior of the lump will remain wholly dry whenever the porosity is less than this critical p_0. We shall examine three schemes (A, B, C) for estimating $P(p)$ by Monte Carlo methods. Both Schemes A and B call for a prohibitive computing time; but it is worth describing them in rough and brief terms, partly because they are the "obvious" methods which first occur to one, and partly because they lead up naturally to the successful Scheme C.

II. Schemes for Estimating $P(p)$ by Monte Carlo Methods

A. Scheme A

Strictly speaking, the length of the side of the block should be very large in comparison with the length of a pore. However, it will probably be adequate to suppose that there are only 200 atoms along the edge of the block. The block therefore has 8 million atoms in an array $200 \times 200 \times 200$, and 24 million bonds. In the straightforward Scheme A, we take a fixed value of p and then decide, by means of random numbers, which of the 24 million bonds are undammed. This fixes a realization of the porous material. Next we scan the 8 million atoms to decide which of them are connected by undammed paths to the wet atoms on the boundary of the block. This can, in principle, be done by an iteration. At any stage of the iteration we examine each atom in turn and wet it if there is already a wet atom at the other end of any of the (six or fewer) undammed bonds leading to the atom under scrutiny; and we repeat the iteration until no more atoms can be wetted thereby. The total number of wet atoms tells us the proportion wet in this realization of the porous material. This proportion is however subject to statistical fluctuations, and to estimate $P(p)$ we shall need to repeat the realization with different random numbers. As a matter of fact, the number of repetitions need not be large, since different parts of the block are likely to behave more or less independently: we shall only need enough repetitions, say half a dozen, to assure ourselves that the observed wet proportion has little variation from realization to realization. This disposes of one fixed value of p. The whole calculation has now to be repeated for other values of p (say 50 different values of p) in the range $0 \leqslant p \leqslant 1$.

Scheme A calls for enormous computing resources. In the first place, the storage requirements are considerable. Throughout the calculation, we have to know the state (wet or dry) of each of 8 million atoms and the state (dammed or undammed) of each of 24 million bonds. Any state can be represented by a single binary digit; but even so we shall need storage for some 800,000 words of 40 bits each. The greater part of the data has therefore to be stored on tape or drums; and, even when it is brought into the arithmetic unit for processing, it will have to be unpacked and repacked. Secondly, at each iteration of the scanning procedure we have to examine the states of six bonds and six nearest-neighbor atoms for each of 8 million atoms, so that 96 million pieces of information have to be processed in each iteration; and, to wet the center of the block, at least 100 iterations will be wanted. Taking into account the repetition of realizations and the range of values of p, the order of magnitude of the number of pieces of information processed is about 10^{12} or 10^{13}. This is a large and lengthy piece of data processing. At a very rough and probably conservative estimate, we have here some 50 years' continuous work for an IBM 704.

B. Scheme B

Scheme B is designed mainly to reduce storage requirements, though it also economizes in computing time. It depends upon the following idea. Since the interior of the material contains many atoms, the proportion of these that are wet will be very nearly equal to the probability that a randomly selected atom is wet; and this selected atom will be wet if and only if there is an undammed path from it to the boundary. Suppose, therefore, that, instead of supplying fluid to all boundary atoms as in Scheme A, we supply fluid to just one source atom deep in the interior and ask what probability there is of wetting at least one atom on the boundary, or equivalently of wetting a large number of other atoms in the course of reaching the boundary. For a more rigorous treatment of this argument see Broadbent and Hammersley, (1957).

It turns out that, if the single source atom succeeds in wetting a few thousand other atoms, it will succeed in wetting a distant boundary. We therefore proceed recursively as follows for each fixed value of p. Suppose that the source atom at the origin has wet n other atoms with known coordinates. We look for an undammed bond from one of these n wet atoms to a dry atom, whereupon we wet this dry atom to obtain a set of $n + 1$ wet atoms for the next stage of the recursion. This single stage of the recursion will require about n operations, the main labor being a check that the added dry atom is not a member of the set of n already wet atoms. We stop the recursion as soon as either a prescribed number N (a few thousand) of atoms are wet or it is not possible to increase n to $n + 1$ because there are no undammed bonds from any of the n wet atoms to a dry atom. For fixed p, we repeat the recursion about 100 times, say, and estimate $P(p)$ to be the percentage of recursions which can be terminated with $n = N$. The number of operations for each value of p will thus be about

$$100 \sum_{n=1}^{N} n \simeq 10^9 \quad \text{or} \quad 10^{10} ; \qquad (1)$$

i.e., about 10^{11} to 10^{12} operations when all values of p are considered. The computing labor is thus about one-tenth of that in Scheme A; but the main improvement is in storage. For we have now only to consider the state of those bonds connected to atoms actually wet in the process. With $N = 5000$, we shall only have to store the state of between 15,000 and 30,000 bonds and the coordinates of 5000 atoms. Allowing 10 binary digits for each of the three coordinates of an atom, the total storage requirement is less than 5000 words of 40 bits each. So the storage required is more than 100 times less than Scheme A. We shall not need recourse to tape storage, and this in turn will substantially reduce the computing time. A rough estimate of the total computing time for Scheme B might be no more than one year's work on an IBM 704, though even this is prohibitive from a practical standpoint.

C. Scheme C

In Scheme B we performed a separate Monte Carlo experiment for each of about 50 values of p. In Scheme C, we shall analyze all values of p simultaneously in parallel. This holds out the possibility of reducing the computing time to about one week.

In Schemes A and B we decided with probability p whether any given bond should be undamned or not. Instead, in Scheme C, we shall allot to each bond independently a number f, chosen at random uniformly between 0 and 1. From the single source atom we shall feed in, not a single fluid as in Scheme B, but an infinite number of different fluids, in fact one fluid for each number g between 0 and 1. The rule to be adopted is that a bond, which has been allotted the random number f, will transmit all those fluids and only those fluids whose numbers g satisfy $g \leqslant f$. To see the effect of this rule, consider how the fluid numbered $g = 1 - p$ will react to some given bond. We have

$$\text{Prob \{bond is undamned for fluid numbered } g = 1 - p\}$$
$$= \text{Prob } \{1 - p = g \leqslant f\} = 1 - \text{Prob } \{f < 1 - p\} = 1 - (1 - p) = p. \quad (2)$$

Hence, to see if the source atom wets N atoms for a particular value of the parameter p under Scheme B, we have only to ask whether the fluid numbered $1 - p$ wets N atoms under Scheme C. If under Scheme C the fluid numbered $1 - p$ wets N atoms, then the foregoing rule ensures that all fluids with numbers $g \leqslant 1 - p$ will also wet N atoms. Hence what we need to know is the number γ of the largest numbered fluid which succeeds in wetting N atoms. If $\gamma \geqslant 1 - p$, then and only then would the source atom wet N atoms for the particular value p under Scheme B. Hence

$$\text{Prob } \{\gamma \geqslant 1 - p\} = P(p), \quad (3)$$

$$\text{Prob } \{1 - \gamma \leqslant p\} = P(p), \quad (4)$$

and we see that the desired solution of the percolation problem is the cumulative distribution function of the random variable $1 - \gamma$.

The determination of γ could proceed as illustrated by Table I.

TABLE I

DETERMINATION OF γ

Line	Atom coordinates			Bond displacements						Critical value
	x	y	z	$+100$	-100	$0+10$	$0-10$	$00+1$	$00-1$	
1	0	0	0	*0.983*	0.609	0.001	0.146	0.420	0.518	0.983
2	+1	0	0	0.502	*0.710*	0.598	0.185	*0.733*	0.702	0.733
3	+1	0	+1	*0.812*	0.409	*0.956*	*0.721*	0.594	*0.003*	
4	+2	0	+1	0.349	*0.805*	0.440	0.053	0.452	0.224	
5	+1	+1	+1	0.212	0.499	0.346	*0.783*	0.287	0.171	
6	+1	−1	+1	...	...	...	...	...	...	0.721

This table gives the first 6 lines out of a total of N lines. The three Euclidean coordinates of successive atoms to be wet are given under the heading "atom coordinates." The six columns of "bond displacements" represent the six bonds leading from the atoms on the left: at the head of each column are the direction cosines $[(+1, 0, 0),$ etc.] of these six bonds. The numbers within these six columns are the values of f associated with these bonds. As Table I is set out, it would seem at first sight that an error has been committed: for the f-values relate to bonds, and Table I seems to indicate that certain bonds can possess more than one value of f. For example, the bond $(+1, 0, 0)$ from the atom $(0, 0, 0)$ has the f-value 0.983; but this bond is the same as the bond $(-1, 0, 0)$ from the atom $(+1, 0, 0)$, and therefore should also have f-value 0.983 instead of the f-value 0.710 shown in the second line of Table I. However, we shall see presently that the computing procedure automatically disregards the value 0.710, and effectively ensures that each bond receives a unique value of f. The calculation proceeds as follows:

In line 1, we start with the source atom $(0, 0, 0)$, and we select six random values of f (0.983, 0.609, ..., 0.518) for the six bonds leading from this source atom. The largest of these, 0.983, is the first critical value and is entered in the right-hand column. Fluids with $g \leqslant 0.983$ (and only these fluids) can escape from the source atom by way of the bond with direction cosines $(+1, 0, 0)$; and these fluids will enter the atom with coordinates $(+1, 0, 0)$, which is at this stage written in line 2 under atom coordinates. The number 0.983 in line 1 is also italicized at this stage to show that it has been used. Next we select six more random numbers $(0.502, ..., 0.702)$ and enter these in line 2. The second of these, 0.710, is italicized because it corresponds to a bond $(-1, 0, 0)$ which would lead back to an atom $(0, 0, 0)$ which is already wet. In the first two lines there are no f-values, which exceed or equal 0.983, and which are not yet italicized. Hence we search for the largest number in these first two lines that is not yet italicized, which is 0.733; we italicize it and we reduce the critical value in the right-hand column to this value. Evidently, fluids with $g \leqslant 0.733$ can escape from the atom $(+1, 0, 0)$ along the bond $(0, 0, +1)$ to the atom $(+1, 0, +1)$, which is now entered in line 3. All fluids with $g > 0.733$ are confined to the first two atoms listed. In the set of six f-values now written into line 3, we may italicize 0.003 because it leads to a wet atom $(+1, 0, 0)$. There are two possible escape bonds, with values of $f = 0.956$ and 0.812, each not less than the last written critical value (0.733), and these two bonds lead to the atoms now entered in lines 4 and 5. We italicize 0.812 and 0.956 in line 3, and also the two entries (0.805, 0.783) in lines 4 and 5 which lead to atoms already wetted. Of the numbers not yet italicized, the largest is 0.721; and accordingly the critical value is reduced to 0.721 and atom $(+1, -1, +1)$ is entered in line 6. The calculation proceeds in this fashion until N lines of Table I are completed; and at this stage γ is the last entry in the column of critical values.

Table I is recalculated about 100 times with, of course, fresh random numbers f at each recalculation; and the resulting 100 values of $1 - \gamma$ give an empiric distribution for $P(p)$.

The total storage requirements are $9N$ numbers, which is reasonable. The amount of calculation depends mainly upon checking which numbers need to be italicized. In any line there is one obvious number, namely the one which leads back again to the atom, which wets the atom in the line in question. To check the remaining five entries, we have to scrutinize the coordinates in all the preceeding lines. Thus for 100 recalculations of the table, we want about $250N^2$ comparisons of pairs of sets of three coordinates. When $N = 8000$, and when we assume that the machine will take 60 μsec to locate and compare a set of three coordinates in the store with a set already held in the arithmetic unit,[1] we obtain an estimated computing time of 270 hours = 11 days.

[1] For the basis of this estimate, it is assumed that three coordinates are packed into a single word of the store and that the computing cycle will require a compare instruction (36 μ sec) together with an instruction to close the cycle and give an increment to the index register (24 μ sec).

288   J. M. HAMMERSLEY

III. Choice of Machine

At the time this calculation was carried out, two machines were available, an IBM 704 and a Ferranti "Mercury." For straightforward calculations the 704 is the faster and it has a larger store. On the other hand, it has only three index registers against seven in the Mercury; furthermore, one of the Mercury's index registers is able to modify all

TABLE II

SUMMARY OF MACHINE OUTPUT

γ	ν	γ	ν	γ	ν	γ	ν
0.783	105*	0.756	86	0.734	3	0.676	8
0.777	1169	0.756	121	0.734	29	0.672	4
0.775	106*	0.756	254	0.734	76	0.666	1
0.775	1344	0.756	292*	0.732	2	0.658	2
0.773	1318	0.754	181	0.732	5	0.654	2*
0.771	1263	0.754	195	0.732	22	0.648	4
0.771	1514	0.752	265*	0.730	1	0.646	1
0.771	1897	0.752	704*	0.730	16	0.646	5*
0.770	248	0.752	1355	0.730	21	0.645	9*
0.768	160	0.750	17	0.729	1	0.643	4
0.768	820	0.750	33	0.729	3	0.641	17*
0.768	1550	0.750	45	0.725	2	0.633	1
0.766	93	0.750	60	0.725	5	0.633	1
0.766	1347	0.750	94	0.725	25	0.629	4
0.766	1362	0.748	8	0.719	10	0.627	1
0.766	1753	0.748	32	0.719	71	0.621	1
0.764	3	0.748	38	0.717	3	0.619	2
0.764	83	0.748	196*	0.711	1	0.619	3
0.764	196	0.748	1354	0.709	21	0.615	1
0.762	14	0.746	160	0.707	7	0.611	1
0.762	40	0.744	33	0.707	41*	0.598	1
0.762	60	0.744	97	0.705	3	0.596	1
0.760	33	0.742	1	0.703	3	0.590	1
0.760	43	0.742	18	0.701	11	0.574	1
0.760	44	0.742	145	0.699	6	0.572	3
0.760	212	0.740	6	0.695	2	0.564	2
0.758	168	0.740	7	0.691	8	0.559	1
0.758	185	0.740	16	0.691	20	0.527	1
0.758	1202	0.740	150	0.686	6	0.525	1
0.758	1400	0.738	31	0.682	7	0.516	3
0.756	3	0.738	33	0.680	7	0.514	1
0.756	81	0.736	250	0.676	1	0.486	1

the other index registers, whereas this is not possible directly on the 704. The net effect of this is that the Mercury can be faster than the 704 if the program is sufficiently complicated. In Scheme C, the computing time depends upon N^2. If, by an elaborate program, we can make the computing time proportional to N^α, where $\alpha < 2$, it will probably pay to use Mercury. At any rate, this was the decision taken; thus the Mercury was programmed to handle a calculation with any present value of $N \leqslant 8192$. We began with 11 calculations, each with $N = 4096$. It appeared from these calculations that the critical value had a small, though just appreciable, chance of being reduced between the 1000th and the 2000th lines of Table I, but a negligible chance of being reduced after the 2000th line. Consequently, in the remaining 117 calculations N was set to 2048. These remaining calculations showed that the chance of reduction between the 1000th and 2000th lines was not as small as the 11 preliminary calculations had suggested. However, we doubt if stopping at $N = 2048$ has seriously affected the final results. Table II shows, for each of the 128 calculations, the final value of γ together with ν, the line number at which this final γ was first attained. Entries marked with an asterisk are the 11 calculations with $N = 4096$. Each calculation with $N = 2048$ took about 7 minutes; and each calculation with $N = 4096$ took about 20 minutes. The fact that 7 minutes is one-third and not one-quarter of 20 minutes reflects roughly the way in which a power law N^α with $\alpha < 2$ was achieved. The total computing time for the 128 calculations on Mercury was about 17 hours. The corresponding time for the IBM 704 would have been about 28 hours; thus, in the final analysis it might have been better to use the 704, inasmuch as the computing time would not have been appreciably greater and the programming effort would have been considerably less. However, if it had been necessary to use $N = 8192$ throughout, the Mercury computing time would have been about 4 days for 100 calculations against about 11 days for the 704. In any case, the foregoing comparisons of computing times should not be taken too seriously; the times given for Mercury are the actual times recorded during computation, whereas the 704 times are only estimates. These estimates are based merely upon the bare skeleton of a program. They are underestimates to the extent that they do not take into account the time that would have been spent in red-tape operations, initial settings, input, output, etc.; and they are overestimates to the extent that the bare skeleton makes no allowance for programming refinements, such as more efficient search routines. It is almost impossible to guess whether these effects of over- and underestimation cancel one another out, or whether one outweighs the other. However, some programs for the

Monte Carlo solution of a number of percolation processes now being written for the IBM 7090 at the Bell Telephone Laboratories, Murray Hill, should permit judgment on these issues.*

IV. Details of the Mercury Calculation

The main problem to be overcome in using Mercury is its small high-speed store of only 1024 words of 40 bits each. There is a larger drum store, but in order to economize in computing time, transfers to and from the drum must be kept to a minimum. For these reasons the layout of Table I was altered considerably. In the altered version, each line consisted of six entries denoted by x, y, z, ξ, π, ϵ. Here x, y, z, are the three coordinates of the atom, and ξ, π, and ϵ are variables which will be defined presently. These six variables in the nth line of the table will be denoted by x_n, y_n, z_n, ξ_n, π_n, ϵ_n.

In the original procedure of Table I, we underlined all entries f corresponding to bonds which led to atoms already wet. In the modified version we proceed as follows. Suppose that we have reached the nth line of the table, and that atom $A_n = (x_n, y_n, z_n)$ has just been wet from some atom B earlier in the table. Let $\delta_0{}^n$ denote the bond from A_n to B; and let $\delta_1{}^n, \delta_2{}^n, ..., \delta_5{}^n$ denote the remaining bonds from A_n counting cyclically on from $\delta_0{}^n$: [e.g., if $\delta_0{}^n$ is $(0, 0, +1)$, then $\delta_1{}^n$ is $(0, 0, -1)$, $\delta_2{}^n$ is $(+1, 0, 0)$, ..., $\delta_5{}^n$ is $(0, -1, 0)$]. Let $f_i{}^n$ be the f-value in the nth line of the table and in the column corresponding to $\delta_i{}^n$ ($i = 1, 2, ..., 5$). There is zero probability that any two of the f-values are equal; so we may suppose that $f_{i_1}^n < f_{i_2}^n < ... < f_{i_5}^n$, where $i_1 i_2 ... i_5$ is a permutation of $12 ... 5$. Since the f-values are independent and identically distributed, $i_1 i_2 ... i_5$ will be a uniformly random permutation of $12 ... 5$. Let c denote the current critical value. If $f_{i_5}^n < c$, then we cannot escape (i.e., wet any other atom) from A_n unless or until c is reduced. If $f_{i_5}^n \geq c$, we attempt to escape from A_n along $\delta_{i_5}^n$: the attempt succeeds or fails according to whether $\delta_{i_5}^n$ leads to a hitherto dry atom or not. If the attempt succeeds, then the coordinates of the atom reached via $\delta_{i_5}^n$ is added to the (x, y, z) list. Next we test to see whether or not $f_{i_4}^n \geq c$. If so, we attempt to escape from A_n along $\delta_{i_4}^n$; and, if the attempt succeeds, we add the atom reached to the (x, y, z) list. Similarly we proceed with $f_{i_3}^n, f_{i_2}^n$, and $f_{i_1}^n$. If there is no atom from which we can escape in this way, then c must be reduced to the largest $f_{i_5}^m$, where m ranges over wet atoms. Note that if, for example, $f_{i_4}^n < c$

---

* Note added in proof: see Frisch *et al.* (1961, 1962), Vyssotsky *et al.* (1961).

there is no point in attempting an escape along $\delta^n_{i_j}$ for $j \leqslant 4$, unless or until c has been reduced. The difference between the original and the modified procedure is that in the latter, we only italicize f-values if they correspond to bonds along which escape is attempted. This economizes in computing time.

We now define ξ_n to be the largest f_i^n which has not yet been italicized in the modified scheme. We shall see presently that π_n and ϵ_n keep a record of δ_0^n, the random permutation $i_1 i_2 ... i_5$ for line n, and the number of attempted escapes from A_n. Suppose that this record of π_n and ϵ_n is available to us. Then it is not necessary to generate and store all the f-values of Table I: the single value ξ_n will suffice. For suppose that we know that k escapes have already been attempted from A_n. Then $\xi_n = f^n_{i_{5-k}}$, where $k = 0, 1, ..., 4$. In order to attempt the next escape from A_n, we shall test to find out whether $\xi_n \geqslant c$; and, if so, we shall attempt to escape along $\delta^n_{i_{5-k}}$. If $k < 4$, we shall need to replace ξ_n by $f^n_{i_{5-k-1}}$ after attempting this escape. This new value, say ξ_n', is, however, the largest of a sample of $5 - k - 1$ numbers chosen uniformly and independently at random from the range 0 to ξ_n; i.e.,

$$\xi_n' = \xi_n \max(\rho_1, \rho_2, ..., \rho_{5-k-1}), \quad (5)$$

when the ρ_j are independent rectangular variates in $(0, 1)$. If $k = 4$, we shall not need a further value of ξ_n after the present one, and we may as well take

$$\xi_n' = 0 \quad \text{if } k = 4. \quad (6)$$

We can thus begin the calculation by taking ξ_n to be the largest of a sample of 5 rectangular variates, and then replacing ξ_n by ξ_n' according to (5) or (6) each time an escape is attempted from A_n. In this way we shall economize in storage of f-values. The procedure involves more computation than if we had stored all f-values; but the additional computation is proportional to N, and may be neglected in comparison with other parts of the computation where the labor will be more nearly proportional to N^2. There is a slight modification to the above procedure when $n = 1$. Then we take ξ_1 initially equal to the largest of a sample of size 6 from the rectangular distribution. After the first escape from $A_1 = (0, 0, 0)$ has occurred, we replace ξ_1 by the second largest of this sample of 6. Thereafter, the calculation proceeds as before.

Next we shall describe the variables π_n and ϵ_n. The latter takes 7 values: $\epsilon_n = 6$ initially and remains at this setting until A_n is wet for the first time, at which stage ϵ_n is reset to one of the values 0, 1, ..., 5 to represent the direction δ_0^n to the atom B which originally wets A_n;

thereafter ϵ_n remains at this reset value. An exception to this rule occurs for $n = 1$ and $n = 2$. A_1 is always set to $(0, 0, 0)$ and A_2 is always set to $(+1, 0, 0)$; ϵ_1 always represents the direction from A_1 to A_2 and ϵ_2 always represents the direction from A_2 to A_1. The variable π_n has 326 possible values, corresponding to the 326 different ordered subsets of five objects (including the empty subset). The five objects can be regarded as the integers 1, 2, 3, 4, 5. $\pi = 0$ represents the empty set. The remaining subsets are placed in the following order: (i) if a and b are the final digits of the subsets α and β, and if $a > b$, then α proceeds β; (ii) if $a = b$, and if α has more members than β, then α proceeds β; (iii) if rules (i) and (ii) fail to decide precedence, then precedence is by lexicographical order (e.g., the next subset after 2415 is 4125). We take $\pi = k$ to represent the kth subset in this order of precedence. We define the descendent of a subset α to be the subset obtained by discarding the last digit of α: the descendent of the empty set is defined to be itself. If π is the number representing the subset α, we write $D(\pi)$ for the number representing descendent of α. For example, we have the values shown in Table III.

TABLE III

TYPICAL EXAMPLES OF π AND $D(\pi)$

Subset	π	$D(\pi)$
41253	143	35
4125	35	247
412	247	323
41	323	130
4	130	0
—	0	0

Those subsets which are permutations of 12345 have π-values of the form $65r + s$, where $r = 0, 1, ..., 4$ and $s = 1, 2, ..., 24$. More generally, the rules for the construction of the values of π are such that, when π is divided by 65, the quotient determines the final digit in the subset represented by π, and the remainder determines the number of digits in the subset. (The empty set follows special rules of its own for technical reasons which need not concern us here.)

The high-speed memory contains a look-up table, in which $D(\pi)$ is stored in address π. The program also contains a short subroutine for decoding π_n, i.e., for dividing a π-value by 65, and determining the final digit and the number of members in the corresponding subset. The final digit of the subset is used to count cyclically on from $\delta_0{}^n$, which is stored in ϵ_n, and the number of members of the subset is used to construct k in the recalculation of ξ_n by Eqs. (5) and (6). After every attempted escape, π_n is replaced by $D(\pi_n)$ by way of the look-up table. If π_n should become zero, then all escape routes from A_n must have been attempted, and the zero value of π_n is used to inhibit any further attempted escapes from A_n. The whole calculation is initiated with $\pi_n = 65r + s$ where r is randomly chosen from $r = 0, 1, ..., 4$ and s is randomly chosen from $s = 1, 2, ..., 24$.

The basic program should now be apparent from the simplified flow diagram shown in Fig. 1. In this flow diagram, primed quantities denote new values; e.g., "$c' = c_0$" means "replace c by c_0." $(x_n{}^*, y_n{}^*, z_n{}^*)$ is

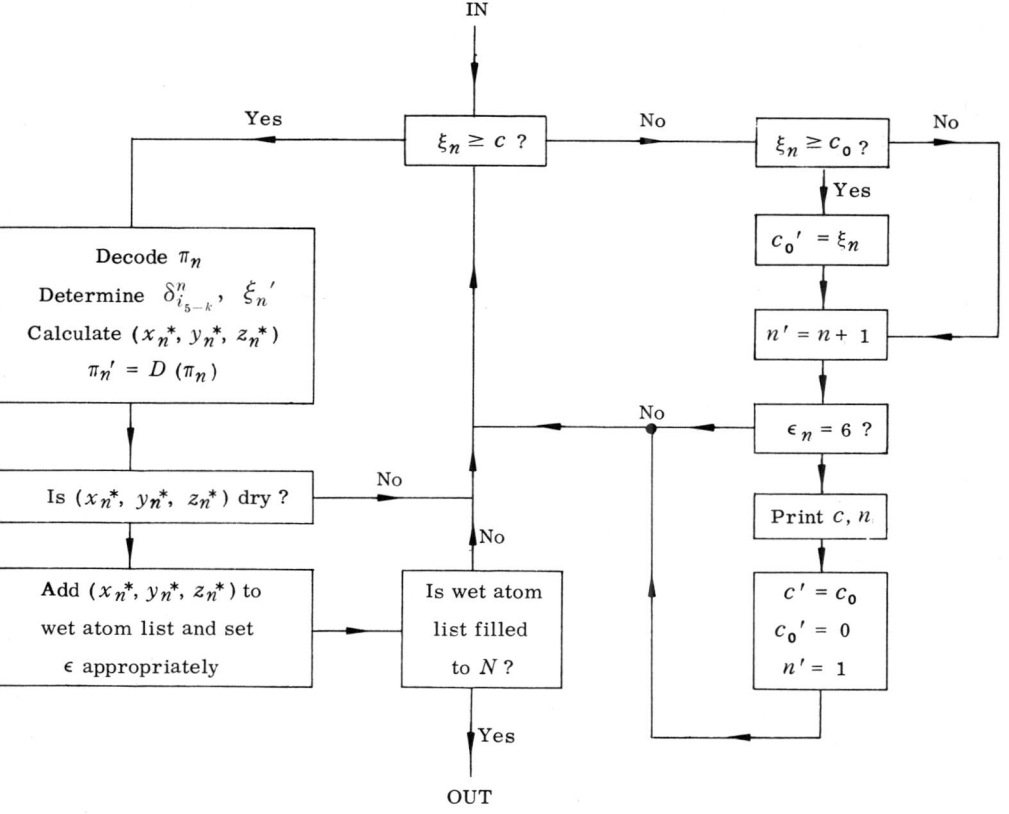

Fig. 1. Simplified flow diagram.

the atom reached from (x_n, y_n, z_n) via $\delta^n_{i_{5-k}}$. If this atom turns out to be dry, it is added to the list of wet atoms and ϵ is set for this new wet atom in the opposite direction to $\delta^n_{i_{5-k}}$. c_0 is a variable which becomes the new critical value c whenever c has to be reduced; thus it is the largest ξ_n (in the wet atom list) which is less than c. The test "$\epsilon_n = 6$?" is used to discover whether the wet atom list has been fully scanned for escape routes. The printed output gives successive points at which c is reduced and the corresponding values of c. Thus γ is the last c-entry to be printed out.

The wet atom list was divided into pages of 32 wet atoms each; and the maximum and minimum values of x_n, y_n, z_n in each page were stored. This defines a rectangular block of three-dimensional space containing all wet atoms in the page. To test to see whether $(x_n{}^*, y_n{}^*, z_n{}^*)$ is dry, the program first looks for blocks containing $(x_n{}^*, y_n{}^*, z_n{}^*)$ and then compares $(x_n{}^*, y_n{}^*, z_n{}^*)$ with all wet atoms in the corresponding pages. This reduces the computing time quite substantially, and is the main reason why the computing time increases more slowly than N^2. The disadvantage of this scheme, however, is that some blocks are very large (especially when they contain atoms wet just before and just after a change in the value of c, or those derived from large blocks); and such blocks are habitually searched in detail. For the IBM 7090 program, referred to previously, and alternative and more efficient search and storing procedure is proposed; this should cut computing time quite considerably.

The complete history of all $(x_n, y_n, z_n, \xi_n, \pi_n, \epsilon_n)$ for all n was stored on the drums. Three partial copies or feeders of this were kept in the high-speed store to deal with (i) the tail of the wet atom list, (ii) the current comparison of ξ_n with c, and (iii) the current comparison of $(x_n{}^*, y_n{}^*, z_n{}^*)$ with the contents of a particular block. The complete list of maxima and minima defining blocks was permanently held in the high-speed store. Fairly elaborate programming was needed to ensure that the drum history was brought up to date from the partial feeders whenever necessary, but *only* when necessary, since it severely wastes computing time to bring the drum history up to date at unnecessary moments. The program could also cross-feed between the three feeders when they overlapped. None of the relevant details appear in the simplified flow diagram.

Virtually the whole calculation (in fact, excepting only the generation of pseudorandom numbers, print out, and the transfer of information to and from the drums) was performed in the index registers of the machine. This effectively trebled the computing speed and quadrupled the storage. In order to speed up the testing of inequalities in the index registers, the values of ξ were restricted to the range $\xi = 0(1)511$: this, however, gives adequate accuracy in γ, since the calculation is entirely free from rounding errors. Use of the index registers for (x, y, z) implied that the calculation was formally on a three-dimensional torus. In order to prevent this formal possibility from occurring in practice, an alarm was set to halt the whole calculation should any point on a antipodal meridian of the source ever become wet. The value of N is, however, sufficiently small that it is most unlikely that the alarm will be sounded when, x, y, z each occupy 10-bit registers. In the actual calculation, the alarm remained unactivated (except of course in test programs deliberately rigged to check the correct functioning of the alarm). The functioning of the main program was tested by printing out the first 256 lines of Table I together with the random numbers used to generate it, and working through these lines with paper and pencil.

The demands on core storage were such that only 320 instructions could be allocated to the main program. Slow or infrequently used routines (e.g., printing routines and initial settings) could naturally be kept elsewhere and brought in under the chapter-changing sequence. But, nevertheless, it remained quite a hard task to compromise between computing speed and number of instructions in the main program.

V. Monte Carlo Refinements and Numerical Results

Let $P_m(p)$ denote the probability that the source atom will wet at least m other atoms when p is the probability of a bond being undamned. Then it is easy to prove that

$$\lim_{m \to \infty} P_m(p) = P(p). \tag{7}$$

In order to see the truth of (7), let us arrange the bonds of the cubic crystal in some definite order $B_1, B_2, ...$, as we may since there are only countably many bonds. Let B be the space of vectors $b = (b_1, b_2, ...)$, where $b_j = 0$ or 1 according as B_j is dammed or undammed. Consider some fixed value of p. This induces a fixed probability measure μ on the subsets of B. Let β_m be the subset such that the source atom wets at least m other atoms, and let β be the subset such that the source atom wets infinitely many other atoms. Evidently β is the intersection of the monotone decreasing sequence $\beta_1, \beta_2, ...$ Hence,

$$P(p) = \mu(\beta) = \lim_{m \to \infty} \mu(\beta_m) = \lim_{m \to \infty} P_m(p); \tag{8}$$

and this relation holds for every fixed p. This establishes (7). It is worth noticing that the foregoing argument is not specific to the cubic crystal. It is valid in any crystal with a countable set of bonds.

For the particular case of the cubic crystal, elementary probability arguments give

$$P_1(p) = 1 - (1-p)^6, \qquad (9)$$

and

$$P_2(p) = 1 - (1-p)^6 - 6p(1-p)^{10}. \qquad (10)$$

For increasing $m > 2$, it becomes increasingly laborious to find explicit expressions for $P_m(p)$; thus one is forced to use a Monte Carlo method to estimate $P_N(p)$, where N is large enough to ignore the difference between $P_N(p)$ and $P(p)$ [cf., (7)]. A more explicit form of Eq. (4) is

$$P_N(p) = \text{Prob}\,(\gamma_N \geqslant 1 - p) \qquad (11)$$

where γ_N is the highest numbered fluid that wets at least N other atoms besides the source atom. The Monte Carlo output is a record of the line numbers n at which the critical value c is reduced, together with the corresponding values of c. It therefore provides γ_m for all $m \leqslant N$.

Next write γ_{mN} for the final critical value that would have been attained at the Nth line if, at the mth line, c had been 1. Since γ_{mN} depends only on ξ_n with $n > m$, whereas γ_m depends only on ξ_n with $n \leqslant m$, we see that γ_{mN} is independent of γ_m. Also

$$\gamma_N = \min\,(\gamma_m, \gamma_{mN}); \qquad (12)$$

and therefore

$$\begin{aligned} P_N(p) &= \text{Prob}\,(\gamma_m \geqslant 1-p)\,\text{Prob}\,(\gamma_{mN} \geqslant 1-p) \\ &= P_m(p)\,P_{mN}(p) \end{aligned} \qquad (13)$$

where

$$P_{mN}(p) = \text{Prob}\,(\gamma_{mN} \geqslant 1 - p). \qquad (14)$$

Let $\hat{P}_n(p)$ denote the Monte Carlo estimate of $P_n(p)$, i.e., the proportion of calculations in which $\gamma_n \geqslant 1 - p$. The Monte Carlo work does not provide γ_{mN}; but we could estimate $P_{mN}(p)$ from the equation

$$\hat{P}_N(p) = \hat{P}_m(p)\,\hat{P}_{mN}(p). \qquad (15)$$

Table IV
Summary of Final Results

p	$P_1(p)$	$\hat{P}_1(p)$	$P_2(p)$	$\hat{P}_2(p)$	$P_2{}^*(p)$	$\hat{P}_N(p)$	$P_N{}^*(p)$
0.20	0.738	0.711	0.609	0.586	0.608	0.000	0.000
0.21	0.757	0.727	0.637	0.609	0.634	0.000	0.000
0.22	0.775	0.734	0.664	0.633	0.668	0.008	0.008
0.23	0.792	0.758	0.690	0.664	0.694	0.070	0.073
0.24	0.807	0.781	0.714	0.687	0.710	0.203	0.211
0.25	0.822	0.789	0.737	0.696	0.725	0.359	0.380
0.26	0.836	0.805	0.758	0.719	0.747	0.477	0.503
0.27	0.849	0.820	0.779	0.766	0.793	0.570	0.580
0.28	0.861	0.836	0.797	0.781	0.804	0.609	0.621
0.29	0.872	0.844	0.815	0.789	0.815	0.641	0.662
0.30	0.882	0.844	0.831	0.789	0.825	0.687	0.724
0.31	0.892	0.844	0.846	0.805	0.851	0.719	0.756
0.32	0.901	0.851	0.860	0.812	0.860	0.742	0.786
0.33	0.910	0.859	0.873	0.820	0.869	0.766	0.816
0.34	0.917	0.867	0.885	0.828	0.876	0.773	0.826
0.35	0.925	0.875	0.896	0.859	0.908	0.789	0.823
0.36	0.931	0.883	0.906	0.867	0.914	0.836	0.874
0.37	0.937	0.898	0.915	0.883	0.921	0.852	0.883
0.38	0.943	0.914	0.924	0.898	0.926	0.875	0.900
0.39	0.948	0.930	0.931	0.922	0.940	0.906	0.915
0.40	0.953	0.930	0.938	0.922	0.945	0.906	0.922
0.41	0.958	0.953	0.945	0.945	0.950	0.930	0.930
0.42	0.962	0.953	0.951	0.945	0.954	0.930	0.936
0.43	0.966	0.961	0.956	0.953	0.958	0.945	0.948
0.44	0.969	0.961	0.961	0.961	0.969	0.953	0.953
0.45	0.972	0.969	0.965	0.969	0.972	0.961	0.957
0.46	0.975	0.969	0.969	0.969	0.975	0.961	0.961
0.47	0.978	0.969	0.972	0.969	0.978	0.961	0.964
0.48	0.980	0.984	0.976	0.984	0.980	0.977	0.969
0.49	0.982	0.992	0.978	0.992	0.982	0.992	0.978
0.50	0.984	0.992	0.981	0.992	0.984	0.992	0.981
0.51	0.986	0.992	0.983	0.992	0.986	0.992	0.983
0.52	0.988	1.000	0.985	1.000	0.988	1.000	0.985
0.53	0.989	1.000	0.987	1.000	0.989	1.000	0.987
0.54	0.991	1.000	0.989	1.000	0.991	1.000	0.989
0.55	0.992	1.000	0.990	1.000	0.992	1.000	0.990
0.56	0.993	1.000	0.991	1.000	0.993	1.000	0.991
0.57	0.994	1.000	0.992	1.000	0.994	1.000	0.992
0.58	0.995	1.000	0.993	1.000	0.995	1.000	0.993
0.59	0.995	1.000	0.994	1.000	0.995	1.000	0.994

An improved estimate of $P_N(p)$ would be

$$P_N^*(p) = P_m(p)\, \hat{P}_{mN}(p). \tag{16}$$

From (15) and (16) we obtain the ratio estimate

$$P_N^*(p) = [P_m(p)/\hat{P}_m(p)]\, \hat{P}_N(p), \tag{17}$$

which can be used when $P_m(p)$ is known theoretically.

Table IV quotes the values of $P_1(p)$ and $P_2(p)$ calculated from (9) and (10), and also the values of $\hat{P}_2(p)$ and $\hat{P}_N(p)$ obtained from the Monte Carlo work. From these, the values of $P_2^*(p)$ and $P_N^*(p)$ are calculated by using (17) with $m = 1$, $N = 2$, and $m = 2$, $N = 2048$ or 4096, respectively. It will be seen that $P_2^*(p)$ is a more accurate estimate of $P_2(p)$ than $\hat{P}_2(p)$ is. Similarly, it may be presumed that $P_N^*(p)$ is a better estimate of $P(p)$ than $\hat{P}_N(p)$ is.

In retrospect, this author would have preferred to estimate $P_{mN}(p)$ directly, by the appropriate modification of the Monte Carlo experiment, and to insert this into (16). The new calculations at the Bell Telephone Laboratories will allow this to be done.

References

Broadbent, S. R., and Hammersley, J. M., (1957). Percolation processes. Crystals and mazes. *Proc. Cambridge Phil. Soc.* **53**, 629-641.

Frisch, H. L., Hammersley, J. M., Sonnenblick, E., and Vyssotsky, V. A. (1961). Critical percolation probabilities (site problem), *Phys. Rev.* **124**, 1021-1022.

Frisch, H. L., Gordon, S. B., Vyssotsky, V. A., and Hammersley, J. M. (1962). Monte Carlo solution of bond percolation processes in various crystal lattices, *Bell Syst. Tech. J.* **41**, 909-920.

Frisch, H. L., Hammersley, J. M., and Welsh, D. J. A. (1962). Monte Carlo estimates of percolation probabilities for various lattices, *Phys. Rev.* **126**, 949-951.

Hammersley, J. M., (1957a) Percolation processes. The connective constant. *Proc. Cambridge Phil. Soc.* **53**, 642-645.

Hammersley, J. M., (1957b) Percolation processes. Lower bounds for the critical probability. *Ann. Math. Statist.* **28**, 790-795.

Hammersley, J. M., (1959) Bornes superieures de la probabilité critique dans un processus de filtration. *Probabilites et ses applications. Proc. 87th Intern. Colloq. Centre National de la Recherche Scientifique, Paris*, 17-37.

Hammersley, J. M. (1961). Comparison of atom and bond percolation processes. *J. Math. Phys.* **2**, 728-733.

Vyssotsky, V. A., Gordon, S. B., Frisch, H. L., and Hammersley, J. M. (1961). Critical percolation probabilities (bond problem), *Phys. Rev.* **123**, 1566-1567.

Author Index

Numbers in italics indicate the pages on which the full references are listed.

Agu, B. N. C., 169, 180, *213*
Alder, B. J., 44, *65*
Anderson, E. C., 132, *134*
Archard, G. D., 137, *213*
Ashkin, J., 97, *133*
Atchison, W. F., 219, 220, 222, 225, 226 *243*

Barfield, W. D., 47, *65*
Bartlett, J. H., 207, *213*
Beach, L. A., 90, *133*
Bell, G. I., 10, 27, 41, *42*, 68, *88*
Benoit, H., 218, *243*
Berger, M. J., 90, *133*, 138, *213*
Bethe, H. A., 97, *133*, 137, 148, 181, 184, 185, *213*, 278, *280*
Bichsel, H., 200, *213*
Birkhoff, R. D., 136, *213*
Blanchard, C. H., 145, *213*
Blunck, O., 136, 146, 178, 205, *213*
Börsch-Supan, W., 205, *213*
Bothe, W., 136, 172, *213*
Breitenberger, E., 93, *133*, 137, *213*
Broadbent, S. R., 281, 284, *298*
Brown, R. T., 209, *213*
Bueche, F., 219, *243*
Burdett, T. A., 169, 180, *213*
Butcher, J. C., 165, *213*
Buys, W. L., 175, 176, *214*

Carlson, B. C., 202, *214*
Carlson, B. G., 10, 41, *42*, 47, *65*, 68, *88*, 137, *214*
Case, K. M., 5, *42*
Chandrasekhar, S., 5, *42*, 47, *65*, 218, *243*
Charpak, G., 170, *214*
Chipman, D., 277, *280*
Collins, F. C., 219, *243*
Cook, J. M., 109, *133*
Cormack, D. V., 154, 191, *214*
Coveyou, R. R., 109, *133*

Davis, D. H., 68, *88*
Davison, B., 5, 10, 32, *42*, 68, *88*
Davisson, C. M., 90, *133*
de Hoffmann, F., 5, *42*
Doggett, J. A., 90, *133*, 207, *214*
Domb, C., 219, 225, *243*
Doob, J. L., 255, *280*
Drawbaugh, D. W., 118, *133*

Ehrman, J. R., 246, *280*
Engelkemeir, D., 102, *133*
Erpenbeck, J. J., 219, 220, 226, 237, *243*
Eyges, L., 188, *214*

Fano, U., 136, 145, 193, 206, *213*, *214*
Feller, W., 255, *280*
Fickett, W., 245, *280*
Fixman, M., 218, 219, *243*
Fleischmann, W., 148, *214*
Flory, P. J., 217, 218, 219, *243*
Fosdick, L. D., 246, *280*
Frank, H., 169, *214*
Friedman, B., 219, *243*
Frisch, H. L., 219, *243*, 290, *298*
Frye, G. M., 80, 85, *88*

Gammel, J. H., 80, 85, *88*
Gans, P. J., 231, *243*
Goad, W., 68, *88*
Gordon, S. B., 290, *298*
Goudsmit, S., 136, 149, *214*
Grimley, T. B., 219, *243*
Grodstein, G. W., 98, 106, *133*
Guth, E., 137, *214*
Guttman, L., 246, 279, *280*

Hammersley, J. M., 225, *243*, 281, 284, 290, *298*
Handscomb, D. C., 246, *280*

AUTHOR INDEX

Hansen, G. E., 27, *42*
Heath, R. L., 128, *133*
Hebbard, D. V., 138, 152, 184, 201, *214*
Heitler, W., 97, 102, *133*
Hermans, J. J., 219, *242*
Higasimura, T., 138, 152, 164, *214*
Hiley, B. J., 225, *242*
Hill, T., 218, *243*
Hiller, L. A., Jr., 219, 220, 222, 225, 226, *243*
Hough, P. V. C., 113, *133*
Howerton, R. J., 85, *88*
Hughes, D. J., 3, *42*

Isaacson, L. M., 219, 220, *243*

Jacobson, J. D., 245, *280*
James, H. M., 218, *243*
Johnston, R., 68, *88*

Kahn, H., 103, 108, 111, 118, *133*, 158, *214*, 250, *280*
Kanter, H., 170, *214*
Kaplan, E. L., 68, *88*
Kaufman, B., 248, 265, *280*
Keller, H. B., 48, 57, *65*
King, G. W., 219, *243*
Kinosita, K., 138, 152, 164, *214*
Koch, H. W., 147, *214*
Kuhn, H., 218, *243*
Kuhn, W., 218, *243*

Landau, L., 136, 146, 205, *214*
Lazar, N. H., 94, 95, 128, *133*
Lee, C. E., 13, 23, 27, 32, 41, *42*
Lehmer, D. H., 222, *243*
Leisegang, S., 136, 146, 178, 205, *213*
Leiss, J. E., 138, 152, 187, 188, *214*
Lewis, H. W., 136, 151, 153, 196, *214*
Lin, S. R., 209, *213*

MacCallum, C., 138, *214*
McGinnies, R. T., 192, 193, *214*
McGowan, F. K., 93, 94, 128, *133*, *134*
Managan, W. W., 102, *133*
Marcer, P. J., 219, 220, 229, *243*
Mather, R., 196, 197, 198, *214*

Mathur, V. S., 148, *214*
Matsukawa, E., 169, 180, *213*
Mazur, J., 240, *243*
Meister, H., 137, 185, 186, *214*
Messel, H., 165, *213*
Metropolis, N., 245, 262, *280*
Meyer, A., 102, *134*
Miller, W., 172, *214*
Miller, W. F., 90, *133*
Mills, C. B., 80, *88*
Molière, G., 136, 148, 206, *214*
Montroll, E. W., 219, *243*, 246, 273, 277, *280*
Moran, H. S., 101, 132, *134*
Morton, K. W., 68, *88*, 225, *242*
Mott, N. F., 207, *214*
Mott, W. E., 93, *134*
Motz, J. W., 147, *214*
Murray, R. B., 102, *134*

Nelms, A. T., 204, *214*
Newell, G. F., 246, 273, 277, *280*
Nigam, B. P., 148, 207, *214*

Onsager, L., 248, 265, *280*

Parker, F. R., 245, *280*
Penner, S., 138, 152, 178, 188, *214*
Peele, R. W., 132, *134*
Penny, S. K., 118, *134*
Placzek, G., 5, *42*

Robinson, C. S., 138, 152, 187, 188, *214*
Roesch, W. C., 137, *214*
Rohrlich, F., 202, *214*
Rose, M. E., 137, 184, 185, *213*
Rosen, L., 80, 85, *88*
Rosenbluth, A. W., 219, *243*, 245, *280*
Rosenbluth, M. N., 219, *243*, 245, *280*
Rossi, B., 136, 137, 148, 151, 172, 205, *214*
Rubin, R. J., 219, 220, *243*

Salzburg, Z. W., 245, *280*
Saunderson, J. L., 136, 149, *214*
Schneider, D. O., 154, 191, *214*
Schwartz, R. B., 3, *42*
Scott, W. T., 136, *214*

Segrè, E., 196, 197, 198, *214*
Seliger, H. H., 170, 172, 176, 181, *214*
Sherman, N., 207, 209, *213*, *214*
Sidei, T., 138, 152, 164, *214*
Smith, L. P., 137, 184, 185, *213*
Snow, W. J., 90, *133*
Snyder, H. S., 136, *214*
Sonnenblick, E., 290, *298*
Spencer, L. V., 98, *134*, 136, 150, 189, 193, 207, 210, *214*
Stelson, P. H., 93, 94, 128, *133*, *134*
Sternheimer, R. M., 204, *214*
Stockmayer, W. H., 218, 219, *243*
Stone, P. M., 32, *42*
Sundaresan, M. K., 148, 207, *214*
Suoninen, E., 278, *280*
Sutton, R. B., 93, *134*
Suzor, F., 170, *214*
Sykes, M. F., 219, 225, *243*
Szegö, G., 13, *42*

Taussky, O., 103, *134*
Taylor, W. J., 218, *243*
Teller, A. H., 245, *280*
Teller, E., 245, *280*
Ter Haar, D., 246, *280*
Todd, J., 103, *134*
Trump, J. G., 169, *214*

Uehling, E. A., 200, *214*

Van de Graaf, R. J., 169, *214*
Van Dilla, M. A., 132, *134*
von Holdt, R., 47, *65*

von Neumann, J., 108, 109, *134*
Vyssotsky, V. A., 290, *298*

Wachspress, E. L., 32, *42*
Wainwright, T., 44, *65*
Wall, F. T., 219, 220, 222, 225, 226, 231, 237, 240, *243*
Wang, M. C., 137, *214*
Warren, B. E., 277, 278, *280*, *282*
Watson, R. E., 207, *213*
Welsh, D. J. A., 290, *298*
Wendroff, B., 48, 57, *65*
Wentzel, G., 137, *214*
Westphal, K., 146, 205, *213*
Weymouth, J. W., 137, *215*
Wheeler, D. J., 219, 220, 222, 225, *243*
Wick, G. C., 47, *65*
Willard, H. B., 94, 95, 128, *133*
Williams, E. J., 136, *215*
Wilson, P. R., 138, 152, 184, 201, *214*
Wilson, R., 187, *215*
Windwer, S., 231, *243*
Wolff, C., 98, *134*
Wood, W. W., 245, *280*
Wright, G. T., 93, *134*
Wu, T., 148, 207, *214*

Yang, C. N., 151, 200, 201, *215*, 249, 279, *280*

Zachariasen, F., 47, *65*
Zerby, C. D., 101, 102, 118, 132, *134*
Zimm, B. H., 218, 219, *243*

Subject Index

Albedo, 168
Adjoint solution, 27
Analogue calculation, 120
Azimuthal angle, 45, 75, 105, 147
   averaging, 108, 152

Boltzmann distribution, 245, 249, 259
   transport equation, 117
Born approximation, 101, 148
Boundary conditions, 37, 50, 142, 186
   effects 10, 191
   periodic 249, 267
Bremsstrahlung 90, 95, 100, 115, 146, 187

Central limit theorem 84, 250, 259
Computing machines
   Ferranti-Mercury 281, 288
   ILLIAC 223, 233, 246, 260
   IBM-704 103, 112, 125, 138, 157, 208, 284, 288
   IBM-7090 58, 82, 163, 290
   LARC 58
Computing time 41, 68, 85, 107, 125, 159, 163, 225, 229, 249, 259, 264, 267, 275, 281, 284, 287, 289, 294
Convergence 40, 57, 213, 265, 273
Conservation laws 17, 25, 38, 44, 155
Coordinates 5, 147, 286
   cylindrical polar 23, 105
   Eulerian 140
   Lagrangian 140
   rectangular Cartesian 17, 105, 150, 220
   spherical 9, 80
Cross-sections 2, 73, 77, 85, 106, *see* also Fission, Transport
   Bhabha 204
   capture 29
   Moller 202
   pair production 95, 98, 113
   radiation 45
   transfer 31
Curie point 278

Difference equation 17, 20, 33, 39, 47, 50, 56

Diffusion equation 10
   method 68, 135, 185
Dimensions 10, 32, 69, 109, 150, 247, 279, 281
Direction cosines 7, 12, 15, 53, 74, 106, 113, 286
   random 109, 117
Distances, mean square end-to-end 218, 236
Distribution, *see* also Gaussian, Temperature
   angular 7, 31, 45, 54, 110, 145, 174, 182, 208, 211
   binomial 251
   cumulative 158, 213, 286
   Goudsmit-Saunderson 149, 161, 207
   Molière 148, 196, 206

Energy spectrum 67, 93, 183, 192
Ergodic class 257
Error analysis 40, 58, 82, 138, 199, 263, 266, 271
   round-off 213, 295
   statistical 10, 61, 84, 87, 117, 119, 132, 171
   truncation 57
Euler formula 14
Existence of solution 4, 68, 256

Ferromagnetism 247
   antiferromagnetism 247, 272
Fission
   critical size 67, 81
   cross-section 29
   neutron 31
   spectrum 3, 80
Flow diagram 56, 70, 104, 111, 161, 223, 228, 232, 279, 293
Fluctuation 140, 146, 177, 193, 261
   statistical 136, 196, 258, 283
Flux
   electrons 191
   neutron 4, 7, 18, 25
   radiation 57
Fortran 69, 138, 158, 208

SUBJECT INDEX

Gamma-ray 90
Gaussian
    abscissas 9
    distribution 92, 147, 151, 196
    quadrature 9, 13, 47
Godiva assembly 85

Ising lattice 246
    order parameter 247, 269, 276
Isotropy 53, 80, 94, 101, 147, 172
    anistropy 31
Iteration procedure 56, 253, 271, 283
    direct 39
    inner 37
    outer 40
    power 34
    scaled 38

Lattice 249
    body-centered cubic 246, 275
    cubic 239, 265, 282, 296
    square 219, 236, 261
    tetrahedral 220, 230
Line broadening 93

Markov
    chain 256
    process 255
Mean free path 41, 63, 185, 195, 204
Memory Storage 58, 107, 115, 125, 158, 163, 224, 229, 233, 284, 290, 295
Mesh 7, 17, 24, 48
Monte Carlo calculations 37, 46, 51, 58, 90, 117, 140, 193, 220, 282, 296
    conditional 118

Percolation process 281
Phase transition 247, 261, 266, 279
Photoelectric effect 97
Planck function 45
Polymers 217
Precision-double 150
Probability
    collision 19, 28
    conditional 255
    critical 282
    transition 138, 245, 255

Programs 160
    assembly 232

Radiation
    annihilation 95, 102, 115, 132
    intensity 45
Radius of gyration 234
    critical 82, 87
Random numbers 53, 56, 62, 73, 76, 78, 109, 262, 283
    generator 103, 157, 222, 267
    mid-square process 262
Random variable 84, 104, 107, 111, 158, 250, 286
Random walk 136, 138, 218
Recursion formula 20, 35, 83, 150, 210, 267, 284
Rejection technique 55, 104, 111, 158

Sampling
    biased 105, 122, 159, 229
    correlated 159, 170
    importance 142, 250
    optimum 254
    random 115, 137, 157, 188
    systematic 105, 118
    technique 106, 226, 250
Scattering
    angular 30, 147
    anistropic 31, 79
    Compton 97, 110
    Coulomb 135, 165, 209
    elastic 78
    inelastic 79, 99, 206
    isotropic 31
    Mott 149, 207
    multiple 136, 209
    Rayleigh 99
    Rutherford 142, 173, 207
    self 30
    single 141, 149, 207
    Thomson 45, 54
Significant figures 36
S_n method 5, 47, 63, 68, 80, 137
Source 2, 26, 29, 180
    boundary 20, 38
    function 49
    perpendicular 178, 189, 197

strength 169, 191
surface 21, 51
volume 51
Stability 57
Standard deviation 76, 84, 87, 124, 170, 260, 263, 271
Steady state solution 32, 40, 44, 47
Stochastic matrix 255
Symmetry 11, 27
axial 92

Temperature
critical 247, 265

distribution 44, 59, 62
Transport
calculations 4, 139, 151
cross-section 30
electron 136
neutron 1, 67
equation 44, 189, 193
linear 2
non-linear 43
radiation 43, 90

Variance 57, 83, 105, 170, 251, 254, 259
Volume, excluded 218